COMPUTATIONAL BIOCHEMISTRY

COMPUTATIONAL BIOCHEMISTRY

By
Manju Yadav
Lecturer
Department of Zoology
M.M.H. College
Ghaziabad (U.P.)
(India)

D P H

DISCOVERY PUBLISHING HOUSE PVT. LTD.
NEW DELHI-110 002

Published by:

DISCOVERY PUBLISHING HOUSE
4383/4B, Ansari Road, Darya Ganj
New Delhi-110 002 (India)
Phone: +91-11-23279245; 23253475; 43596065
Mobile: +91 9811179893 / +91 9871656464
E-mail: discoverybooksindia@gmail.com
orderdphbooks@gmail.com
namitwasan9@gmail.com
web: www.discoverypublishinggroup.com

First Published: 2009

Reprinted: 2025

ISBN: 978-81-8356-463-2

Computational Biochemistry

Printed at:
Infinity Imaging Systems
Delhi

Preface

The present title "*Computational Biochemistry*" is an exciting new field bringing together life scientists, mathematicians, computer scientists and engineers to explore a new and deeper understanding of biological systems. The present text has been compilated for computer science and mathematics graduate and level undergraduate students. It will also be of interest to molecular biologists interested in bioinformatics.

The rationable of the book is to present algorithmic ideas in computational biology and to show how they are connected to molecular biology and to biotechnology. To achieve this goal, the book has a substantial "computating biology without formulas" component that presents biological motivation and computational ideas in a simple way. This simplified presentation of biology and computing aims to make the book accessible to computer scientists entering this new areas and to biologists who do not have sufficient background for more involved computational techniques. Every chapter has an introductory section that describes both computational and biological ideas without any formulas. Thus book concentrate on computational ideas rather than details of the algorithms and makes special efforts to prevent there ideas in a simple way.

To make the work more comprehensive and informative, the author has consulted many authoritative books, research journals, abstracts, monographs etc. He is grateful to all those great scholars whose work are cited or substantially reproduced.

There can be no claim to originality except in the manner of treatment and much of the information has been obtained from the books and scientific journals available in the different libraries.

The author expresses his thanks to his friends and colleagues whose continue inspirations have initiated him to bring out this book.

The author expresses his gratitude to Mr. Wasan and staff of M/s Discovery Publishing House Pvt. Ltd. for their whole hearted cooperation in the publication of this book.

In the mean time, the author will remain sincerely responsible for any shortcomings of the book and be grateful to the readers for their suggestions and constructive criticism for the continuous betterment of the book. He takes this opportunity to appeal to the readers to send their suggestions straightaway to his Publisher.

Author

Contents

CHAPTER

INTRODUCTION

Biostatistics is the statistical analysis that evaluates the reliability of biochemical data objectively is Presented. Statistical programs are introduce. The applications of spreadsheet(Excel) and database (Access) software packages to analyze and organize biochemical date are described.

BIOSTATISTICAL METHODS

Many investigations in biochemistry are quantitative.Thus, some objective methods are necessary to aid the investigators in presenting and analyzing research data. *Statistics* refers to the analysis and interpretation of data with a view toward objective hypothesis testing. *Descriptive statistics* refers to the process of organizing and summarizing the data in a way as to arrive at an orderly and informative presentation. However, it might be desirable to make some generalizations from these data. *Inferential statistics* is concerned with inferring characteristics of the whole from characteristics of its parts in order to make generalized conclusions.

Numerical Data

All numerical data are subject to uncertainty for a variety of reasons; but because decisions will be made on the basis of analytical data, it is important that this uncertainty be quantified in some way. Variation between replicate measurements may be due to a variety of causes, the most predictable being random error that occurs as a cumulative result of a series of simple, indeterminate variations. Such error gives rist to results that will show a normal distribution about the mean. The number(*n)* of measurements(x_i or $x_{i)}$ falling within the range of a particular group is known as *frequency.* The measurement occurring with the greatest frequency is konwn as the mode. The middle measurement in an ordered set of data is typically defined as *median.* That is, there are just as many observations larger than the median as there are smaller. The average of all measurements is known as the *mean,* and in theory, to determine this value(μ), many replicates are required. In practice, when the number of replicates is limited, the calculated mean($\bar{x}$ or $\bar{x}$) is an acceptable approximation of the true value.

The sum of all deviations from the mean—that is, $\Sigma(x_i - \bar{x})$ — will always equal zero. Summing the absolute values of the deviations from the mean results in a quantity that expresses dispersion about the mean. This quantity is divided by n to yield a measure known as the *mean deviation* or *standard error of the mean(*SEM), which expresses the confidence in the resulting mean value:

$$\text{SEM} = \Sigma|(x_i - \bar{x})|/n$$

An approach to eliminate the sign of the deviations from the mean is to square the deviations. The sum of the squares of the deviations from the mean is called the *sum of squares*(ss) and is defined as

$$\text{Populations SS} = \Sigma(X_i - \mu)^2$$

$$\text{Sample} = \Sigma(X_i - \bar{x})^2$$

As a measure of variability or dispersion, the sum of squares considers how far the X_i's deviate from the mean. The mean sum of squares is called the *variance*(or *mean squared deviation*), and it is denoted by σ^2 for a population:

$$\sigma^2 = \Sigma(X_i - \mu)^2/N$$

The best estimate of the population variance is the sample variance, s^2:

$$s^2 = \Sigma(X_i - \bar{X})^2/(n-1) = [\Sigma X_i^2 - \{\Sigma X_i\}^2/n]/(n-1)$$

Dividing the sample sum of squares by the degree of freedom(n – 1) yields an unbiased estimate. If all observations are equal, then there is no variablity and $S^2 = 0$. The sample variance becomes increasingly large as the amount of variability or dispersion increases.

The most acceptable way of expressing the variation between replicate measurements is by calculating the *standard deviation(s)* of the data:

$$s = [\Sigma(\bar{x} - x)^2 / n - 1]^{1/2}$$

where x is an individual measurement and n is the number of individual measurements. An alternative, more convenient formula to use is

$$s = [\{\Sigma x^2 - (\Sigma x)^2/n\}/(n - 1)]^{1/2}$$

The calculation of standard deviation requires a large number of replicates. For any number of replicates less than 30, the value for s is only an approximate value and the function(n – 1) known as the degree of freedom(DF) is used rather than(n).

In addition to random errors derived from samplings, systematic errors are peculiar to each particular method or system. They cannot be assessed statistically. A major effect of systematic error known *as bias* is a shift in the position of the mean of a set of readings relative to be the original mean.

Analytical methods should be precise, accurate, sensitive, and specific. The precision or reproducibility of a method is the extent to which a number of replicate measurements of a sample agree with one another and is expressed numerically in terms of the standard deviation of a large number of replicate determinations. Statistical comparison of the relative precision of two methods used the variance ratio(F_ϕ) or the F test.

$$F_\phi = s_1^2/s_2^2$$

The basic assumption, or null hypothesis(H_0), is that there is no significant difference between the variance(s^2) of the two sets of data. Hence, if such a hypothesis is true, the ratio of two values for variance will be unity or almost unity. The values for s_1 and s_2 are calculated from a limited number of replicates and, as a result, are only approximate values. The values calculated for F will vary from unity even if the null hypothesis is true. Critical values for F (F_{crit}) are available for different degrees of freedom; and if the test value for F exceeds F_{crit} with the same degrees of freedom, then the null hypothesis can be rejected.

Accuracy is the closeness of the mean of a set of replicate analysis to the true value of the sample. Often, it is only possible to assess the accuracy of one method relative to another by comparing the means of replicate analyses by the two methods using the t test. The basic assumption, or null hypothesis, made is that there is no significant difference between the mean value of the two sets of data. This is assessed as the number of times the difference between the two means is greater than the standard error of the difference(t value.)

$$t_\phi = (\bar{x}_i - \bar{x}_2) / [\{\Sigma(x_i - \bar{x}_i)^2 + \Sigma(x_2 - \bar{x}_2)^2\}^{1/2}] / n(n-1)$$

The critical value of the t test can be abbreviated as $t_{\alpha(2), v}$, where $\alpha(2)$ refers to the two-tailed probability of α and $v = n - 1$(degree of freedom). For the two-tailed t test, compare the calculated t value with the critical value from the t distribution table. In general, if $|t| \geq t_{\alpha(2), v}$, then reject the null hypothesis. When comparing the means of replicate determinations, it is desirable that the number of replicates be the same in each case.

The sensitivity of a method is defined as its ability to detect small amounts of the test substance. It can be assessed by quoting the smallest amount of substance that can be detected. The specificity is the ability to detect only the test substance. It is important to appreciate that specificity is often linked to sensitivity. It is possible to reduce the sensitivity of a method with the result that interference effects become less significant and the method is more specific.

Variance

We need to become familiar with the topic of *analysis of variance,* often abbreviated ANOVA, in order to test the null hypothesis(H_0): $\mu_1 = \mu_2 = ... = \mu_k$, where k is the number of experimental groups, or samples. in the ANOVA, we assume that $\sigma_1^2 = \sigma_2^2 = ... = \sigma_k^2$, and we estimate the population variance assumed common to all k groups by a variance obtained using the pooled sum of squares(within-groups SS) and the pooled degree of freedom(within-groups DF):

$$\text{within-groups SS} = \sum_{i=1}^{k}\left[\sum_{i=1}^{n}(X_{ij} - \bar{X}_i)^2\right]$$

and

$$\text{within-groups DF} = \Sigma(n_i - 1)$$

These two quantities are often referred to as the *error sum of squares* and the *error degrees of freedom,* respectively. The former divided by the latter is a statistical value that is the best estimate of the variance, σ^2, common to all k populations:

$$\sigma^2 = \sum_{i=1}^{k}\left[\sum_{i=1}^{n}(X_{ij} - \bar{X}_i)^2\right] \Big/ \sum_{i=1}^{k}(n_i - 1)$$

The amount of variability among the k groups is important to our hypothesis testing. This is referred to as the *group sum of squares and* can be denoted as

$$\text{among- groups SS} = \sum_{i=1}^{k} n_i(\bar{X}_i - \bar{X})^2$$

and the *groups degrees of freedom is*

$$\text{among-groups DF} = k - 1$$

We also consider the variability present among all N data, that is,

$$\text{Total SS} = \sum_{i=1}^{k}\sum_{j=1}^{n}(X_{ij} - \bar{X})^2$$

and

$$\text{Total DF} = N - 1$$

In Summary, each deviation of an observed datum from the grand mean of all data is attributable to a deviation of that datum from its group mean plus the deviation of that group mean from the grand mean, that is.

$$(X_{ij} - \bar{X}) = (X_{ij} - \bar{X}_i) + (\bar{X}_i - \bar{X})$$

Further more, the sums of squares and the degree of freedom are additive,

total SS = group SS + error SS $= \sum_{i=1}^{k}\sum_{j=1}^{n} X_{ij}^2 - (\Sigma\Sigma X_{ij})^2/N$

total DF = group DF + error DF

Computationally,

$$\text{total SS} = \sum_{i=1}^{k}\sum_{j=1}^{n} X_{ij}^2 - C$$

where

$$C = (\Sigma\Sigma X_{ij})^2/N$$

and

$$\text{Groups SS} = \sum_{i=1}^{k}\left[\left(\sum_{j=1}^{n} X_{ij}\right)^2 / n_i\right] - C$$

$$\text{error SS} = \sum_{i=1}^{k}\sum_{j=1}^{n} X_{ij}^2 - \sum_{i=1}^{k}\left[\left(\sum_{j=1}^{n} X_{ij}\right)^2 n_i\right] = \text{total SS} - \text{groups SS}$$

Dividing the group SS or the error SS by the respective degree of freedom results in a variance referred to as *mean squared deviation from the mean(mean square,* MS):

$$\text{groups MS} = \text{groups SS/groups DF}$$

$$\text{error MS} = \text{error SS/error DF}$$

Table 1.1 summarizes the single factor ANOVA calculations. The test for the equality of means is a one-tailed variance ratio test, where the groups MS is placed in the numerator so as to inquire whether it is significantly larger than the error MS:

$$F = \text{groups/error MS}$$

The critical value for this test is $F_{\alpha(1),(k-1),(N-k)}$. If the calculated F is at least as large as the critical value, then we reject H_0.

It has become uncommon for ANOVA with more than factors to be analyzed on a computer, owing to considerations of time, ease, and accuracy. it will presume that established computer programs will be used to perform the necessary mathematical manipulation of ANOVA.

Table 1.1. Single Factor ANOVA Calculations

Source of Variation	*Sum of Squares, SS*	*Degree of Freedom, DF*	*Mean Square, MS*
Total $[X_{ij} - \bar{X}]$	$\sum_{i=1}^{k}\sum_{j=1}^{n} X_{ij}^2 - C$	$N - 1$	
Group(*i.e.,* among group) $[X_i - \bar{X}]$	$\sum_{i=1}^{k}\left[\left(\sum_{j=1}^{n} X_{ij}\right)^2 n_i\right] - C$	$k - 1$	Groups SS/groups DF
Error(*i.e.,* within groups) $[X_{ij} - \bar{X}_I]$	Total SS – groups SS	$N - k$	Error SS/error DF

Note: $C = (\Sigma\Sigma X_{ij})^2/N$; $N = \sum_{i=1}^{k} n_i$; *k* is the number of groups n_i is the number of data in groups *i*.

Linear Regression and Correlation

The relationship between two variables may be one of dependency. That is, the magnitude of one of the variable(the dependent variable) is assumed to be determined by the magnitude of the second vriable(the independent variable). Sometimes, the independent variable is called the predictor or regressor variable, and the dependent variable is called the response or criterion variable. This dependent relationship is termed *regression.* However, in many types of biological data, the relationship between two variables is not one of dependency. In such cases, the magnitude of one of the variables changes with changes in the magnitude of the second variable, and the relationship is *correlation.* Both simple linear regression and simple linear correlation consider two variables. In the simple regression, the one variable is linearly dependent on a second variable, whereas neither variable is linearly dependent on a Second variable, whereas neither variable is functionally dependent upon the other in the simple correlation.

It is very convenient to graph simple regression data, using the abscissa(X axis) for the independent variable and the ordinate(Y axix) for the dependent variable. The simplest functional relationship of one variable to another in a population is the simple linear regression:

$$Y_i = \alpha + \beta X_i$$

Here, α and β are population parameters(constants) that describe the functional relationship between the two variables in the population. However, in a population the data are unlikely to be exactly on a straight line, thus Y may be related to X by

$$Y_i = \alpha + \beta X_i + \varepsilon_i$$

where ε_i is referred to as an error or *residual.*

Generally, there is considerable variability of data around any straight line. Therefore, we seek to define a so-called "best-fit" line through the data. The criterion for "best-fit" normally utilizes the concept of *least squares.* The criterion of least squares considers the vertical deviation of each point from the line($Y_i - Y_i$) and defines the best-fit line as that which results in the smallest value of the sum of the squares of these deviations for all values of Y'_i with respect to Y_i. That is, $\Sigma^n_{i=1}(Y_i - Y'_i)^2$ is to *be minimum* where *n* is the number of data point. The sum of squares of these deviations is called the *residual sum of squares*(or the *error sum of squares*). Because it is impossible to possess all the data for the entire population, we have to estimate parameters α and β from a sample of *n* data, where *n* is the number of pairs of X and Y values. The calculations required to arrive at such estimates and to execute the testing of a variety of important hypotheses involve the computation of sums of squared deviations from the mean. This requires calculation of a quanitity referred to as the sum of the cross-products of deviations from the mean:

$$\Sigma xy = \Sigma(X_i - \bar{X})[Y_i - \bar{Y}] = \Sigma X_i Y_i - \{[\Sigma X_i)(\Sigma Y_i)]\}/N$$

The parameter β is termed the *regression coefficient,* or the *slope* of the best-fit regression line. The best estimate of β is

$$b = \Sigma xy/\Sigma x^2 = \{\Sigma(X_i - \bar{X})(Y_i - \bar{Y})\}/\Sigma(X_i - \bar{X})^2$$

$$= [\Sigma X_i Y_i - \{(\Sigma X_i)(\Sigma Y_i)\}/n]\ [\Sigma X^2_i - (\Sigma X_i)^2\}/n]$$

Although the denominator in this calculation is always positive, the numerator may be either positive, negative, or zero. The regression coefficient expresses what change in Y is associated, on the average, with a unit change in X.

A line can be defined uniquely, by stating, in addition to β any one point on the line, conventionally on the line where X = 0. The value of Y in the population at this point is the parameter α, which is called the Y *intercept*. The best estimate of α is

$$a = \bar{Y} - b\,\bar{X}$$

By specifying both *a,* and *b,* a line is uniquely defined. Because *a* and *b* are calculated using the criterion of least squares, the residual sum of squares from this line is the smallest. Certain basic assumptions are met with respect to regression analysis.

1. For any value of X there exists in the population a normal distribution of Y values and, therefore, a normal distribution of ε's.
2. The variances of these population distributions of Y values(and of ε's) must all be equal to one another, that is, homogeneity of variances.
3. In the population, the mean of the Y's at a given X lies on a straight line with all other mean Y's at the other X's; that is, the actual relationship in the population is linear.
4. The values of Y are to hvae come at random from the sampled population and are to be independent of one another.
5. The measurements of X are obtained without error. In practice, it is assumed that the errors in X data are at least small compared with the measurement errorn in Y.

For ANOVA, the overall variability of the dependent variable termed the *total sum of squares* is calculated by computing the sum of squares of deviations of Y_i values from $\bar{Y}$:

$$\text{total SS} = \Sigma y^2 = \Sigma(Y_i - \bar{Y})^2 = \Sigma Y_i^2 - (\Sigma Y_i)^2/n$$

Then, one determines the amount of variability among the Y_i values that results from there being a linear regression; this is termed the *linear regression sum of squares.*

$$\text{regression SS} = (\Sigma\, xy)^2/\Sigma\, x^2 = \Sigma(Y_i' - \bar{Y})^2$$

$$= [\Sigma\, X_i Y_i - \{(\Sigma\, X_i)(\Sigma\, Y_i)\}/n]^2/[\Sigma\, X_i^2 - (\Sigma\, X_i)^2/n]$$

The value of the regression SS will be equal to that of the total SS only if each data point falls exactly on the regression line.The scatter of data points around the regression line is defined by the *residual sum of squares,* which is calculated from the difference in the total and linear regression sums of squares:

$$\text{residual SS} = \Sigma(Y_i - Y_i')^2 = \text{total SS} - \text{regression SS}$$

The degrees of freedom associated with the total variablity of Y_i values are $n - 1$. The degrees of freedom associated with the variability among Y_i's due to regression are always 1 in a simple linear regression. The residual degrees of freedom are calculable as

$$\text{esidual DF} = \text{total DF} - \text{regression DF} = n - 2$$

Once the regression and residual mean squares are calculated(MS = SS/DF), the null hypothesis may be tested by determining

$$F = \text{regression MS/residual MS}$$

This calculated F value is then compared to the critical value, $F_{\alpha(1),\nu_1\nu_2}$, where υ_1= regression DF = 1, and υ_2 = residual DF = $n - 2$. The residual mean square is often written as $s^2_{y,x}$, a representation denoting that is it is the variance of Y after taking into account the dependence of Y on X. The square root of this quantity – that is, $S_{Y.X}$, – is called the *standard error of estimate*(occasionally termed the *standard error of the regression*).

The proportion of the total variation in Y that is explained or accounted for by the fitted regression is termed the *coefficient of determination*, r^2, which may be thought of as a measure of the strength of the straight line relationship:

$$r^2 = \text{regression SS/total SS}$$

The quantity r is the *correlation coefficient* which is calculated as

$$r = \Sigma xy/(\Sigma x^2\ \Sigma y^2)^{1/2}$$

Table 1.2. ANOVA Calculations of Simple Linear Regression

Source of variation	*Sum of Squares, SS*	*DF*	*Mean Squars, Ms*
Total $[Y_i - \bar{Y}]$	Σy^2	$n - 1$	
Linear regression $[Y'_i - \bar{Y}]$	$(\Sigma xy)^2 / \Sigma x^2$	1	Regression SS/ regression DF
Residual $[Y_i - Y'_{i}]$	Total SS- regression SS	$n - 2$	Residual SS/ residual DF

The quantity is readily computed as

$$r = [\Sigma XY - (\Sigma X \Sigma Y)/n] / [\{\Sigma X^2 - (\Sigma X^2)/n\}\{\Sigma Y^2 - (\Sigma Y)^2/n\}]^{1/2}$$

A regression coefficient, b, may lie in the range of $\infty \geq r \geq -\infty$, and it expresses the magnitude of a change in Y associated with a change in X. But a correlation coefficient, r, is unitless and $1 \geq r \geq -1$. Thus, the correlation coefficient is not a measure of quantitative change of one variable with respect to the other, but it is a measure of intensity of association between the two variables.

Multiple Regression and Correlation

If we have simultaneous measurements for more than two variables where number of variables = m), and one of the variables is assumed to be dependent upon the others, then we are dealing with a *multiple regression* analysis.

$$Y_j = \alpha + \beta_1 X_{1j} + \beta_2 X_{2j} + \beta_3 X_{3j} + \ldots + \beta_m X_{mj}$$

The underlying assumptions are as follows:

1. The observed values of the dependent variable(Y) have come at random from a normal distribution of Y values at the observed combination of independent variables.
2. All such normal distributions have the same variance.
3. The values of Y are independent of each other.
4. The error in measuring the X's is small compared to the error in measuring Y.

If none of the variables is assumed to be functionally dependent on any other, then we are dealing with *multiple correlation*, a situation requiring the assumption that each of the variables exhibits a normal distribution at each combination of all the other variables.

The computational procedures required for most multiple regression and correlation analyses are difficult, and demand computer capability to perform the necessary operation. A computer program for multiple regression and correlation analysis will typically include an analysis of variance(ANOVA) of the regression.

Table 1.3. ANOVA Calculations of Multiple Regression

Source of variation	*Sum of Squares, SS*	*DF*	*Mean Squars, Ms*
Total	$\Sigma(Y_j - \bar{Y}_j)^2$	$n - 1$	
Regression	$\Sigma(Y'_j - \bar{Y}_j)^2$	m	Regression SS/regression DF
Residual	$\Sigma(Y_i - Y'_i)^2$	$n - m - 1$	ResidualSS/residual DF

Note: n = total number of data points(*i.e.*, total number of Y values) and m = number of independent variables in the regression

If we assume Y to be functionally dependent on each of the X's, then we are dealing with multiple regression. If no such dependence is implied as the case of multiple correlation, then any of the $M = m + 1$ variables could be designated as Y for the purpose of utilizing the computer program. In either situation, we Can test the interrelationship among the variables as

$$F = \text{regression MS/ residual MS} = \{R^2/(1 - R^2)\}(\text{regression DF})$$

The ratio is the *cofficient of determination* for a multiple regression or correlation. In the

$$R^2 = \text{regression SS/total SS}$$

regression situation, it is an expression of the proportion of the total variability in Y attributable to the dependence of Y on all the X's as defined by the regression model fit to the data. In the case of correlation, R^2 may be considered to be amount of variability in any one of the M variables that is accounted for by correlating it with the other M – 1 variables. The square root of the coefficient of determination is referred to as the *multiple correlation coefficient.*

$$R = (R^2)^{1/2}$$

The square root of the residual mean square is the *standard error of estimate* for the multiple regression:

$$S_{Y\cdot 1,\,2,\,3,\,\ldots\, m} = (\text{residual MS})^{1/2}$$

The subscript (Y·1, 2, 3, ... m) refers to the mathematical dependence of variable Y on the independent variables 1 through m.

SPREADSHEET APPLICATION

A spreadsheet is a collection of data entries: text, numbers, or a combination of both(alphanumeric) arranged in rows and columns used to display, manipulate, and analyze data. Microsoft Excel is a spreadsheet software package that allows you to do the following:

- Manipulate data through built-in or user-defined mathematical functions.
- Interpret data through graphical displays or statistical analysis.
- Create tables of numeric, text, and other formates.
- Graph tabulated input in various formates.
- Perform various curve-fitting procedures, either a built-in or a user- defined and-on Solver.
- Write user-defined macros for applications routines to automate or enhance a spreadsheet for a particular purpose.

The user is referred to the *Microsoft Excel User's Guide* or online Help for information.

Launching Microsoft Excel brings you into the workspace of Excel and a new workbook. A work book is a collection of related spreadsheets organized in rows(number headings) and columns (letter headings). The intersection of a row and a column is a cell that is addressed by row and column headings. To select a group of cells, click on the first cell and drag down to the last cell.

This collection of cells is known as a range. Inserting cells, rows, and columns is done through either the Insert or shortcut Edit menu. Excel inserts the new column to the left of the highlighted column. New rows are inserted above the highlighted row. At the bottom of the spreadsheets are sheet tabs representing the worksheets of the workbook. Click on these tabs one at a time to move from one sheet to another within the same workbook. The Excel commands are either in the form of a drop-down menu or in the form of icon buttons grouped as toolbars.

Mathematical Operations

There are three types of data that can be entered in the cells of the spreadsheet; number, date/ time, and text. For multiple entries of serial number with constant increment, enter the initial value into the first cell, highlight the cells, and select Edit → Fill → Series. To fill multiple entries of the same value into a range, enter the value into the first cell, move the pointer to the right-hand corner of the active cell to activate fill handle(a bold crosshair), and drag it through the range. Data are edited with the usual *cut, copy* and *paste* operations in the Edit menu and the decimal places of scientific numbers is controlled via dialog box in Format → Cells.

The fundamental operation of a spreadsheet is performing calculations on data. Excel performs mathematical operations through formula and functions. Formulas are written by using the formula bar and beginning with an equal sign(=). Functions can be either user-defined or bulit-in excel functions. Because the formulas and functions are acting on cells of the spreadsheet, the variables used therein are the cell references of those cells. This is where an understanding of the difference between relative and absolute (Preceded by a $ Sign) references is important. Excel's built-in functions are accessed through the Function Wizard tool(f_x) found on the standard toolbar. To use a wizard, just follow the instructions

in the dialog boxes, moving through them one step at a time. Excel has many mathematical, statistical, and scientific functions. These have the general syntex: = Function Name(arguments). For example, to calculate the mean and standard deviation, type Mean: in B1 and S.D: in C1, select B2 and C2, and enter the formulas = *average(range)* and = *stdev(range)*. The calculated Mean and Standard Deviation are Placed in the cells B2 and c2, respectively.

Application

Statistical functions are selected from two menus within Excel.The First approach is through the Function Wizard.Click the Function Wizard, f_x to activate the Function Wizard dialog box; select Statistical under the Function Category list and select Statisstical Functions under theFunctions Name(fefer to Help for the definition and usage of a function). The second approach is form Tools → Data Analysis. If Data Analysis is not present when the Tools command is selected, this means that the Analysis Tool Pak was not loaded during Excel installation. The Analysis Tools Pak can be loaded from Tools → Add-Ins.

To perform F test or *t* test;

- Enter the data into a workbook.
- Select Tools in the menu bar and select Data Analysis.
- From the box, select F- Test or *t*- Test Two-Sample for Variances and click OK.
- Enter variable 1 and variable 2(in absolute format, *e.g.*, $C$4:$C$10).
- Under the output options, select New Worksheet Ply to report the results on a new worksheet.

A report is generated, which contains all the required information to interpret the data.

To perform ANOVA:

- Enter the data into a workbook.
- Select Tools in the menu bar and select Data Analysis.
- Select ANOVA: Single Factor from the list of options to bring up the dailog box.
- Enter the input range covering the entire data set(in absolute format, *e.g.*,$B$2:$D$6)
- Check whether the data are group in columns or rows.
- Alpha can be set(default is 0.05).
- Select the output range(select new Worksheet Ply to report on a new worksheet).
- Click on the Ok button.

The comparison between the F- test result(F) and the critical value(F_{crit}) Provide a decision concerning the similarity/ difference of the groups.

Regression Analysis

Traditional approaches to experimental data processing are largely based on linearization and /or graphical methods. However, this can lead to problems where the model describing the data is inherently nonlinear or where the linearization process introduces data distortion. In this case, nonlinear curve-fitting techniques for experimental data should be applied.

Excel Provides some built-in tools for fitting models to data sets. By far the most common routine methods for experimental data analysis is linear regression, from which the best-fit model is obtained by minimizing the least-squares error between the *y*-test data and an array of predicated *y* data calculated according to a linear equations with common *x* values. Linear regression can be accessed in Excel by LINEST function or *via* Data Analysis tool.

LINEST allows for detailed multiple linear regression:

- Select a blank sheet and enter the data in arrays.
- Activate a free cell.
- Click the Function Wizard icon and select Statistical and LINEST.
- Enter the array for the *know –y*'s, for example,(A1:A10), and *known–x's*, for example,(B1:B10),(C1:C10),(D1:D10), but leave the *const* and *stats* boxes blank. Click Ok.

The LINEST result is returned at the assigned(activated) free cell. To display the Slope and Intercept:

- Enter LINEST {m,b} in the new cell.
- Highlight the range of cells corresponding to 2 × *n* matrix, where *n* is the number of parameters, *x*. Select LINEST from the Function Wizard and enter the imput as before.
- Do not press Return, but instead click the mouse in the formula bar and place the cursor at the end of the entry.
- Pres CTRL + SHIFT + ENTER. The values of the slope(coefficients) and the intercept have been entered into the highlighted cell range.

The detailed linear regression analysis is obtained via the Tools menu us follows:

- Select Tool → Data Analysis → Regression. This opens the Regression dialog box.
- Enter the cell ranges for the array of *y* and *x* values using absolute format –for example, $A $2:$A$17 in the appropriate boxes.
- Check Confidence Level(e.g., 95%).

- Enter the output range where you want the regression analysis report to be copied (check New Worksheet ply for reporting on a new worksheet).
- Check the options to display the plots(e.g, Residual plots and Line Fit plots) and click OK.

The detailed statistical report includes the slope and intercept coefficients for the "best-fit" line, the standard error in these coefficients, and a listing of the residues. The goodness of fit is evaluated from the high correlation coefficient, R^2(R^2 = 1.00 for a perfect fit),and residuals evenly scattered about zero along the entire range.

Perhaps the most common situation involving graphing scientific data is to generate a linear regression plot with y error bars. In most situations, the rror in the x data is regarded as being so much smaller than that of the y data that it can be effectively ignored. Excel allows several methods for generating error bars where *custom error* by which the experimental standard deviation of several estimates of a value can be used. This is accomplished by obtaining the mean values, =AVERAGE { }, and the standard deviations, = STADEV { }, for experimental data. Generate the regression line for the mean values and add error bar via Insert → Error Bars, given by the standard deviations.

Statistical Packages

A number of custom-made statistical software packages are available commercially. The student is urged to be familiar with at least one of them because of their versatility and effixiency.

SPSS: The statistical analysis software, SPSS, is a common statistical package available in most of university networks, and the student could learn its use by following the tutorial session of the SPSS program. Open and click Spsswin. exe to start the SPSS program. Go to Help and select the Spss Tutorial. From the Main menu page, follow the session. The biochemical data for statistical analysis can be entered directly by starting File, and then choose New and Data. However, it can be advantageous for the student to prepare data files with Excel(filename. xls) in advance. In this case, start File and then choose Open and Data Click File type to select Excel(filename. xls). from the menu bar, select Statistics to initiate the data analysis.

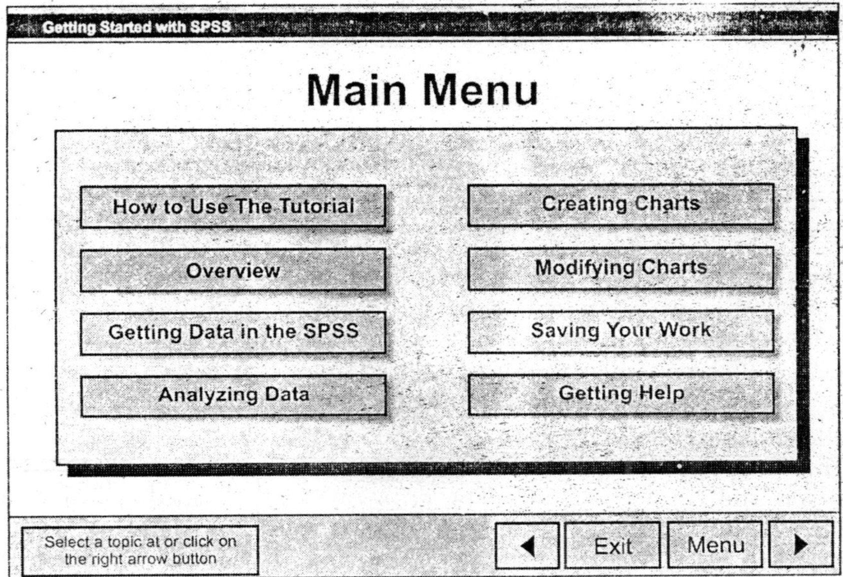

Fig. 1.1. SPSS Home page.

SyStat: SyStat is a stand-along statistical package of SPSS Inc, that performs comprehensive statistical analysis. The user's guide, SYSTAT *Getting Started,* should be consulted. There are three types of windows(main data and graph windows), each with its own meanus. The main window displays prints and saves results from statistical analyses. The data window(opened by the Data command of the File menu from the main window) allows entering, editing, and viewing of data. The graph window(opened from Graph menu or by double-clicking a graph button form the main window) provides facilities for editing graphs.

The new data can be entered directly(via File → New → Data) or imported from various spreadsheets(via File → Open → Data). Selecting Microsoft Excel and entering filename. *xls* imports the Excel data file. To define variable names, double click the variable heading to bring up the variable properties box(for a numeric data type, the string variable ends with $). A statistical analysis is initiated by selecting the analytical options form Statistic menu of the data window. The analytical results are displayed in the output pan by selecting it from the organization pan of the main window. The data file is saved from the data window as filename.syd, and the analysis results(outputs) are saved from the main window as filenames.syo.

To carry our linear regression analysis as an example, select "Regression, correlation, least squares curve-fitting, nonparametric correlation," and then select any one of the methods(*e.g.*, Least squares regression line, Least squares straight line). Enter number of data points to be analyezed, then data, x_i and y_i. Click the Calculate Now button. The analytical results, *a*(intercept), *b*(slope), *f* degrees of freedom), and r(correlation coefficient) are returned.

DATA MANAGEMENT

The data are the unevaluated, independent entries that can be either numeric or non-numeric –for example, alphabetic or symbolic. Information is ordered data which are retereved according to a user's need. If raw data are to be effectively processed to yield information, they must firts be organized logically into files. In computer files, a field is the smallest unit of data which contains a single fact about an item in the table; and a collection of records of the same type is called a file, which contains all facts about a topic in the table. Adatabase system is a computerized information system for the management of data by means of a genera-purpose software Package called a database management system(DBMS). A *database* is a collection of interrelated files created with a DBMS. The content of a database is ontained is obtained by combining data from all the different sources in an organization so that dta are available to all users such that redundant dta can be eliminated or at least minimized.

Figure 1.6 illustrates the relationship among the components of a typical database. Two files are interrelated(logically linked) if they contain common data types that can be at interpreted as being equivalent. Two files are also interrelated(physically linked) if the value of a data type of onefiles contains the physical address of a record in the other file. A relational database stores information in a collection of files, each containing data about one subject. Because the files are logically linked, you can use information from more than one file at a time. The Microsoft Access database is a collection of data and objects related to a particular topic (Hutchinson and Giddeon, 2000). The data represent the information stored in the database, and the objects help users define the structure of that information and automate the data manipulating tasks. Access supports SQL(Structured Query Language) to create, modify, and manipulate records in the table to facilitate the process. It is a table-oriented processing.

Launching Microsoft Access and click New dtabase from File menu brings you the Database Window with five tabs representing the object types that Access supports:

Table is used to define the data structure for different related topics.

Query is used to select and retrieve data stored in table(*s*).

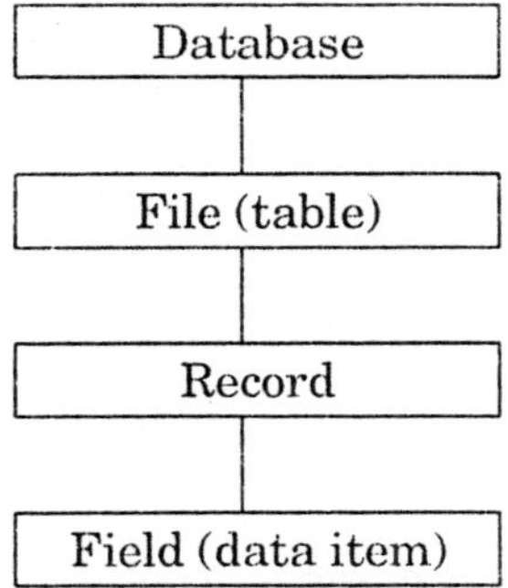

Fig. 1.2. Relationship among components of databases.

Form used to create user-interface screens for entering, displaying, and editing data.

Report is used to present information in a specified format and organization.

Macro is used to automate tasks.

Mudule is used as a repository of declarations and procedures to design advanced Access applications.

To build database file(table), activate Table tab, then:

(*i*) Click New and select Design View option.

(*ii*) Define field name, data type, and optional description as a remark for the field.

(*iii*) Highlight the unique field, right click, and assign the Primary key to the field.

(*iv*) Save as filename.mdb.

Molecular structures can be entered as objects:

(*i*) Activate Table tab, and select Design view.

(*ii*) Insert a row where the structures would appear in the field, and type the Field name.

(*iii*) Select OLE *object* in the Data type column.

The database file as a table object can be opened from the activated *Table* tab for manipulation. For examples, fields (columns) can be added from the Datasheet view by clicking the right edge or the existing column and then selecting Column form Insert menu. This brings a new column (labeled Field 1) or columns(repeated selection of Column option). Double click the new column to enter the field name. Records (rows) can be added also form the Datasheet view after clicking New record button on the tool bar.

Data can be imported(linked) from Microsoft Excel in two ways.

From Excel,

(*i*) Select MS Access Form to create form in New database.

(*ii*) Select columns to be included by transferring them from available fields to Selected fields.

(*iii*) Set layout and enter filename to Finish.

(*iv*) Open the Table after the import process automatically launching Access in the Access Window.

From Access

(*i*) Activate Microsoft Access window by choosing New database from File menu and entering filename.

(*ii*) Select Get external data from File menu.

(*iii*) Select Import or Link table.

(*iv*) Select Microsoft Excel for file type and enter filename in the import(link) dialog box.

(*v*) Follow the selections to Finish.

(*vi*) Repeat the Import to build the database containing more than one table for organizing data by relationship via Relationships button.

Alternatively, you can import/export data between Access and Excel by Copy and Paste operations, provided that matching records/fields(Access) and ragne of cells (Excel) are selected in the exchange.

To create interacting screen via activated *Form* tab:

(*i*) Click New and select Form Wizard option.

(*ii*) Pick data fields by transferring available fields to selected fields.

(*iii*) Choose form types: columnar, tabular, datasheet, or justified.

(*iv*) Click New object button on the tool bar and select Auto Form.

The form is saved and appended to the database file bdname.mdb. Similar steps are followed to create a report after invoking Report Wizard option from the activated *Report* tab. The Wizard provides options for grouping level, sorting order, layout, and style.

To retrieve specified data via *Query* object:

(*i*) Activate Query tab and click New.

(*ii*) Select Design view from New query dialog box.

(*iii*) Add table(s) from where records and fields are to be retrieved after 'Show Table" dialog box is removed.

(*iv*) Click(pick) and drag the fields to the design query grid.

(*v*) Specify Criteria for the retrieval by keying in the expression or right-click the grid and select build to bring Expression builder.

(*i*) Select appropriate relation operator with the expression.

(*ii*) Microsoft Access uses the operators: +, --, /, *, &, = , >, <, <> And, Or, Not, Like and().

(*iii*) Save the query file which is appended to database file and can be retrieved (opened) from *Query* tab.

WORKSHOPS

Use spreadsheet(Excel) or a statistical program to solve the following problems(1–8) after reviewing pertinent topics discussed in biochemistry texts

1. The ionization of cysteine proceeds by the following main sequences with $pk_1 = 1.92$, $pk_2 = 8.33$ and $pk_3 = 10.78$

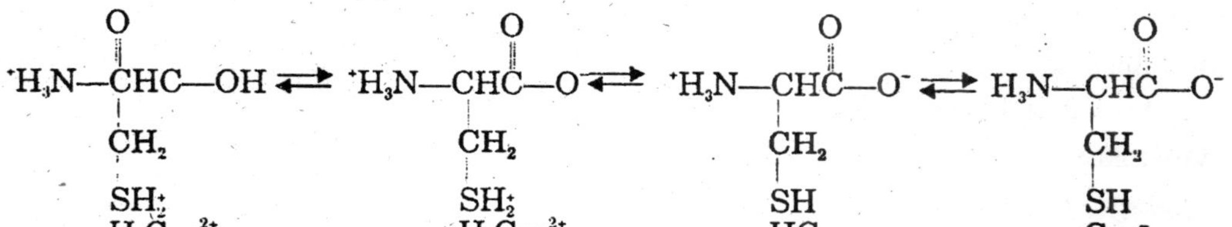

$$^{+}H_3N{-}\underset{\underset{SH_2^{+}}{|}}{\underset{CH_2}{|}}{CH}{-}\overset{O}{\overset{\|}{C}}{-}OH \rightleftharpoons {}^{+}H_3N{-}\underset{\underset{SH_2^{+}}{|}}{\underset{CH_2}{|}}{CH}{-}\overset{O}{\overset{\|}{C}}{-}O^{-} \rightleftharpoons {}^{+}H_3N{-}\underset{\underset{SH}{|}}{\underset{CH_2}{|}}{CH}{-}\overset{O}{\overset{\|}{C}}{-}O^{-} \rightleftharpoons H_2N{-}\underset{\underset{SH}{|}}{\underset{CH_2}{|}}{CH}{-}\overset{O}{\overset{\|}{C}}{-}O^{-}$$

H_3Cys^{2+} $\quad$ H_2Cys^{2+} $\quad$ HCys $\quad$ Cys^{-}

The ratio of each pair of ionic forms for a given p^H can be calculated by using the Henderson-Hasselbalch requation:

$$[H_2cys^{+}]/[H_3\,Cys^{2+}] = 10^{(pH-1.92)}$$
$$[HCys]/[H_2Cys^{+}] = 10^{(pH-8.33)}$$
$$[Cys^{-}]/[HCys] = 10^{(pH-10.78)}$$

$[H_3Cys^{2+}]$ expressed as a fraction of the total concentration may be calculated from the above ratio by using the partition equation:

$$1.00 = [H_3Cys^{2+}]\{1 + ([H_2Cys^{+}])/[H_3Cys^{2+}])\}$$
$$+([HCys]/[H_2Cys^{+}])([H_2Cys^{+}]/[H_3Cys^{2+}])$$
$$+[Cys^{-}]/[HCys^{+}])([HCys]/[H_2Cys^{+}]/[H_3Cys^{2+}])$$

This partition equation is solved for $[H_3Cys^{2+}]$, and the concentrations of the other ionic forms can be calculated from $[H_3Cys^{2+}]$ according to

$$[H_2Cys^{+}] = 10^{(pH-1.92)}[H_3Cys^{2+}]$$
$$[HCys] = 10^{(pH-8.33)}[H_2Cys^{+}]$$
$$[Cys^{-}] = 10^{(pH-10.78)}[HCys]$$

Find the isoelectric point and the concentrations of all ionic species from 1.0M cysteine solution at $p^H = 7.00$

2. The molecular weight is an intrinsic property of biomolecules. Sedimentation velocity is one of physicochemical methods which have been employed to determine molecular weights of many biomacromolecuoles. The molecular weight(M) of a biomacromolecule can be computed according to the Savelberg equation:

$$M = \{RTs\}/\{D(1-\bar{v}\rho)\}$$

where the symbols have the following meanings: R, gas constant(8.314×10^7 erg mol^{-1} deg^{-1}); T, absolute temperature; *s*, sedimentation coefficient; *d*, diffusion coefficient; $\bar{v}$, partial specific volume of protein; and ρ, density of the solvent. The density(ρ) of water at 20°C is 0.998

and the partial specific volume is taken to be 0.735 g cm^{-3}. Construct the spreadsheet table by supplementing the following information of biomacromolecules/particles with their molecular/particle weights.

	$S_{20.w}$($\times 10^{-13}$ sec)	$D_{20.w}$($\times 10^{-7}$ cm^2/sec)	$\bar{v}$ (cm^3/g)
Bacteriophage T7	453	0.603	0.639
Bushy stunt virus	132	1.15	0.740
Catalase, horse liver	11.3	4.10	0.730
chymotrypsinogen, bovine	2.54	9.50	0.721
Ferricytochrome c, bovine heart	1.91	13.20	0.707
Fibrinogen, G, human	7.9	2.02	0.706
Immunoglobin G, human	6.9	4.00	0.739
Lysozyme, hen's egg white	1.91	11.20	0.741
Myoglobin, horse heart	2.04	11.30	0.741
Ribosome	82.6	1.52	0.61
R Nase, bovine pancreas	2.00	13.10	0.707
Serum albumin, bovine	4.31	5.94	0.734
Tobacco mosaic virus	192	0.44	0.730
Urease, jack bean	18.6	3.46	0.730

3. The free change(ΔG) of a chemical reaction is related to its reaction quotient(Q, ratio of products to reactants) by

$$\Delta G = \Delta G^0 + Rt \text{ In } Q$$

where ΔG^0 is the standard free energy change and R is the gas constant(8.3145 J mol^{-1} deg^{-1}). At equilibrium(ΔG = 0), the standard free energy is related to the equilibrium constant (K_{eq}) by

$$\Delta G^0 = Rt \text{ In } K_{eq}$$

Biochemists have adopted a modified standard state(1 atm for gases and pure solids or 1 M solutions at p^H 7.0.) using symbols ΔG^0 for the standard free energy change, K'_{eq} for the equilibrium constant, and so on, such that

$$\Delta G^0 = \Delta G^0 + RT \text{ In } [H^+] \quad \text{for a reaction in which } H^+ \text{ is prduced, and}$$

$$\Delta G^0 = \Delta G^0 - RT \text{ In } [H^+] \quad \text{for a reaction in which } H^+ \text{ is consumed}$$

The free energy change in biochemical systems becomes

$$\Delta G' = \Delta G^{0\prime} + RT \text{ In } Q'$$

and

$$\Delta G^{0\prime} = -RT \text{ In } K'_{eq}$$

For coupled reaction system in which a product of one reaction is a reactant of the subsequent reaction, the overall free energy change is the sum of the free energy changes of all coupled reactions:

$$\Delta G^{0\prime} \text{ (overall)} = \Sigma \Delta G_i^{0\prime}$$

The equilibrium constants for the enzymatic reactions of the tricarboxylic acid cycle at p^H 7.0 are given below:

Reaction	*Enzyme*	K_{eq}
1.	Citrate synthase	3.2×10^5
2.	Aconitase	0.067
3.	Isocitrate dehydrogenase	29.7
4.	α-Ketoglutrate dehydrogenase complex	1.8×10^5
5.	Succiny I-CoA synthetase	3.8
6.	Succinate dehydrogenase	0.85
7.	Fumerase	4.6
8.	Malate dehydrogenase	6.2×10^{-6}

Calculate the standard free energy for each step and the overall ΔG° for one pass around the cycle, that is,

$$\text{Acetyl-CoA} + 3NAD^+ + [FAD] + GDP + H_3PO_4 + 2H_2O$$
$$\rightarrow \text{CoASH} + 3NADH + [FADH_2] + GTP + 2CO_2\ 3H^+$$

4. For biochemical reactions undergoing oxidation-reduction (redox reaction), the free energy change is related to the electromotive force(emf) or redox potential(ΔE′) of the reaction:

$$\Delta G' = -nF\Delta E'$$

where n denotes number of electrons involved in the reaction and F is the faraday(1 F = 96,494 C Mol^{-1} = 96,494 J V^{-1} mol^{-1}). The Nernst equation,

$$\Delta E' = \Delta E'_0 - (RT/nF) \text{ in } Q'$$

describes the redox potential of a redox reaction with its standard redox potential ($\Delta E'_0$) which can be calculated from the standard reduction potentials (E'_o) of its component half-reactions

$$\Delta E'_0 = E'_0(\text{acceptor}) - \Delta E'_0(\text{donor})$$

and related to the standard free energy changes by

$$\Delta E'_0 = -\Delta E'/nF = -(RT/nF) \text{ In } K_{eq}$$

From the standard reduction potentials of half-reactions listed below, construct the plausible electron transfer system for the oxidation of ethanol to acetaldehyde and water.

Half-Reaction	$\varepsilon^{0\prime}$ *(volts)*
$\frac{1}{2}O_2 + 2H^+ + 2e^- = H^2O$	0.825
Cytochrome $a_3(Fe^3+) + e^-$ = cytochrome $a_3(Fe^{2+})$	0.385
Cytochrome $a(Fe^3+) + e^-$ = cytochrome $a(Fe^{2+})$	0.29
Cytochrome $c(Fe^3+) + e^-$ = cytochrome $c(Fe^{2+})$	0.254
Cytochrome $c_1(Fe^3+) + e^-$ = cytochrome $c_1(Fe^{2+})$	0.22
Cytochrome $b(Fe^3+) + e^-$ = cytochrome $b(Fe^{2+})$	0.077
Ubiquinone + $2H^+ + 2e^-$ = ubiquinol	0.045
FAD + $2H^+ + 2e^-$ = $FADH_2$(in flavoprotein)	–0.04
Acetaldehyde + $2H^+ + 2e^-$ = enthanol	– 0.197
$NAD^+ + H^+ + 2e^-$ = NADH	–0.315

Claculate $\Delta E^{0\prime}$ and $G^{0\prime}$ for each redox step x step, and identify the sites that may couple with phosphorylation(ATP formation) assuming $\Delta G^{0\prime} = -30.5$ kJ mol^{-1} for the hydrolysis of ATP to ADP.

5. Amino acid analyses of acetylcholine esterase(AchE) and acetylcholine receptor(Ach R) or *electrophorus electicus* have been performed by various investigators.

	Ala	*Cys*	*Glu*	*Lys*	*Tyr*
Ach E	5.5	1.1	9.4	4.3	3.8
	6.2	1.6	10.4	4.6	3.6
	5.6	1.2	9.2	4.5	3.5
	5.8	1.3	9.8	4.4	3.4
	6.0	1.4	10.3	4.6	3.8
	5.8	1.5	10.0	4.5	3.9
Ach R	7.5	1.8	12.8	5.7	5.0
	5.4	1.7	9.0	6.3	3.8
	5.9	1.4	10.5	5.8	4.2
	6.4	1.7	10.9	6.0	4.4
	6.9	1.9	11.4	6.2	4.7
	7.0	1.6	9.8	5.9	4.5

The compositions(moles %) of selected amino acids are compared in the Table. Perform ANOVA to deduce whether AChE and ACh R are identical or different protein molecule(s) based on the amino acid compositions.

6. In an early work on the compositions of DNA, Chargaff(1955) noted that the ratios of adenine to thymine and of guanine to cytosine are very close to unity in a very large number of DNA samples. This observation has been used to support the helical structure of DNA proposed by Watson and Crick(1953). From the base compositions of DNA and RNA give in the following tables, deduce the statistical significance for the statement that the base ratio(A/T(U) and G/C are unity for DNA but vary for RNA.

The base Composition(%) of Some DNA

Tissue	*Adenine*	*Guanine*	*Cytosine*	*Thymine*
Calf thymus	27.3	22.7	21.6	28.4
Calf thymus	28.2	21.5	22.5	27.8
Beef spleen	27.9	22.7	20.8	27.3
Beef spleen	27.7	22.1	21.8	28.4
Beef liver	28.8	21.0	21.7	29.0
Beef pancreas	27.8	21.9	20.9	28.5
Beef kidney	28.3	22.6	22.0	28.2
Beef sperm	28.7	22.2	21.0	27.2
Sheep thymus	29.3	21.4	21.0	28.3

Sheepn liver	29.3	20.7	21.1	29.2
Sheep spleen	28.0	22.3	21.0	28.6
Sheep sperm	28.8	22.0	19.8	27.2
main thymus	30.9	19.9	19.9	29.4
Man liver	29.2	19.5	20.4	30.3
Man spleen	30.9	21.0	18.4	29.4
Man sperm	27.8	22.2	22.5	31.6
Herring sperm	29.7	20.8	20.4	27.5
Salmon Sperm	31.2	19.1	19.2	29.1
abacia lixuld sperm	13.4	37.12	19.2	30.5
Sarcian lutea	27.3	22.7	37.1	12.4
Wheet germ	31.3	18.7	22.8	27.1
Yeast	31.3	20.5	17.1	32.9
Pneumococcus type III	29.8	20.5	18.0	31.6
Vaccinia virus	29.5	20.6	20.0	29.9

The Base Composition(%) of some RNA

Tissue	*Adenine*	*Guanine*	*Cytosine*	*Thymine*
Bakers' yeast	25.1	30.1	20.1	24.7
Bakers' yeast	25.7	28.1	21.8	24.4
Escherichia coli	27.0	27.5	23.0	22.5
Rabbit liver	19.3	32.6	28.2	19.9
Rat liver	19.1	33.5	26.6	20.8
Rat liver	19.0	34.0	30.2	16.8
Sheep liver	20.4	39.4	25.8	14.4
Man liver	11.4	44.4	31.6	12.6
Calf liver	21.6	40.6	23.3	14.5
Calf liver	19.5	35.0	29.1	16.4
Calf pancreas	14.2	48.6	23.6	13.4
Calf spleen	17.9	35.2	31.6	15.3
Calf thymus	18.5	43.9	23.6	12.0
Tobacco mosaic virus	29.8	25.4	18.5	26.3
Cucumbervirus	28.7	25.5	18.2	30.6
Tomato bushy stunt virus	27.6	27.6	20.4	24.4
Turnip yellow mosaic virus	22.6	17.2	38.0	22.2
Southern bean mosaic virus	25.9	25.9	23.0	25.2
Potato X	34.4	21.4	22.8	21.4
Sarcina lutea	16.7	28.4	32.9	22.0

7. Heating DNA solution produces an increase in UV absorption(hyperchromic effect) and a decrease in viscosity, resulting in the disruption of the hydrogen bonds between the two strands of the double helix. A plot of optical absorbance at 260 nm versus temperature is known, as melting curve, and the temperature corresponding to the midpoint of the increase in absorance is refereed to as the melting temperature, T_m. A sample of DNA purified from rabbit liver nuclei is dissolved in a solution containing 0.15 M NaCI and 0.015 M sodium citrate. The change in absorabance at 260 nm is followed while the temperature is increased from 55°C to 98° C as shown:

T,°C:	60	65	70	75	80	82.5	85	87.5	90	92.5	95	98
A_{260}:	1.00	1.00	1.01	1.02	1.06	1.09	1.12	1.15	1.18	1.21	1.23	1.24

Estimate T_m of the sample by the melting curve. Perform regression analysis to deduce the correlation between T_m and the base composition of DNA by comparing the following data of DNAs from other sources:

DNA Sample		*Base Composition(%)*			
	T_m*(°C)*	*A*	*G*	*C*	*T(HMC for T_2)*
Poly(DAT)	65.6	50.0	0	0	50.0
T_2 bacteriophage	78.0	18.0	32.0	32.0	18.0
Yeast	84.0	32.9	17.1	17.1	32.9
Calf thymus	86.7	28.2	21.8	21.8	28.2
Salmon sperm	87.1	29.1	20.9	20.9	29.1
Rabbit liver	?	28.0	22.0	22.0	28.0
Escherichia coli	88.3	25.0	25.0	25.0	25.0
S. *marcescens*	93.5	21.0	29.0	29.0	21.0
Micrococcus lysodeikticus	95.4	16.0	34.0	34.0	16.0
Mycobacterium phlei	96.2	14.5	36.5	36.5	14.5
Poly(DGC)	105.8	0	50.0	50.0	0

8. Many biochemical and toxicological properties of compounds X_i depend on solute-solvent interaction can be rationalized in terms of the linear solvation-energy relationship(LSER)

$$X_i = X_0 + m(V/100) + s\pi^* + b\beta + a\alpha = X_m + s\pi^* + b\beta + a\alpha$$

where V, π^*, β, and α are the solute van der Waals molar volume, the solovatochromic parameters, a measure of basicity(hydrogen acceptor), and a measure of acidity(hydrogen donor) of compounds, X_i, respectively. The toxicity of various chemical solvents toward organisms expressed as log(1/C), where C is the LD_{50} molar concentration, can be correlated with LSER parameters:

Solvent	*Log(1/c)*	π^*	β	α
n Hexane	4.06	−0.08	0	0
Methanol	0.41	0.60	0.62	0.98
Ethanol	0.76	0.54	0.77	0.86
1-Propanol	1.20	0.51	0.83	0.80

Solvent	*Log(1/c)*	$\pi*$	β	α
2-Propanol	1.08	0.46	0.95	0.78
1-Butanol	2.40	0.46	0.88	0.79
tert-butanol	1.71	0.41	1.01	0.62
Ethylene glycol	0.14	0.85	0.52	0.92
Chololform	2.87	0.76	0	0.34
Diethlyl ether	1.77	0.27	0.47	0
Acetone	0.78	0.72	0.48	0.07
2-Butanone	1.26	0.67	0.48	0.05
Acetophenone	2.79	0.90	0.49	0.
Ethyl formate	1.30	0.61	0.46	0
Ethyl acetate	1.68	0.55	0.45	0
Butyl acetate	2.49	0.46	0.44	0
Ethyl propionate	2.10	0.50	0.42	0
Dimethy lacetamide	0.47	0.88	0.76	0
Acetonitrile	0.69	0.85	0.31	0.15
Benzene	2.85	0.59	0.10	0
p- Xylene	3.64	0.43	0.12	0
Anisole	2.90	0.73	0.22	0
Pyridine	1.69	0.87	0.64	0
Quinoline	2.97	0.64	0.64	0

Perform the regression analyses to assess the contribution of LSER parameters for LD_{50}(log 1/c) of the sample organism.

9. General constituents and approximate energy values of common foods are given below. Apply Access to design a database such that the information can be retrieved according to sources of macronutrients for food types(*e.g.*, protein sources or carbohydrate sources for cereal, fish fruit, dairy, meat/poultry, vegetable).

Food	*Water (%)*	*Protein (g)*	*Lipid (g)*	*Carbolydrate, g*		*Mineral (G)*	*Energy (cal)*
				Total	*Fiber*		
Apple	84.8	0.2	0.6	14.1	1.0	0.3	56
Asparagus	91.7	2.5	0.2	5.0	0.7	0.6	26
Avocado	71.8	1.8	14.0	7.4	1.5	1.1	149
Bacon	19.3	8.4	69.3	1.0	0	2.0	665
Banana	75.7	1.1	0.3	22.2	0.5	0.8	85
Barley	11.1	8.2	1.0	78.8	0.5	0.9	349
Bean, cooked red	69.0	7.8	0.5	21.4	1.5	1.3	118
Beef, T-bone	47.5	14.7	37.1	0	0	0.7	397
Buleberry	85.0	0.7	0.5	13.6	1.5	0.2	35

Food	*Water (%)*	*Protein (g)*	*Lipid (g)*	*Carbolydrate, g*		*Mineral (G)*	*Energy (cal)*
				Total	*Fiber*		
Broccoli	89.1	3.6	0.3	5.9	1.5	1.1	32
Butter	15.5	0.6	81.0	0.4	0	2.5	716
Cabbage	92.4	1.3	0.2	5.4	0.8	0.7	24
Cantaloupe	91.2	0.7	0.1	7.5	0.3	0.5	30
Carrot	88.2	1.1	0.2	9.7	1.0	0.8	42
Cauliflower	91.0	2.7	0.2	5.2	1.0	0.9	27
Celery	94.1	0.9	0.1	3.9	0.6	1.0	17
Cherry	83.7	1.2	0.3	14.3	0.2	0.5	58
Cheese, brick	40.0	23.2	30.0	1.9	0	4.9	370
Rice, white	12.0	6.7	0.4	80.4	0.3	0.5	363
Rye, whole-grain	11.0	12.1	1.7	73.4	2.0	1.8	334
Salmon, Atlantic	63.6	22.5	13.4	0	0	1.4	217
Sardine	70.7	19.3	8.6	0	0	2.4	160
Sausage, pork	38.1	9.4	50.8	Trace	0	1.7	498
Shrimp	78.2	18.1	0.8	1.5	Trace	1.4	91
Soybean, green	69.2	10.9	5.1	13.2	1.4	1.6	134
Soybean, milk	92.4	3.4	1.5	2.2	0	0.5	33
Spinach	90.7	3.2	0.3	4.3	0.6	1.5	26
Strawaberry	89.9	0.7	0.5	8.4	1.3	0.5	37
Sweet potato	70.6	1.7	0.4	26.3	0.7	1.0	114
Tomato	93.5	1.1	0.2	4.7	0.5	0.5	22
Trout, rainbow	66.3	21.5	11.4	0	0	1.3	195
Tuna	71.0	25.0	3.6	0	0	1.4	139
Turkey, light meat	73.0	24.6	1.2	0	0	1.2	116
Veal, lean loin	69.0	19.2	11.0	0	0	1.0	181
Walnut	3.5	14.8	64.0	15.8	2.1	1.9	651
Watermelon	92.6	0.5	0.2	6.4	0.3	0.3	26
Wheat, all-purpose flour	12.0	10.5	1.0	76.1	0.3	0.4	364
Yam	73.5	2.1	0.3	23.2	0.9	1.0	101
Yogurt	88.0	3.0	3.4	4.9	0	0.7	62

10. Vitamins are essential nutrients that are required for animals, usually in the minute amounts, in order to ensure healthy growth and development, Because the animal cannot synthesize them in amount adequate for its daily needs, vitamins must be supplied in the diet. The composition(per 100g food) of some vitamins in our foods are given on next page :

Food	Vit A (I.U)	Vit D (μg)	Vit E (μg)	Thiam (*mg*)	Rflv (*mg*)	Niac (*mg*)	F.*a* (*mg*)	P.*a*. (μg)	Pyd*x* (*mg*)	Vit C (*mg*)
Apple	65		0.74	0.03	0.02	0.1	0.5		20	4.5
Banana	190		0.40	0.05	0.06	0.7	9.6		320	10
Bean, dry	20		3.60	0.55	0.15	2.3	117	850		
Beef	60	0.33	0.63	0.07	0.15	3.8	12	1,100	275	
Broccoli	2,500			0.10	0.23	0.9	33.9	1,400		113
Cabbage	130	0.005		0.05	0.05	0.3	24		205	51
Carrrot	11,000	0.075	0.45	0.06	0.05	0.6	8		170	8
Cauliflower	60			0.11	0.10	0.7	29.1	920		78
Cheese	1,150	0.83		0.02	0.70		28.5	680	98	
Chicken	60		0.25	0.05	0.09	10.7	3.0	715	7.5	
Egg	1,180	6.6*	2.0	0.11	0.30	0.10	5.1	2,700	35	
Grapefruit	45		0.26	0.04	0.02	0.2	2.7		21	36
Halibut	440	3,500*		0.07	0.07	8.3			110	
Herring	110	8.2		0.02	0.15	3.6				3
Lamb			0.77	0.16	0.22	5.1	1.9	600	300	
Lettuce	1,200		0.50	0.05	0.07	0.04	29		71	7
Milk, whole	150	0.11		0.03	0.17	0.1	11	290	82	1
Orange	200		0.24	0.10	0.04	0.4	5.1	340	37	55
Peach	1,330			0.02	0.05	1.09	2.3		16	7
Pea	575		2.1	0.32	0.13	2.9	20	820	46	17
Peanut			22*	1.14	0.13	17.2	56.6	2,500	300	
Pork			0.71	0.77	0.19	4.1	5.1	920	510	
Potato			0.06	0.10	0.04	1.5	66	525	405	20
Rice				0.07	0.03	1.6			395	
Salmon	230	7.4		0.12	0.03	6.7			590	
Sardine, can	220	34.5		0.04	0.16	3.4			280	
Soyabean	385		140*	0.77	0.20	0.18	3.2	185	960	14
Spinach	8,100	0.005		0.10	0.24	0.6	132		60	51
Strwaberry	40			0.03	0.07	0.6	5.3		44	59
Sweet potato	14,000		4.0	0.10	0.06	0.6	12	940		22
Tomato	900		0.36	0.06	0.04	0.7	9			23
Tuna, can	80	5.8		0.05	0.12	11.9			440	
Wheat flour				0.55	0.12	4.3	38	400	490	

Note 1: Group B vitamins are thiamine(Thiam), riboflavin(Rflv), niacin(niac), folic acid(F.*a*) pantothenic acid(P.*a*). and pyridoxins(P*ydx*).

	Vit A	Vit D	Vit E	Thiam	Rflv	Niac	F.a	P.a.	Pydx	Vit C
Food	(I.U)	(μg)	(μg)	(*mg*)	(*mg*)	(*mg*)	(*mg*)	(μg)	(*mg*)	(*mg*)

Note 2: Values with * relating to vitamin D are for Egg yolk and halibnut liver oil, and those * relationg to vitamin E are for peranut oil and soyabeam oil.

Apply Access to build a database and create queries to retrieve records of(*a*) fishes that would supply recommended minimum daily supply of vitamin D(5 *mg*),(*b*) fruits that would supply recommended daily allowance of vitamin C(50 mg), and(*c*) foods of plant origin that would supply recommended daily allowance of vitamin A(800 I.U.).

CHAPTER 2

POTENTIAL ENERGY AND FORCE FIELDS

Central to the success of any computational approach to the study of chemical systems is the quality of the mathematical model used to calculate the energy of the system as a function of its structure. For smaller chemical systems studied in the gas phase, quantum mechanical (QM) approaches are appropriate. The success of these methods was emphasized by the selection of John A. People and Walter Kohn as winners of the 1998 Noble prize in chemistry. These methods, however, are typically limited to systems of approximately 100 atoms or less, although approaches to treat large systems are under development Systems of biochemical or biophysical interest typically involve macromolecules that contain 1000-5000 or more atoms plus their condensed phase environment. This can lead to biochemical systems containing 20,000 atoms or more. In addition, the inherent dynamical nature of biochemicals and the mobility of their environments require that large number of conformations, generated via various methods, be subjected to energy calculations. Thus, an energy function is required that allows for 10^6 or more energy calculations on systems containing on the order of 10^5 atoms.

Empirical energy functions can fulfill the demands required by computational by studies of biochemical and biophysical systems. The mathematical equations in empirical energy functions include relatively simple terms to describe the physical interactions that dictate the structure and dynamic properties of biological molecules. In addition, empirical force fields use atomistic models, in which atoms are the smalls particles in the system rather than the electrons and nuclei used in quantum mechanics. These two simplifications allow for the computational speed required to perform the required number of energy calculations on biomolecules in their environments to bee attained, and, more important, via the use of properly optimized barameters in the mathematical models the required chemical accuracy can be achieved. The use of empirical energy functions was initially applied to small organic molecules, where it was referred to as molecular mechanics and more recently to biological systems.

FUNCTIONS OF POTENTIAL ENERGY

A Potential energy function is a mathematical equation that allows for the potential energy, v, of a chemical system to be calculated as a function of its three-dimensional (3D) structure, R. The equation includes terms describing the various physical interactions that dictate the structure and properties of a chemical system. The total potential energy of a chemical system with a defined 3D structure, $V(R)_{total}$ can be separated into terms for the internal, V(R)internal and external, $V(R)_{external,}$ potential energy as described in the following equations.

$$V(R)_{total} = V(R)_{internal} + V(R)_{external} \quad ...(1)$$

$$V(R)_{internal} = \sum_{bonds} K_b(b-b_0)^2 + \sum_{angles} K_\theta(\theta-\theta_0)^2 + \sum_{dihedrals} K_\chi[1+\cos(n\chi-\sigma)] \quad ...(2)$$

and

$$V(R)external = \sum_{\substack{nonbonded \\ atompairs}} \left(\varepsilon_{ij} \left[\left(\frac{R_{\min,ij}}{r_{ij}} \right)^{12} - \left(\frac{R_{\min,ij}}{r_{ij}} \right)^{6} \right] + \frac{q_i q_j}{\varepsilon_D r_{ij}} \right) \quad ...(3)$$

The internal terms are associated with covalently connected atoms, and the external terms represent the noncovalent or nonbonded, interactions between atoms. The external terms are also referred to as interaction, nonbonded or intermolecular terms.

Beyond the form of Eqs. (1)-(3), which is discussed below, it is important to emphasize the difference between the terms associated with the 3D structure, R, being subjected to the energy calculation and the parameters in the equations. The terms obtained from the 3D structure are the bond lengths, b_j the valence angles, θ; the dihedral or torsion angles, χ; and he distances between the atoms, $r_{ij.}$ A diagrammatic representation of two hypothetical molecules in Figure 2.1 allows for visualization of terms. The values of these terms are typically obtained from experimental structures generated from X-ray crystallography or NMR experiments, from modeled structures (*e.g.*, from homology modeling of a protein; or a structure generated during a molecular dynamics (MD) or Monte Carlo (MC) simulation. The remaining terms in Eqs. (2) and (3) are referred to as the parameters. These terms are associated with the particular type of atom and the types of atoms covalently bound to it. For example, the parameter q, the partial atomic charge, of a sodium cation is typically set to + 1, while that of a chloride anion is set to – 1. Another example is a C—C single bond versus a C = C double bond, where the former may have bond parameters of b_0 = 1.53 Å, K_{b} = 225 kcl/(mol. Å$^{2)}$ and the latter b_0 = 1.33 Å, k_b = 500 kcal/(mol. Å^2) Thus, different parameters allow for different types of atoms and different molecular connectivities to be treated using the same from of Eqs. (2) and (3). Indeed, it is the quality of the parameters, as judged by their ability to reproduce experimentally, and quantum-mechanically determined target data (e.g., information on selected molecules that the parameters are adjusted to

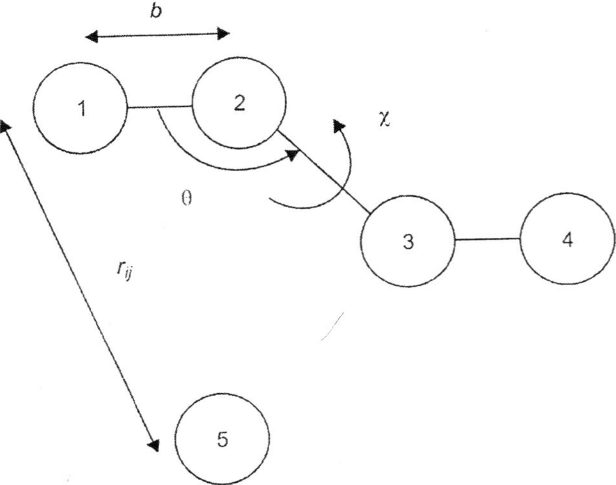

Fig. 2.1. Hypothetical molecules to illustrate the energetic terms included in Eqs. (1)-(3). Molecule A comprises atoms 1-4, and molecule B comprises atom 5. Internal terms that occur in molecule A are the bonds, b, between atoms 1 and 2, 2 and 3, and 3 and 4; angles θ, involving atoms 1-2-3 and atoms 2-3-4, and a dihedral or torsional angle, χ, described by atoms 1 2-3-4. Bonds can also be referred to as 1,2 atom pairs or 1, 2 interactions; angles as 1, 3 atom pair or 1,3 interactions; and dihedrals as 1,4 atom pairs or 1, 4 interactions. Molecule B is involved in external interactions with all four atoms in molecule A, where the different interatomic distances, r_{ij}, must be known. Note that external interactions (both van der Waals and Coulombic) can occur between the 1, 2, 1, 3 and 1, 4 pairs in molecule A. However, external interactions involving 1, 2 and 1, 3 interactions are generally not included as part of the external energies (i.e., 1, 2 and 1, 3 exclusions), but 1,4 influence are. Often the 1,4 external interaction energies are scaled (i.e,. 1,4 scaling) to diminish the influence of these external interactions on geometries, vibrations, and conformational energetics. It should also be noted that additional atoms that could be present in molecule A would represent 1,5 interactions, 1,6 interactions, and so on, an d would also interact with each other via the external terms.

reproduce) that ultimately determines the accuracy of the results obtained from computational studies of biological molecules. The mathematical form of Eqs. (2) and (3) represents a compromise between simplicity and chemical accuracy. Both the bond-twitching and angle-bending terms are treated harmonically, which effectively keeps the bonds and angles near their equilibrium values. Bond and angle parameters include b_0 and θ_0, the equilibrium bond length and equilibrium angle, respectively. K_b and K_θ are the force constants associated with the bond and angle terms, respectively.

The use of harmonic terms is sufficient for the conditions under which biological computations are performed. Typically MD or MC simulations are performed in the vicinity of room temperature and in the absence of bond-breaking or bond-making events; because the bonds and angles stay close to their equilibrium values at room temperature, the harmonic energy surfaces accurately represent the local bond and angle distortions. It should be noted that the absence of bond breaking is essential for simulated annealing calculations performed at elected temperatures. Dihedral or torsion angles represent the rotations that occur about a bond, leading to changes in the relative positions of atoms 1 and 4. These terms are oscillatory in nature (e.g., rotation about the C—C bond in ethane changes the structure from a low energy staggered conformation to a high energy eclipsed conformation, then back to a low energy staggered conformation, and so on), requiring the use of a sinusoidal function to accurately model them.

In Eq. (2), the dihedral term includes parameters for the force constant, K_χ; the periodicity or multiplicity, n; and the phase, δ. The magnitude of K_χ dictates the height of the barrier to rotation, such that $K\chi$ associated with a double bond would be significantly larger that for a single bond. The periodicity, n, indicates the number of cycles per 360^0 rotation about the dihedral. In the case of an sp^3-sp^3 bond, as in ethane, n would equal 3, while the $sp^2 - sp^2$ C=C bond in ethylene would have $n = 2$. The phase, δ, dictates the location of the maxima in the dihedral energy surface allowing for the location of the minima for a dihedral with $n = 2$ to be shifted from 0° to 90° and so on. Typically, δ is equal to 0 or 180, although recent extensions allow any value from 0 to 360 to be assigned to δ. Finally, each torsion angle in a molecule may be treated with a sum of dihedral terms that have different multiplicities, as well as force constant and phases [*i.e.*, the peptide bond can be treated by a summation of 1-fold ($n = 1$) and 2-fold ($n = 2$) dihedral terms with the 2-fold term used to model the double-bonded character off the C—N bond and the 1-fold term used to model the energy difference between the cis and trans conformations].

The use of a summation of dihedral terms for a single torsion angle, a Fourier series, greatly enhances the flexibility of the dihedral 3 term, allowing for more accurate reproduction of experimental and QM energetic target data Equation (3) describes the external or nonbond interaction terms. These term may be considered the most important of the energy terms for computational studies of biological systems. This is because of the strong influence of the environment on the properties of macromolecules as well as the large number of nonbond interactions that occur in biological molecules themselves (*e.g.*, hydrogen bonds between Watson-Crick base pairs in DNA, peptide bond-peptide bond hydrogen bonds involved in the secondary structures of proteins, and dispersion interactions between the aliphatic portions of lipids that occur in membranes).

Although the proper treatment of nonbond interactions is essential for successful biomolecular computations, it has been shown that the mathematical model required to treat these terms accurately can be relatively simple. Parameters associated with the external terms are the well depth, ε_{ij}, between atoms *i* and *j*; the minimum interaction radius, $R_{min,ij}$; and the partial atomic charge, q_i. Also included is the dielectric constant, ε_D, which is generally treated as equal to 1, the permittivity of vacuum, although exceptions do exist.

The term in square brackets in Eq. (3) is used to treat the van der Waals (VDW) interactions. The particular form in Eq. (3) is referred to as the Lennard-Jones (LJ) 6-12 term. The $1/r^{12}$ term represents the exchange repulsion between atoms associated with overlap of the electron clouds of the individual atoms (*i.e.*, the pauli exclusion principle). The strong distance dependence of the repulsion is indicated by the 12th power of this term. Representing London's dispersion interactions or instantaneous dipole-induced dipole interactions is the $1/r^6$ term, which is negative, indicating its favorable nature. In the LJ 6-12 equation there are two parameters; The well depth, ε_{ij}, indicates the magnitude of the favorable London's dispersion interactions between two atoms *r, j;* and $R_{min,\ ij}$ is the distance between atom *i* and *j* at which the minimum LJ energy occurs and is related to the VDW radius of an atom integration. Typically, ε_{ij} and $R_{min,i}$ are not determind for every possible interaction pair, *i, j;* but reather e_i and $R_{mir,\ i}$ parameters are determined for the individual atom types (*e.g.*, sp^2 carbon versus sp^3 carbon) and then combining rules are used to create the *ij* cross terms. These combining rules are generally quite simple, being either the arithmetic mean [*i.e.*, $R_{min,ij} = (R_{min,i} + R_{min,j})/2$] or the geometric mean [*i.e.*, $\varepsilon_{ij} = (\varepsilon_i\ \varepsilon_j)^{1/2}$]. The use of combining rules greatly simplifies the determination of the ε_i and $R_{min,i}$ parameters.

In special case the use of combining rules can be supplemented by specific *i, j* LJ parameters, referred to as off-diagonal terms, to treat interactions between specific atom types that are poorly modeled by the use of combining rules. The final term contributing to the external interactions is the electrostatic or Coulombic term. This term involves the interaction between partial atomic charges, q_i and q_j on atoms *i* and *j* divided by the distance, r_{ij}, between those atoms with he appropriate dielectric constant taken into account.

The use of a charge representation for the individual atoms, or monopoles, effectively includes all higher order electronic interactions, such as those between dipoles and quadrupoles. Combined, the Lennard-Jones and Coulombic interactions have been shown to produce a very accurate representation of the interaction between molecules, including both the distance and angle dependencies of hydrogen bonds once the 3D structure of a molecule and all the parameters required for the atomic and molecular connectivities are known, the energy of the system can be calculated via Eqs. (1)-(3). First derivatives of the energy with respect to position allow for determination of the forces acting on the atoms, information that is used in the energy minimization or MD simulations. Second derivatives of the energy with respect to position can be used to calculate forces constants acting on atoms, allowing the determination of vibrational spectra via normal mode analysis.

Extended-Atom Models

Always a limiting factor in computational studies of biological molecules is the ability to treat systems of adequate size for the required amount of simulation time or number of conformation to be sampled. One method to minimize the size of the system is to use extended-atom models versus all-atom models. In extended-atom models he hydrogens are not explicitly represented but rather are treated as part of the nonhydrogen atom to which they are covalently bound. For example, an all-atom model would treat a methyl group as four individual atoms (a carbon and three hydrogens), whereas in an extended atom model the methyl group would be treated as a single atom, with the LJ parameters and charges adjusted to account for the omission of the hydrogens.

Although this approach could be applied for all hydrogens it was typically used only for nonpolar (aliphatic and aromatic) hydrogens; polar hydrogens important for hydrogen bonding interactions were treated explicitly. Extended-atom models were most widely applied for the simulation of proteins in vacuum, where the large number of nonpolar hydrogens yields a significant decrease in the number of atoms compared to all-atom models. However, as more simulations were performed with explicit solvent representation, making the proportion of nonpolar hydrogens in the system much smaller, with ever-increasing computer resources the use of extended-atom models in simulations has decreased. Extended-atom models, however, are still useful for applications where a large sampling of conformational space is required.

Furthering the Potential Energy

The potential energy function presented in Eqs. (2) and (3) represents the minimal mathematical model that can be used for computational studies of biological systems. Currently, the most widely used energy functions are those included with the CHARMM AMBER, and GROMOS programs. Two extensions beyond the terms in Eqs. (2) and (3) are often used to treat out-of-plane distortions. Such as those that occur with aromatic hydrogens (*i.e.*, Wilson wags). Historically, the improper term was also used to maintain the proper chirality in extended-atom models of proteins (*e.g.*, without the H_α hydrogen, the chirality of amino acids is undefined).

Some force fields also contain a Urey-Bradly term that treats 1, 3 atoms (the two terminal atoms in an angle;) with a harmonic bond-stretching term in order to more accurately model vibrational spectra. Beyond the extensions mentioned in the previous paragraph, a variety of terms are included in force fields used for the modeling of small molecules that can also be applied to biological systems. These types of force fields are often referred to as Class II force fields, to distinguish then from the class I force fields such as AMBER, CHARMM, and GROMOS discussed above. For example, the bond term in Eq. (2) can be expanded to include cubic and quartic terms, which will more accurately treat the anharmonicity associated with bond stretching.

Another extension is the addition of cross terms that express the influence that stretching of a bond has on the stretching of an adjacent bond. Cross terms may also be used between the different types of terms such as bond angle or dihedral angle terms, allowing for the influence of bond length on angle bending or of angle bending on dihedral rotations, respectively, to be more accurately modeled. Extensions may also be made to the interaction portion of the force field [Eq. (3)]. These may include terms for electronic polarizability (see below) or the use of $1/r^4$ terms to treat ion-dipole interactions associated with interactions between, for example, ions and the peptide backbone. In all case the extension of a potential energy function should, in principle, allow for the system of interest to be modeled with more accuracy. The gains associated with the additional terms, however, are often significant only in specific cases (*e.g.*, the use of a $1/r^4$ term in the study of specific cation-peptide interactions), making their inclusion for the majority of calculations on biochemical systems unwarranted, especially when those terms increase the demand on computational resources.

Substitutes

The form of the potential energy function in Eqs. (1)-(3) was developed based on a combination of simplicity with required accuracy. However, a number of other forms can be used to treat the different terms in Eqs. (2) and (3). One alternative form used to treat the bond is referred to as the More potential. This term allows for bond-breaking events to occur and includes anharmonicity in the bond-stretching surface near the equilibrium value. The ability to break bonds, however, leads to forces close to zero at large bond distances. Which may present a problem when crude modeling techniques are used to generate structures.

A number of variations in the form of the equation to treat the VDW interactions have been applied. The $1/r^{12}$ term used for modeling exchange repulsion overestimates the distance dependence of the repulsive wall, leading to the use of an $1/r^9$ term or exponential repulsive terms. A more recent variation is the buffered 14-7 form which was relected because of its ability to reproduce interactions between rare gas atoms. Concerning electrostatic interactions, the majority of potential energy functions employ the standard coulombic term shown in Eq. (3), with one variation being the use of bond dipoles rather than atom-centered partial atomic charges. As with the extensions to the force fields discussed above, the alternative forms discussed in this paragraph generally do not yield significant gains in accuracy for biomolecular simulations performed in condensed phase environments at room temperature. Although for specific situations they may.

FROM POTENTIAL ENERGY TO FORCE FIELDS

Equations (1)-(3) in combination are a potential energy function that is representative of those commonly used in biomolecular simulations. As discussed above the form of this equations, his adequate to treat the physical interactions that occur in biological systems. The accuracy of that treatment, however, is dictated by the parameters used in the potential energy function, and it is the combination of the potential energy function and the parameters that comprises a force field. In the remainder of this chapter we describe various aspects of force fields including their derivation (i.e., optimization of the parameters), those widely available, and their applicability.

Overview of Available Force Fields

Currently there variety of force fields that may, in principle, be used for computational studies of biological systems. Of these force fields however, only a subset have been designed specifically for biomolecular simulation. As discussed above, the majority of biomolecular simulations are performed with the *CHARMM*, *AMBER*, and *GROMOS* packages. Recent publication of new *CHARMM* and *AMBER* force fields allows for these to be discussed in detail. Although the forms of the potential energy functions in *CHARMM* and *AMBER* are similar, with *CHARMM* including the additional improper and Urey-Bradley terms (see above). Significant philosophical and parameter optimization differences exist.

The latest versions of both force fields are all atom representations, although extended-atom representations are available to date, a number of simulation studies have been performed on nucleic acids and proteins using both *AMBER* and *CHARMM*. A direct comparison of crystal simulations of bovine pancreatic trypsin inhibitor show that the two force fields behave similarly, although differences in solvent-protein interactions are evident side-by-side tests have also been performed on a DNA duplex, showing both force fields to be in reasonable agreement with experiment although significant, and different, problems were evident in both cases. It should be noted that as of the writing of this chapter revised versions of both the *AMBER* and *CHARMM* nucleic acid force fields had become available.

Several simulation of membranes have been performed with the *CHARMM* force field for both saturated and unsaturated lipids. The availability of both protein and nucleic acid parameters in *AMBER* and *CHARMM* allows for protein-lipid and DNA-lipid simulations can also be performed with *CHARMM*. A number of more general force fields for the study of small molecules are available that can be extended to biological molecules. These force fields have been designed with the goal of being able to treat a wide variety of molecules, based on the ability to transfer parameters between chemical systems and the use of additional terms (*e.g.* cross term in their potential energy functions. Typically, these fields have been optimized to treat small molecules in the gas phase, although exceptions do exist. Such force fields may also be used for biological simulations; however, the lack of emphasis on properly treating biological systems generally makes them inferior to those discussed in the previous paragraphs. The optimized potential for liquid simulation (OPLS) force fiels was initially developed for liquid and hydration simulations on a variety of organic compounds. This force field has been extended to proteins nucleic acid bases, and carbohydrates, although its widespread use has not occurred. Some of the most widely used force fields for organic molecules are MM3 and its predecessors. An MM3 force field for proteins has been reported; however, it too has not been widely applied to date. The consistent force field (CFF) series of force fields have also been developed to treat a wide selection of small molecules and include parameters for peptides. However, those parameters were developed primarily on the basis of optimization of the internal terms. A recent extension of CFF, *COMPASS*, has been published that concentrates on producing a force field suitable for condensed phase simulations, although no field to which significant effort was devoted to allow for its application to a wide variety of compounds is the Merck Molecular Force Field (MMFF), During the development of MMFF, a significant effort was placed on optimizing the internal parameters to yield good geometries and energetics of small compounds as well as the accurate treatment of nonbonded interactions. This force field has been shown to be well behaved in condensed phase simulations of proteins; however, the results appear to be inferior to those of the *AMBER* and *CHARMM* models.

Two other force fields of note are UFF and *DREIDING* These force fields were developed to treat a much wider variety of molecules, including inorganic compounds, than the force fields mentioned previously, although their application to biological systems has not been widespread. It should also be noted that a force field for a wide variety of small molecules, *CHARM* (note the small "m," indicating the commercial version of the program and parameters), is available and has been applied to protein simulations with limited success. Efforts are currently under way to extend the *CHARMm* small molecule force field to make the nonbonded parameters consistent with those of the *CHARMM* force fields, thereby allowing for a variety of small molecules to be included in computational studies of biological systems. Although the list of force fields discussed in this subsection is by no means complete, it does emphasize the wide variety of force fields that are available for different types of chemical systems as well as differences in their development and optimization.

Free Energy Force Fields

All of the force fields discussed in the preceding sections are based on potential energy functions. To obtain free energy information when using these force fields, statistical mechanical ensembles must be obtained via various simulation techniques. An alternative approach is to use a force field that has been optimized to reproducer free energies directly rather than potential energies. For example, a given set of dihedral parameters in a potential energy function may be adjusted to reproduce a QM-determined torsional potential energy surface for a selected model compound. In the case of a free energy force field, the dihedral parameters would be optimized to reproduce the experimentally observed probability distribution of that dihedral in solution. Because the experimentally determined probability distribution corresponds to a free energy surface, a dihedral energy surface calculated using this force field would correspond to the free energy surface in solution. This allows for calculations to be performed in vacuum while yielding results that. In principle, correspond to the free energy in solution. The best Known of the free energy force fields is the Empirical conformational Energy Program for peptides (ECEPP) ECEPP parametrs (both internal and external) were derived primarily on the basis of crystal structures of a wide variety of peptides. Such an approach yields significant savings in computational costs when sampling large number of conformations; however, microscopic details of the role of solvent on the biological molecules are lost.

This type of approach is useful for the study of protein folding as well as protein-protein or protein-ligand interactions an alternative to obtaining free energy information is the use of potential energy functions combined with methods to calculate the contribution of the free energy of solvation. Examples include methods based on the solvent accessibilities of atoms continuum electrostatics-based models and the generalized Born equation. With some of these approaches the availiability of analytical derivatives allows for their use in MD simulation; however, they are generally most useful for determining solvation contributions associated with previously generated conformations.

Practical Utilization

Clearly, the wide variety for force fields requires the user to carefully consider those that are available and choose that which is most appropriate for his or her particular application. Most important in this selection process is a knowledge of the information to be obtained from the computational study. If atomic details of specific interactions are required, then all-atom models with the explicit inclusion of solvent will be necessary. For example, experimental results indicate that a single point mutation in a protein increases its stability. Application of an all-atom model with explicit solvent inMD simulations would allow for atomic details of interaction of the two side chains with the environment to be understood, allowing for more detailed interpretation of the experimental data.

The use of free energy perturbation techniques would allow for more quantitative data to be obtained from the calculation, although this approach requires proper treatment of the unfolded states of the proteins, which is difficult. In other cases, a more simplified model, such as an extended-atom force field with the solvent treated implicitly via the use of an R-dependent dielectric constant, may be appropriate. Examples include cases in which sampling of a large number of conformation of a protein or peptide is required. In these case the use of the free energy force fields may be useful. Another example is a situation in which the interaction of a number of small molecules with a macromolecule is to be investigated. In such a case it may be appropriate to treat both the small molecules and the macromolecule with one of the small-molecule based force fields, although the quality of the treatment of the macromolecule may be sacrificed. In these cases the reader is advised against using one force field for the macromolecule and a second, unrelated, force field for the small molecules. There are often significant differences in the assumptions made when the parameters were being developed that would lead to a severe unbalance between the energetics and forces dictating the individual macromolecule and small molecule structures and the interactions between those molecules.

If possible, the user should select a model system related to the particular application for which extensive experimental data are available. Tests of different force fields (and programs) can then be performed to see which best reproduces the experimental data for the model system and would therefore be the most appropriate for the application.

EVOLUTION OF EMPIRICAL FORCE FIELDS

As emphasized by the word "empirical" to describe the force fields used for biomolecular computations, the development of these force fields is largely based on the methods and target data used to optimize the parameters in the force field. Decisions concerning these methods and target data are strongly dependent on the force field developer. To a large extent, even the selection of the form of the potential energy function itself is empirical, based on considerations of what terms are and are not required to obtain satisfactory results. Accordingly, the philosophy, or assumptions, used in the development of a force field will dictate both its applicability and its quality. A brief discussion of some of the philosophical considerations behind the most commonly used force fields follows.

Philosophical Views

Step 1 in the development of a force field is a decision concerning its applicability and transferability. The applicability discussed already and can be separated, on one level, into force fields for biological molecules and those for small molecules. Applicability also includes the use of explicit solvent representations (*i.e.,* the solvent molecules themselves are included in the simulations), implicit solvent models, *i.e.,* the solvent is included in a simplified, continuum-based fashion, the simplest being the use of a dielectric constant of 78 (for water) versus 1 (for vacuum)], or free energy based force fields.

Transferability is concerned with the ability to take parameters optimized for a given set target data and apply them to compounds not included in the target data. For example, dihedral parameters about a C—C single bond may be optimized with respect to the rotational energy surface of ethane. In a transferable force field those parameters would then be applied for calculations on butane. In a nontransferable force fields, the parameters for the C—C—C—C–– and C—C—C—H dihedrals not in ethane would be optimized specifically by using target data on butane. Obviously, the definition of transferability is somewhat ambiguous, and the extent to which parameters can be transferred is associated with chemical similarity. However, because of the simplicity of empirical force fields, transferability must be treated with care. Force fields for small molecules are generally considered transferable, the transferability being attained by the use of various cross terms in the potential energy function.

Typically, a set of model compounds representing a type of functional group (e.g., azo compounds or bicarbonates) is selected. Parameters corresponding to the functional group are then optimized to reproduce the available target data for the selected model compounds. Those parameters are then transferred to new compounds that contain that functional group but for which unique chemical connectivities are present. A recent comparison of several of the small-molecule force fields discussed above has shown this approach to yield reasonable results for conformational energies; however, in all cases examples exist of catastrophic failures. Such failures emphasize the importance of user awareness when a force field is being applied to a novel chemical system. This awareness includes an understanding of the range of functional groups used in the optimization of the force field and the relationship of the novel chemical systems to those functional groups.

The more dissimilar the novel compound and the compounds included in the target data, the less confidence the user should have in the obtained results. This is also true in the case of bifunctional compounds, where the physical properties of the first functional group could significantly change those of the second group and vice versa. In such cases it is recommended that some tests of the force field be performed via comparison with QM data. Of the biomolecular force fields, AMBER is considered to be transferable, whereas academic CHARMM is not transferable. Considering the simplistic form of the potential energy functions used in these force fields, the extent of transferability should be considered to be minimal, as has been shown recently. As stated above, the user should perform suitable tests on any novel compounds to ensure that the force field is treating the systems of interest with sufficient accuracy. Another important applicability decision is whether the force field will be used for gas-phase (i.e., vacuum) or condensed phase (e.g., in solution, in a membrane, or in the crystal environment) computation.

Owing to a combination of limitations associated with available condensed phase data and computational resources, the majority of force fields prior to 1990 were designed for gas-phase calculations. With small-molecule force fields this resulted in relatively little emphasis being placed on the accurate treatment of the external interaction terms in the force fields. In the case of the biomolecular force fields designed to be used in vacuum via implicit treatment of the solvent environment, such as the CHARMM param 19 and AMBER force fields, care was taken in the optimization of charges to be consistent with the use of an R-dependent dielectric constant. The first concerted effort to rigorously model condensed phase properties was with the OPLS force field. Those efforts were based on the explicit use of pure solvent and aqueous phase computations to calculate experimentally accessible thermodynamic properties.

The external parameters were then optimized to maximize the agreement between the calculated and experimental thermodynamic properties. This very successful approach is the basis for the optimization procedures used in the majority of force fields currently being developed and used for condensed phase simulations. Although while a number of additional philosophical considerations with respect to force fields could be discussed, presentation of parameter optimization methods in the remainder of this section will include philosophical considerations.

It is worth reemphasizing the empirical nature of force fields, which leads to the creators of different ones having a significant impact on the quality of the resulting force field even when exactly the same form of potential energy function is being used. This is in large part due to the extensive nature of parameter space. Because of the large number of different individual parameters in a force field, an extensive amount of correlation exists between those parameters. Thus a number of different combinations of parameters could reproduce a given set of target data. Although additional target data can partially overcome this problem, it cannot eliminate it, making the parameter optimization approach central to the ultimate quality of the force field. It should be emphasized that even though efforts have been made to automate parametrization procedures, a significant amount of manual intervention is generally required during parameter optimization.

Method of Optimization

Knowledge of the approaches and target data used in the optimization of an empirical force field aids in the selection of the appropriate force field for a given study and acts as the basis for extending a force field to allow for its use with new compounds (see below). Presented in Table 2.1 is a list of the parameters in Eqs. (2) and (3) and the type of target data used for their optimization. The information in Table 1 is separated into categories associated with those parameters. It should be noted that separation into the different categories represents a simplification; in practice there is extensive correlation between the different parameters, as discussed above; for example, changes in bond parameters that affect the geometry may also have an influence on $\Delta G_{solvation}$ for a given model compound.

Internal parameters are generally optimized with respect to the geometries, vibrational spectra, and conformational energetics of selected model compounds. The equilibrium bond lengths and angles and the dihedral multiplicity and phase are often optimized to reproduce gas-phase geometric data such as those obtained form QM, electron diffraction, or microwave experiments. Such data, however, may have limitations when they are used in the optimization of parameters for condensed phase simulation. For example, it has been shown that the internal geometry of N-methylacetamide (NMA), a model for the peptide bond in proteins, is significantly influenced by the environment. Therefore, a force field that is being developed for condensed phase simulations should be optimized to reproduce condensed phase geometries rather than gas-phase values. This is necessary because the form of the potential energy function does not allow for subtle changes in geometries and other phenomena that occur upon going from the gas phase to the condensed phase to be reproduced by the force field.

Table 2.1. Types and sources of Target Data Used in the Optimization of Empirical Force Field parameters.

Term	*Target data*	*Source*
Internal		
Equilibrium terms, multiplicity, and phase (b_0, θ_0, n, δ)	Geometries	QM, electron diffraction, mi crowave, crastal survey
Force constants (K_b, K_β) K_χ)	Vibrational spectra, Conformational properties	QM, IR, Raman QM, IR, NMR, crystal survey
External		
VDW terms (ε_i, $R_{min, i}$)	Pure solvent properties [56] ($\Delta H_{vaporization,}$ molecular volume)	Vapor pressure, calorimetry, densities
	Crystal properties ($\Delta H_{sublimation}$ [56] latic pa rameters, non-bond distances)	X-ray and neutron diffraction, vapor pressure, calorimetry
	Interaction energies (dimers, rare gas-model compound, water-model compound)	QM, microwave, mass spectrometry
Atomic charges (q_i)	Dipole moments [57]	QM, dielectric permittivity, Stark effect, microwave
	Electrostatic potentials	QM
	Interaction energies (dimers, water-model compound)	QM, microwave, mass spectro metry
	Aqueous solution ($\Delta G_{solvation}$, $\Delta H_{solvation}$, partial molar volume [58])	Calorimetry, volume variations

QM = quantum mechanics; IR = infrared spectroscopy.

The use of geometric data form a survey of the Cambridge crystal Database (CSD) can be useful in this regard. Geometries from individual crystal structures can be influenced by non-bond interactions in the crystal, especially when ions are present. Use of geometric data from a survey overcomes this limitation by averaging over a large number of crystal structures, yielding condensed phase geometric data that are not biased by interactions specific to a single crystal. Finally, QM calculations can be performed in the presence of water molecules or with a reaction filed model to test whether condensed phase may have an influence on the obtained geometries optimization of the internal force constants typically uses vibrational spectra and conformational energetics as the primary target data.

Vibrational spectra, which comprise the individual frequencies and their assignments, dominate the optimization of the bond and angle force constants. It must be emphasized that both the frequencies and assignments should be accurately reproduced by the force field to ensure that the proper molecular distortions are associated with the correct frequencies. To attain this goal it is important to have proper assignments from the experimental data, often based on isotopic substitution. One way to supplement the assignment data is to use QM-calculated spectra from which detailed assignments in the form of potential energy distributions (PED_S) can be obtained. Once the frequencies and their assignments are known, the force constants can be adjusted to reproduce these values. It should be noted that selected dihedral force constants will be optimized to reproduced conformational energetics, often at the expense of sacrificing the quality of the vibrational spectra. For example, with ethane it is necessary to overestimate the frequency of the C—C torsional rotation in order to accurately reproduce the barrier to rotation. This discrepancy emphasizes the need to take into account barrier heights as well as the relative conformational energies of minima, especially in cases when the force field is to be used in MD simulation studies where there is a significant probability of sampling regions of conformational surfaces with relatively high energies.

As discussed with respect to geometries, the environment can have a significant influence on both the vibrational spectra and the conformational energetics. Examples include the vibrational spectra of NMA [20] and the conformational energetics of dimethylphosphate, a model compound used for the parametrization of oligonucleotides. Increasing the size of the model compound used to generate the target data may also influence the final parameters. An example of this is the use of the alanine dipeptide to model the protein backbone versus a larger compound such as the alanine tetrapeptide optimization of external parameters tends to be more difficult as the quantity of the target data is decreased relative to the number of parameters to be optimized compared to the internal parameters, leaving the solution more undetermined. This increased the problems associated with parameter correlation, thereby limiting the ability to apply automated parameter optimization algorithms. An example of the parameter correlation problem with van der Waals parameters is presented in Table 2.2, where pure solvent properties for ethane using three different sets of parameters are presented (AD MacKerell Jr, M Karplus, unpublished work), As may be seen, all three sets of LJ parameters presented in Table 2.2 yield heats of vaporization and molecular volumes in satisfactory agreement with the experimental data, in spite of the carbon R_{min} varying by over 0.5 A among the three sets.

The presence of parameter correlation is evident. As the carbon R_{min} increases and ε values decrease, the hydrogen R_{min} decreases and ε values increase. Thus, it is clear that special care needs to be taken during the optimization of the non-bond parameters to maximize agreement with experimental data while minimizing parameter correlation. Such efforts will yield a force field that is of the highest accuracy based on the most physically reasonable parameters. Van der Waals or Lennard-Jones contributions to empirical force fields are generally considered to be of less importance than the electrostatic term in contributing to the non-bond interactions in biological molecules. This view, however, is not totally warranted. Studies have shown significant contributions from the VDW term to heats of vaporization of polar-neutral compounds, including over 50% of the mean interaction energies in liquid NMA as well as in crystals of nucleic acid bases, where the VDW energy contributed between 52% and 65% of the mean interaction energies. Furthermore, recent studies on alkanes have shown that VDW parameters have a significant impact on their calculated free energies of solvation. Thus, proper optimization of VDW parameters is essential to the quality of a force field for condensed phase simulation of biomolecules. Significant progress in the optimization of VDW parameters was associated with the development of the OPLS force field. In those efforts the approach of using Monte Carlo calculation on pure solvents to compute heats of vaporization and molecular volumes and then using that information to refine the VDW parameters was first developed and applied. Subsequently, developers of other force fields have used this same approach for optimization of biomolecular force fields.

Van der Waals parameters may also be optimized based on calculated heats of sublimation of crystals, as has been done for the optimization of some of the VDW parameters in the nucleic acid bases. Alternative approaches to optimizing VDW parameters have been based primarily on the use of QM data. Quantum mechanical data contains detailed information on the electron distribution around a molecule, which , in principle, should be useful for the optimization of VDW parametrs. In practice, however, limitations in the ability of QM approaches to accurately treat dispersion interactions make VDW parameters derived solely from QM data yield condensed phase properties in poor agreement with experiment Recent work has combined the reproduction of experimental properties with QM data to optimize VDW parameters while minimizing problems associated with parameter correlation. In that study QM data for helium and neon atoms interacting with alkanes were used to obtain the relative values of the VDW parameters while the reproduction of pure solvent properties was used to determine their absolute values, yielding good agreement for both pure solvent properties and free energies of aqueous solvation.

Table 2.2. Ethane Experimental and Calculated Pure Solvent Properties[a]

Lennard Jones Parameters[b]			
Carbon	*Hydrogen*	*Heat of vaporization[c]*	*Molecular volume*
3.60/0.190	3.02/0.0085	3.50	90.7
4.00/0.080	2.71/0.0230	3.48	90.9
4.12/0.080	2.64/0.0220	3.49	91.8
Experiment		3.56	91.5

[a] Calculations performed using MC BOSS [66] with the CHARMM combination rules. Partial atomic charges (c = - 0.27 and H = 0.09) were identical for all three simulations.

[b] Lennard-Jones parameters are $R_{min/\varepsilon}$ in angstroms and kilocalories per mole, respectively.

[c] Heat of vaporization in kilocalories per mole and molecule volume in cubic angstroms at -89^0C [56].

The reproduction of both experimental pure solvent and free energies of aqueous solvation has also been used to derive improved parameters. From these studies it is evident that optimization of the VDW parameters is one of the most difficult aspects of force field optimization but also of significant importance for producing well-behaved force fields. Development of models to treat electrostatic interactions between molecules represent one of the most central, and best studies areas in force field development. For biological molecules, the computational limitations discussed above have led to the use of the coulombic model included in Eq. (3). Despite its simplistic form, the volume of work done on the optimization of partial atomic charges, as well as the appropriate dielectric constant, has been huge.

The present discussion is limited to currently applied approaches to the optimization of partial atomic charges. These approaches are all dominated by the reproduction of target data from QM calculations, although the target data can be supplemented with experimental data on interaction energies and orientations and molecular dipole moments when such data are available. Method 1 is based on optimizing partial atomic charges to reproduce the electrostatic potential (ESP) around a molecule determined via QM calculations. Programs are available to perform this operation and some of these methodologies have been incorporated into the *GAUSSIAN* suite of programs. A variation of the method, in which the charges on atoms with minimal solvent accessibility are restrained, termed *RESP* has been developed and is the basis for the partial atomic charges used in the 1995 *AMBER* force field. The goal of the ESP approach is to produce partial atomic charges that reproduce the electrostatic field created by the molecule.

The limitation of this approach is that the polarization effect associated with the condensed phase environment is not explicitly included, although the tendency for the HF/6-31G* QM level of theory to overestimate dipole moments has been suggested to account for this deficiency. In addition, experimental dipole moments can be included in the charge-fitting procedure. An alternative method, used in the *OPLS*, *MMFF*, and *CHARMM* force fields, is to base the partial atomic charges on the reproduction of minimum interaction energies and distances between small-molecule dimmers and small molecule-water interacting pairs determined form QM calculations. In this approach a series of small molecule-water (monohydrate) complexes are subjected to QM calculations for different idealized interactions. The resulting minimum interaction energies and geometries, along with available dipole moments, are then used as the target data for the optimization of the partial atomic charges.

Application of this approach in combination with pure solvent and aqueous solvent simulations has yielded offsets and scale factors that allow for the production of charges that yield reasonable condensed phase properties. Advantages of this method are that the use of the monohydrates in the QM calculations allows for local electronic polarization to occur at the different interacting sites, and the use of the scale factors accounts for the multibody electronic polarization contributions that are not included explicitly in Eq. (3). As for the dielectric constant, when explicit solvent molecules are included in the calculation, a value of 1, as in vacuum, should be used because the solvent molecules themselves will perform the charge screening.

The omission of explicit solvent molecules can be partially accounted for by the use of an R-dependent dielectric, where the dielectric constant increases as the distance between the atoms, r_{ij}, increases (e.g., at a separation of 1 Å the dielectric constant equals 1; at a 3 Å separation the dielectric equals 3; and so on). Alternatives include sigmoidal dielectrics; however, their use has not been widespread. In any case, it is important that the dielectric constant used for a computation correspond to that for which the force field being used was designed; use of alternative dielectric constants will lead to improper weighting of the different electrostatic interaction, which may lead to significant errors in the computations.

Interactive Importance

Proper condensed phase simulations require that the non-bond interactions between different portions of the system under study be properly balanced. In bimolecular simulations this balance must occur between the solvent-solvent (e.g., water-water), solvent-solute (e.g., water-protein), and solute-solute (e.g., protein intramolecular) interactions Having such a balance is essential for proper partitioning of molecules or parts of molecules in different environments. For example, if the solvent-solute interaction of a glutamine side chain were over estimated, there would be a tendency for the side chain to move into and interact with the solvent.

The first step in obtaining this balance is the treatment of the solvent-solvent interactions. The majority of biomolecular simulation are preformed using the TIP3P and SPC/E water models. The SPC/E water model is known to yield better pure solvent properties than the TIP3P model; however, this has been achieved by overestimating the water-dimer interaction energy (*i.e.*, the solvent-solvent interactions are too favorable).

Although this overestimation is justifiable considering the omission of explicit electronic polarizability from the force field, it will cause problems when trying to produce a balanced force field due to the need to overestimate the solute-solvent and solute-solute interaction energies in a compensatory fashion. Owing to this limitation, the TIP3P model is suggested to be a better choice for the development of a balanced force field. It is expected that water models that include electronic polarization will allow for better pure solvent properties while having the proper solvent-solvent interactions to allow for the development of balanced force fields. It is important when applying a force field to use the water model for which that particular force filed was developed and tested. Furthermore, extensions of the selected force field must maintain compatibility with the originally selected water model.

Target Data

Throughout this chapter and in Table 2.1 the inclusion of QM results as target data is evident, with the use of such data in the optimization of empirical forces fields leading to many improvements. Use of QM data alone, however, is insufficient for the optimization of parametrs for condensed phase simulations. This is due to limitations in the ability to perform QM calculations at an adequate level combined with limitations in empirical force fields. As discussed above, QM data are insufficient for the treatment of dispersion interactions, disallowing their use alone for the optimization of van der Waals paramenters. The use of HF/6-31G-calculated intermolecular interaction energies for the optimization of partial atomic charges has been successful because of extensive testing of the ability of optimized charges to reproduce experimentally condensed phase values, thereby allowing for the appropriate offsets and scaling factors to be determined. In many cases, results from QM calculations are the only data available for the determination of conformational energetics.

There is a need for caution in using such data alone, as evidenced by recent work showing that the rigorous reproduction of QM energetic data for the alanine dipeptide leads to systematic variations in the conformation of the peptide backbone when applied to MD simulations of proteins. Furthermore, QM data are typically obtained in the gas phase, and as discussed above, significant changes in geometries, vibrations, and conformational energetics can occur in going from the gas phase to the condensed phase. Although the ideal potential energy function would properly model differences between the gas and condensed phases, this has yet to be realized. Thus, the use of QM results as target data for the optimization of force fields must include checks against experimentally accessible data whenever possible to ensure that parameters appropriate for the condensed phase are being produced.

Application to Charmm

Selection of a force field is often based on the molecules of interest being treated by a particular force field. Although many of the force fields discussed above cover a wide range of functionalities, they may not be of the accuracy required for a particular study. For example, if a detailed atomistic picture or quantitative data are required on the binding of a series of structurally similar compounds to a protein, the use of a general force fields may not be appropriate. In such cases it may be necessary to extend one of the force fields refined for biomolecular simulations to be able to treat the new molecules. When this is to be done, the optimization procedure must. be the same as that used for the development of the original force field. In the remainder of this chapter a systematic procedure to obtain and optimize new force field parameters is presented. Due to my familiarity with the CHARMM force field, this procedure is consistent with those parameters. An outline of the parametrization procedure is presented in Figure 2.2. A similar protocol for the AMBER force filed has been published and can be supplemented with information from the AMBER web page.

Selection of Model Compounds

Step 1 of the parametrization process is the selection of the appropriate model compounds. In the case of small molecules, such as compounds of pharmaceutical interest, the model compound may be the desired molecule itself. In other cases it is desirable to select several small model compounds that can then be "connected" to create the final, desired molecule. Model compounds should be selected for which adequate experimental data exist, as listed in Table 2.1. Since in almost all cases QM data can be substituted when experimental data are absent (see comments on the use of QM data, above), the model compounds should be of a size that is accessible to QM calculations using a level of theory no lower than HF/6-31G*. This ensures that geometries, vibrational spectra, conformational energetics, and model compound-water interaction energies can all be performed at a level of theory such that the data obtained are of high enough quality to accurately replace and supplement the experimental data. Finally, the model compounds should be of such a size that when they are connected to create the final molecule, QM calculations of at least the HF/3-21G level (though HF/6-31G is preferable) can be performed to test the linkage.

(1) Model compound selection
(2) Target data
(3) Topology creation and initial parameter selection
  Assign atom types
  Assign connectivity
  Assign partial atomic charges
  Assign initial parameters
(4) Parameter optimization
  Starting geometry

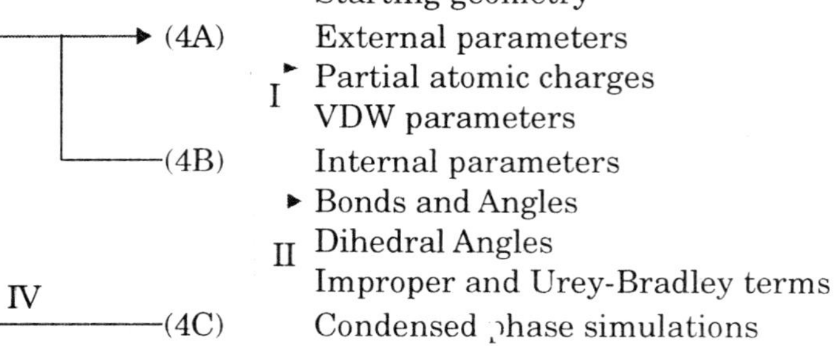

Fig. 2.2. Outline of the steps involved in the preparation of a force field for the inclusion of new molecules and optimization of the associated parameters. Iterative loops (I) over individual external terms. (II) over individual internal terms, (III) over the external and internal terms. In loop (IV) over the condensed phase simulation, both external terms and internal terms are included.

For illustration of the parametrization concepts, methotrexate, the dihydrofolate reductase inhibitor, was selected as a model system. Its structure is shown in Figure 2.3(a) Methotrexate itself is too large for QM calculations at a satisfactory level, requiring the use of smaller model compounds that represent the various parts of methotrexate.

Examples of model compounds that could be used for the parametrization of methotrexate are included as compounds 1-3 in Figure 2.3(*a*) which are, associated with the pteridine, benzene,

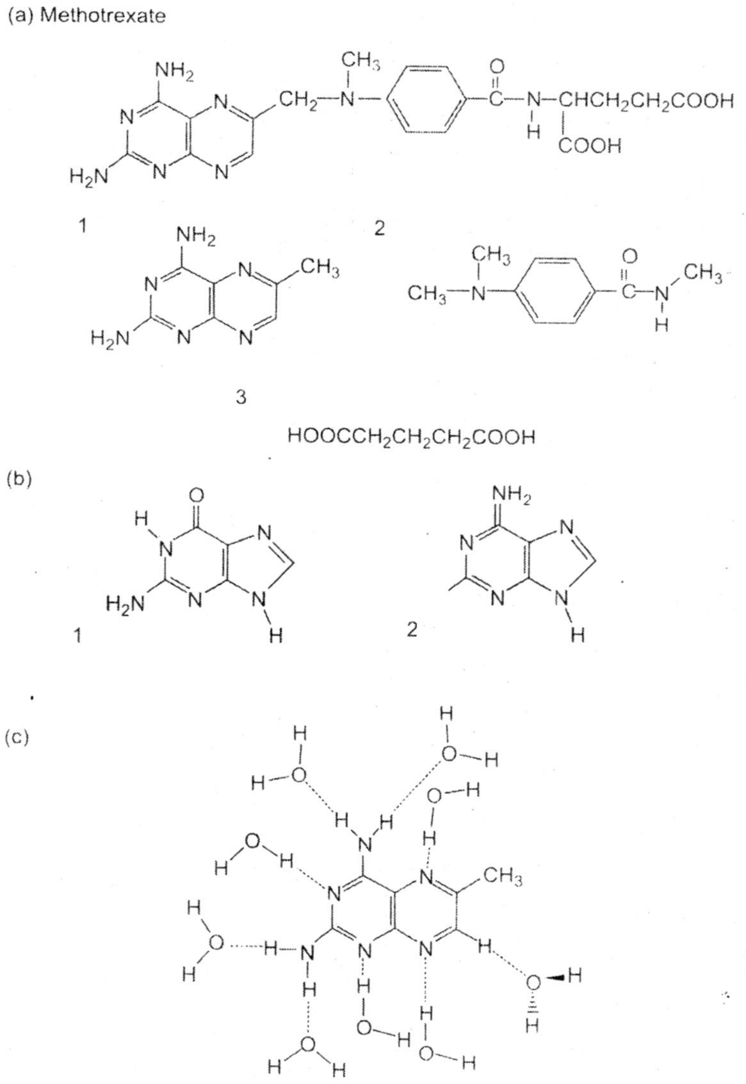

(a) Methotrexate

NH_2 CH_3 O N N CH_2—N— C—N—$CHCH_2CH_2COOH$ H COOH H_2N N N

1 NH_2 N N CH_3 H_2N N N

2 CH_3 O CH_3—N— C—N CH_3 H

3 $HOOCCH_2CH_2CH_2COOH$

(b) H O N N H_2N N N H 1

2 NH_2 N N N N H

(c) H H—O H O—H O—H H H N H O H H N N CH_3 H O H—N N N H O H H H H O—H H O H O—H H H H

Fig. 2.3. Structure of methotrexate and the structures of three model compound that could be used for parameter optimization of methotrexate. (*b*) The structures of (1) guanine and (2) adenine. (*c*) Interaction orientations between model compound of 1 (*a*) and water to be used in the optimization of the partial atomic charges. Note that in the optimozation procedure the water-model compound dimers are treated individually (*e.g.*, monohydrates).

and diacid moieties, respectively. It may be assumed that some experimental data would be available for the pteridine and diacid compounds and that information on the chemical connectivities internal to each compound could be obtained from a survey of the CSD each of these compounds is of such a size that HF/6-31G* calculations are accessible, and at least HF/3-21G calculations would be accessible to the dimers, as required to test the parameters connecting the individual model compounds. An alternative model compound would include the amino group with model 3, yielding glutamic acid; however, that would require breaking the amide bond on compound 2, which would cause the loss of some of the significant chemical characteristics of methotrexate. Of note is the use of capping methyl groups on compounds 1 and 2 with 1 the methyl group will ensure that the remainder of the molecule. The same is true in the case of model compound 2, although in this case the presence of the methyl groups is even more important; the properties of a primary amine, even in an amide, can be expected to differ significantly from those of the secondary amine present in methotrexate. Including the methyl cap ensures that the degree of substitution of the amine, or any other functional group, is the same in the model compound as in the final compound to be used in the calculations.

Target Data Identification

Simultaneous with the selection of the appropriate model compounds is the identification of the target data, because the availability of adequate target data in large part dictates the selection of the model compound. Included in Table 2.1 is a list of the various types of target data and their sources. Basically, the parameters for the new compounds will be optimized to reproduce the selected target data. Thus, the availability of more target data will allow the parameters to be optimized as accurately as possible while minimizing problems associated with parameter correlation, as discussed above. With respect to the types of target data, efforts should be made to identify as many experimental data as possible while at the same time being aware of possible limitations in those data (e.g., countertion contributions in IR spectra of ionic species).

The experimental data can be supplemented and extended with QM data; however, the QM data themselves are limited due to the level of theory used in the calculations as well as the fact that they are typically restricted to the gas phase. As discussed above, target data associated with the condensed phase will greatly facilitate the optimization of a force fields for condensed phase simulations.

Creation of Topology and Initial Parameter Selection

Once the model compounds are selected, the topology information (e.g., connectivity, atomic types, and preliminary partial atomic charges) must be input into the program and the necessary parameters supplied to perform the initial energy calculation. This is initiated by identifying molecules already present in the force field that. Closely mimic the model compound. In the case of model compound 1, in Figure 2.3a, the nucleic acid bases guanine and adenine, shown as compounds 1 and 2, respectively, in Figure 2.3b, would be reasonable starting points. Although going from a 5, 6 to a 6,6 fused ring system, the distribution of heteroatoms between the ring systems is similar and there are common amino substituents.

The initial information for model compound 1 would be taken from guanine (e.g., assign atomic types and atomic connectivity). To this an additional aromatic carbon would be added to the five-membered ring and the atomic types on the two carbons in the new six-membered ring would have to be switched to those corresponding to six-membered rings. For the methyl group, atomic types found on thymine would be used. Atomic types for the second amino group on model compound 1, which is a carbonyl in guanine, would be extracted from adenine. This would include information for both the second amino group and unprotonated ring nitrogen. Completion of the topology for compound 1 in Figure 2.3a would involve the creation of reasonable partial atomic charges. In one approach, the charges would be derived on the basis of analogy to those in guanine and adenine; with the charges on the new aromatic carbon and covalently bound hydrogen set equivalent and of opposite sign, the now methylated aromatic carbon would be set to a charge of zero and the methyl group charges would be assigned a total charge of zero (*e.g.*, C = –0.27, H = 0.09). Care must be taken at this stage that the total charge on the molecule is zero. Alternatively, charges from Mulliken population analysis of an HF/6-31G calculation could act as a starting point. Concerning the VDW parameters, assignment of the appropriate types of atoms to the model compound simultaneously assigns the VDW parameters. At this point the information required by *CHARMM* to create the molecule is present, but the parameters necessary to perform energy calculations are not all available yet.

In the case of *CHARMM*, the program is designed to report missing parameters when an energy calculation is requested. Taking advantage of this feature, missing parameters can be identified and added to the parameters file. The advantage of having the program identify the missing parameters is that only new parameters that are unique to your system will be added. It is these added parameters that will later be adjusted to improve the agreement between the empirical and target data properties for the model compound. Note that no parameters already present in the parameter file should be changed during the optimization procedure, because this would compromise the quality of the molecules that had previously been optimized. It is highly recommended that the use of wild cards to create the needed parameters be avoided, because it could compromise the ability to efficiently optimize the parameters.

Parameter Optimization

Empirical force field calculation during the optimization procedure should be performed in a fashion consistent with the final application of the force field. With recent developments in the Ewald method, particularly the particle mesh Ewald (PME) APPROACH [85], it is possible to perform simulations of biological molecules in the condensed phase with effectively no cutoff of the non-bond interactions. Traditionally, to save computational resources, no atom-atom non-bond interactions beyond a specified distance are included in the calculation; the use of PME makes this simplification unnecessary (i.e., distance based truncation of non-bond interactions).

All empirical calculations in the gas phase (e.g., water-model compound interactions, energy minimizations, torsional rotation surfaces) should be performed with no atom-atom truncation, and condensed phase calculation should be performed using PME. In addition, condensed phase calculation should also be used with a long-tail correction for the VDW interactions. Currently, such a correction is not present in CHARMM, although its implementation is in progress. Other consideration are the dielectric constant. Which should be set to 1 for all calculations, and the 1, 4 scaling factor which should also be set to 1.0 (no scaling). Initiation of the parameter optimization procedure requires that an initial geometry of the model compound be obtained. The source of this can be an experimental, modeled, or QM-determined structure. What is important is that the geometry used represent the global energy minima and that it be reasonably close to the final empirical geometry that will be obtained from the parameter optimization procedure.

External Parameters : The parameter optimization process is initiated with the external terms because of the significant influence those terms have on the final empirical geometries and conformational energetics. Since reasonable starting geometries can readily be assigned from an experimental or QM structure, the external parameters obtained from the initial round of parametrization can be expected to be close to the final values. Alternatively, starting the optimization procedures with the internal terms using very approximate external parameters could lead to extra iterations between the internal and external optimization procedures owing to possibly large changes in the geometries, vibrations, and conformational energetics when the external parameters were optimized during the first or second iteration. It must be emphasized that the external parameters are influenced by the internal terms such that iterations over the internal and external parameters are necessary.

Partial Atomic Charges

Determination of the partial atomic charges requires minimum interaction energies and geometries for individual water molecules interacting with different sites on the model compounds. An example of the different interaction orientations is shown in Figure 2.3c for compound 1, Figure 2.3a. As may be seen, idealized interactions with all the hydrogen bonding sites as well as nonpolar sites are investigated. Note that the procedure is carried out only on the individual monohydrates (*i.e.,* the model compound and one water molecule) and not on a system in which the model compound is interacting with multiple water molecules.

Typically to obtain the QM target data the model compound is geometrically optimized at the HF/6-31G level. Individual water molecules are then placed in idealized orientations and only the interaction distance and in some cases, an additional angle are optimized at the HF/6-31G level while the model compound and water intramolecular geometries are fixed; the water geometry is typically the TIP3P geometry. From this optimization the minimum interaction energy and distance are obtained. The interaction energy is determined on the basis of the difference between the total energy of the model compound-water supramolecular complex and the individual monomer energies. No correction for basis-set superposition error (BSSE) is required in the present approach. At this stage the QM interaction distances and energies are offset and scaled as follows to produce the final target data that will allow for the optimization of partial atomic charges that yield reasonable condensed phase properties.

The offsets and scalings are preformed as follows. The QM distances are decreased by 0.2 Å for polar-neutral interactions., by 0.1Å for hydrogen bonds involving charged species, and not offset for interaction between water and nonpolar sites. Scaling of the interaction energies by 1.16 is performed for all interaction involving polar-neutral compounds and no scaling is performed for the charged compounds. The 1.16 scaling factor is based on the ratio of the TIP3P to HF/6-31G water-dimer interaction energies. The overestimation of the TIP3P water interaction energy partially accounts for the omission of explicit polarizability in the force filed and use of the same 1.16 scaling factor maintains the balance between the solvent-solvent and solute-solvent interaction. In addition to the water-model compound interactions it may also be useful to perform calculations to obtain model compound-model compound interaction information. Such data are most useful for small model compound (*e.g.,* methanol or NMA) and special cases, such as the Watson-Crick base pairs, where specific types of interactions dominate the properties of the system. Once the target data are obtained, the partial atomic charges can then be optimized to reproduce the QM interaction offset distances and scaled energies. Along with reproduction of the QM interaction data the partial charge optimization can also include dipolemoments (magnitude as well. as direction), form either experiment or QM calculations, as target data.

For polar-neutral compound the empirical dipole moments are typically larger than the experimental or QM gas-phase values owing to the need to include electronic polarization effects implicitly in the charge distribution, though no well-defined scale factor is applied. When performing the charge optimization it is suggested that groups of atoms whose sum of partial atomic charges yields a unit charge be used. Typically unit charge groups of three to seven atoms are used; however, in the most recent version of the *CHARMM* nucleic acid parameters the unit charges were summed over the entire bases. The use of unit charge groups allows for the use of the group truncation option in *CHARMM* and simplifies the reorganization of the charge when combining model compounds into larger chemical entities. Aliphatic and aromatic partial atomic charge are a special case. In *CHARMM* all the aliphatic hydrogens are assigned a charge of 0.09, with the carbon charge adjusted to yield a unit charge.

For example, in methane, with hydrogen charges of 0.09, the carbon charge is –0.36; and in methanol, with aliphatic hydrogen charges of 0.09 and charges of –0.66 and 0.43 on the hydroxyl oxygen and hydrogen atoms, the charge on the aliphatic carbon is set to –0.04 to yield a total charge of 0.0. The use of the same charge on all the aliphatic hydrogens, which was set to 0.09 based on the electrostatic contribution to the butane torsional surface [88], is justified by both the free energy of solvation of alkanes and the pure solvent properties of ethane being insensitive to the charge distribution. Concerning the aromatic carbons and hydrogens, charges of 0.115 and -0.115, based on the condensed phase properties of benzene are used on all aromatic functional groups. The only exception are rings containing heteroatoms (*e.g., phridine),* where the carbon-hydrogen pairs adjacent to the heteroatom amy have different charges; these charges would be determined via interactions with water as presented above. At points of substitution of aromatic groups, the carbon charge is set to 0.0 except when substitued by a heteroatom (*e.g.,* tyrorine), where a charge that yields good interaction energies of the substituent with water is assigned to the ring carbon (*i.e..,* to the atom in tyrosine). Beyond its simplicity. The use of the same charges for alkanes and aromatic functional groups during the fitting procedure also allows charges to be systematically altered when small model compounds are connected. This is done by replacing an aliphatic or aromatic hydrogen with the new covalent bond required to connect the two model compounds and moving the charge previously on the hydrogen into the carbon to which it was attached. For example, to connect model compounds 1 and 2 in Figure 2.3a, a hydrogen from each of the terminal methyl groups would be removed, the charges on each of the methyl carbons increased by 0.09 and the carbon-carbon covalent bond created.

Van der Walls Parameters

The van der waals and Lennard-Jones (LJ) parameters are the most challenging of the parameters in Eqs.(2) and (3) to optimize. In the majority of cases, however, direct transfer of the VDW parameters based on analogy with atom already in the force field will yield satisfactory results. This is particularly true in cases where ligands that interact with biological molecules are being parametrized. In the case of methotrexate, all the atoms in the three model compounds are directly analogous to those in DNA (*e.g.,* the pteridine ring) or in proteins (*e.g.,* the benzene ring, the amide group, and the carboxylic acids). In cases where the condensed phase properties of the new molecules themselves are to be studied, formal optimization of the VDW parameters is required. This requires the identification of appropriate experimental condensed phase target data along with QM calculation on rare gas-model compound interaction.

Detailed presentations of the methods to perform such optimizations have been published elsewhere. Note that in cases where optimization of the VDW parameters is required, the majority of the VDW parameters can be transferred directly from the available parameters (*e.g.*, alkane functional groups or polar hydrogens) and only the VDW parameters of one or two atoms (*e.g.*, the sulfur and oxygen in dimethylsulfoxide) actually optimized. If the VDW parameters are included in the optimization procedure it is essential that the partial atomic charges be rechecked for agreement with the target data. If significant changes do occur, further optimization of the charges must be performed, followed by rechecking of the VDW parameters, and so on, in an iterative fashion.

(b) Internal Parameters : Molecular geometries are dominated by the bond and angle equilibrium terms and the dihedral multiplicity and phase, whereas the vibrational spectra and conformational energetics are controlled primarily by the internal term force constant, as presented in Table. 2.1 Typically, the vibrational spectra will be largely influenced by the bond and angle force constants while dihedral force constants requiring iteration over the different internal parameter optimization steps. Furthermore, the external parameters can have a significant influence on the geometries, vibrations, and energetics. Thus, any change in those parameters be rechecked with respect to the target data and additional optimization performed as required.

Bond, Angle and Dihedral Terms

Adjustment of the bond and angle equilibrium values to reproduce the geometric target data is generally straightforward. It should be noted, however, that the minimized structure generally does not have bond lengths and angles that corresponded directly to the equilibrium bond and angle parameters. This is due to the influence of the other terms in the potential energy function on the final geometry. In planar systems (*e.g.*, aromatic rings) the angles should sum to 360° for the three angles around a single atom, to 540° for the five endocyclic angles in a five-membered ring, and to 720° for the six endocyctic angles in a six membred ring to ensure that their is no angular strain at the planar minima.

Initial optimization of the dihedral phase and multiplicity should be performed by assigning only one dehydral parameter to each torsion angle and assigning that dihedral the appropriate values for the type of covalent bond (*e.g.*, an $sp^3 - sp^3$ bond would have a multiplicity of 3 and a phase of 0.0) Note that for nonplanar systems incorrect selection of the dihedral phase and multiplicity will often lead to minima with the wrong conformations. Force constant optimization is initially performed by reproducing experimental or QM vibrational spectra. Quantum mechanical frequencies generally have to be scaled to yield experimentally relevant values. This is best done by comparison with experimental data on the same molecule and calculation of the appropriate scale factor, however, if no experimental data are available, then published scale factors associated with different QM levels of theory should be applied.

As mentioned, already the parameters should be optimized to reproduce the assignments of the frequencies as well as their numerical values. This is best performed by producing a potential energy distribution (PED), where the contributions of the different normal modes (*e.g.*, symmetric methyl sketch, asymmetric stretch of water, and so on) are assigned to the individual frequencies. The module *MOLVIB* in *CHARMM* allows for calculation of PED for the empirically calculated vibrational spectra as well as for vibrational spectra calculated via

QM calculations. Thus, the PED from the empirical vibrational spectra can be compared directly with assignments for experimental data based on isotopic substitution and/or QM-based assignments. Final optimization of the dihedral parameters is performed on the bases of conformational energetics of the model compounds.

Typically, only dihedrals containing all non-hydrogen atoms are used for adjustment of the conformational energies, with dihedrals that include terminal hydrogens optimized on the basis of the vibrational spectra. While experimentally determined conformational energetics are available for a wide variety of molecules, typically data from QM calculations at the HF/6-31G level or higher are used. Use of QM data allows for calculation of entire energy surfaces and also yields the energies in geometry (*i.e.,* bond lengths and angles) as a function of a torsional energy surface, allowing off all minima when serval local minima are present in addition QM data include changes for an additional check of the bond and angle parameters.

Model compound 2 in Figure 2.3 offers a good example. In that molecule at least two torsional energy profiles should be calculated, one for the C_{methyl} —N—$C_{aromatic}$ —$C_{aromatic}$ torsion and a second for the $C_{aromatic}$—$C_{aromatic}$—C=O torsion. Torsional surfaces for the methyl rotations and about the amide could also be investigate although the strong analogy of these groups with previously parametrized functional groups indicates that the assigned parameters should yield reasonable results. Each torsional surface would be calculated at the HF/6-31G level of theory or higher by fixing the selected torsion at a given value and relaxing the remainder of the molecule, followed by incrementing that dihedral by 15° or 30° reoptimizing, and so on, until a complete surface is obtained (*i.e.,* an adiabatic surface.

Note that the energy form the fully optimized structure (global minima), with no dihedral constraints, should be used to offset the entire surface with respect to zero. Using these as target data, the corresponding empirical torsional energy surface is calculated and compared to the QM surface. The dihedral parameters are then optimized to maximize the agreement between the empirical and target surfaces. At this stage it is often helpful to add additional dihedrals with alternate multiplicities (*i.e.,* create a Fourier series) to more accurately reproduce the target surface, although all the dihedral parameters (*i.e.,* the force constant, phase, and multiplicity of a single dihedral term) contributing to a torsional surface should first be adjusted to reproduce the target surface. It should be noted that the empirical energy surface associated with rotation of a torsion angle will contain contributions from all the terms in the potential energy function, both external and internal, and not just from the dihedral term. This again emphasizes the necessity of literating over all the individual optimization steps in order to obtain a final, consistent set of parameters. Following initial optimization of all the bond, angle, and dihedral parameters, it is important to reemphasize that all the target data must be rechecked for convergence via an iterative approach. This is due to the parameter correlation problem.

Even though excellent agreement may be achieved with the geometric target data initially, adjustment of the force constants often alters the minimized geometry such that the bond and angle equilibrium parameters must be readjusted; typically the dihedral multiplicity and phase are not significantly affected. Thus, an iterative approach must be applied for the different internal parameters to ensure that the target data are adequately reproduced.

Improper and Urey-Bradley Terms

When initially optimizing the internal parameters for a new molecule, only the bond, angle, and dihedral parameters should be included. If at the end of the interactive optimization of the internal parameters, agreement with the target data is still not satisfactory, then improper and Urey-Bradley terms can respect to out-of-plane distortions associated with planar groups (Wilson wags). For example, with model compound 2, Figure 2.3 improper terms could be added for the aromatic hydrogens as well as for the amine and amide substituted. Urey-Bradley terms are often helpful for the proper assignment of symmetric and asymmetric stretching modes in, for example, methyl group. They can also be used for optimizing energy surfaces including changes in angles as a function of torsional surfaces. This approach has been used for proper treatment of the conformational energetics of dimethyl phosphate and methylphosphate.

(c) Convergence Criteria : As mentioned several times in the preceding section and as emphasized in Figure 2.2, the parameter optimization approach is an iterative procedure required to ensure satisfactory agreement with all available target data. Iterative cycles must be performed over the partial atomic charges and the VDW parameters (loop I,), over the internal parameters (loop II). and over the external and internal parameter optimization protocols (loop III). Furthermore, in certain cases it may be necessary to introduce another iterative cycle based on additional condensed phase simulations (loop IV), although this is typically not required for the optimization of small-molecule parameters. With any iterative approach it is necessary to have convergence criteria in order to judge when to exit an iterative loop. In the cases of the optimization of empirical force field parameters it is difficult to define rigorous criteria due to the often poorly defined and system-dependent nature of the targe data; however, guidelines for such criteria are appropriate. In the case of the geometries, it is expected that bond lengths, angles, and torsion angles of the fully optimized model compound should all be within 0.02 Å, 1.0°, and 1.0° respectively, of the target data values. In cases where both condensed phase and gas-phase data are available, the condensed phase data should be weighted more than the gas-phase data, although in an ideal situation both should be accurately fit. It should be noted that, because of the harmonic nature of bonds and angles, values determined from energy minimization are generally equivalent to average values from MD simulations, simplifying the optimization procedure. This, however, is less true for torsional angles and for non-bond interactions, for which significant differences in minimized and dynamic average values can exist.

With respect to vibrational data, generally a root-mean-square (RMS) difference of 10 cm^{-1} or an average difference of 5% between the empirical and target data should be considered satisfactory. Determination of these values, however, is generally difficult, owing to problems associated with unambiguous assignment of the normal modes to the individual frequencies. Typically the low frequency modes (below 500 cm-1) associated with torsional deformations and out-of-plane wags and the high frequency modes (above 1500cm-1) associated with bond streching are easy to assign, but significant mixing of different angle bending and ring deformation modes makes assignments in the intermediate range difficult. What should be considered when optimizing force constants to reproduce vibrational spectra is which modes will have the greatest impact on the final application for which the parameters are being developed. If that final application involves MD simulations, then the low frequency modes, which involve the largest spatial displacements, are the most important.

Accordingly, efforts should be made to properly predict both the frequencies and assignments of these modes. For the 500 - 1500 cm^{-1} region, efforts should be made to ensure that frequencies dominated by specific normal modes are accurately predicted and that the general assignment patterns are similar between the empirical and target data. Finally, considering the simplicity of assigning starching frequencies, the high frequency modes should be accurately assigned, although the common use of the SHAKE algorithm to constrain covalent bonds during MD simulations, especially those involving hydrogens, leads to these modes often having no influence on results of MD simulations. With respect to conformational energetics, the most important task is to properly select the target data for the optimization.

As discussed above, these data are typically from QM calculations, with the level of theory depending on the size of the model compound and available computational resources. Typically, the larger the basis set, the better, although the HF/6-31G level has been shown to yield generally satisfactory results, especially for predicting the relative energies of local minima. Calculation of barrier heights, particularly in nonaromatic ring systems that include heteroatoms often requires the use of MP2/6-31G calculations. For molecules that contain multiple local minima that have similar energies, higher levels of theory are often required, and care must be taken that the actual order of the energies of the minima is correct. Emphasis should also be placed on properly reproducing the energetics associated with barriers. This is especially true for barriers of 2 kcal/mol or less, which can be frequently crossed during room temperature MD simulations. It should be noted that many force field optimization and comparison studies omit information on energy barriers; such an omission can lead to poor energetic properties that could cause significant problems in MD simulations. Once a set of target energies have been selected, the parameters should be adjusted to reproduce these as accurately as possible, generally within 0.2 kcal/mol of the target data. The lower energy conformation should be weighted more than the high energy terms, because those conformations are sampled more frequently in MD simulations. If necessary, at a later time, higher level QM calculations can be performed on the model compound to obtain improved target data that can be used to reoptimize the diherdal parameters.

Two final Points concerning the dihedral parameters. First, in optimizing these terms to reproduce the conformational energetics, there is often a decrease in the agreement with the vibrational target data. In such cases, it is suggested that the agreement with the energetic data be maximized. The second point concerns the height of the empirical energy barriers compared with the target data. In the CHARMM force field, emphasis has been placed on making the energy barriers lower than that of the target data rather than higher when ideal agreement cannot be achieved, This will tend to make the molecule more flexible and therefore more sensitive to environmental conditions. Creation of artificially high energy barriers will make the molecule more rigid, possibly locking it in an undesirable conformation with respect to its surrounding environment. The use of intermolecular minimum interaction energies and distances between the model compounds and water makes the assignment of convergence criteria for the partial atomic charges straightforward.

Typically, the energetic average difference (average over all interaction energy pairs) should be less than 0.1 kcal/mol and the rms difference should be less than 0.5 kcal/mol. The small value of the average difference is important because it ensures that the overall solvation of the molecule will be reasonable, while the rms criterion of 0.5 kcal/mol ensures that no individual term is too far off the target data. Emphasis should be placed on accurately reproducing the more favorable interactions, which are favorable interactions, which are expected to be the more important in MD or MC simulations, at the expense of the less favorable interactions. With distances, the rms difference should be less than 0.1 Å ; note that the $1/r^{12}$ repulsive wall leads to the differences generally being larger than the QM values, especially in the analysis of data from MD simulations. For both energies and differences. The criteria presented above are with respect to the target data after they have been offset and scaled (see above). In the case of small-molecule dimers (*e.g.*, watson-crick baseparis), the difficulty is again in selection of the appropriate target data, as with the conformational energetics discussed in the preceding paragraph, rather than the degree of convergence. Again, suitable experimental or QM data must be identified and then the empirical parameters must be optimized to reproduce both set of data as closely as possible.

If problems appear during application of the parameters, then the target data themselves must be reassessed and the parameters reoptimized as necessary. Concerning the VDW parameters, the ability to directly apply previously optimized values makes convergence criteria unnecessary. If VDW parameter optimization is performed based on pure solvent or crystal simulations, then the heats of vaporization or sublimation should be within 2% of experimental values, and the calculated molecular or unit cell volumes should be also. If rare gas-model compound data are used, the references cited above should be referred to for a discussion of the convergence criteria.

(d) Condensed Phase Testing : In the majority of cases where parameters are being optimized for a small compound or series of compounds to use with an already available force field, final testing of the parameters via condensed phase simulations is often not necessary. Rigorous use of crystal data, especially survey data, in force filed optimization is discussed elsewhere. If, however, crystal structures for one or more of the compounds exist or a crystal structure of the compound bound to a macromolecule exists, then additional testing is appropriate and can quickly identify any significant problems with the parameters. These tests are preformed by creating the explicit crystal environment followed by energy minimization and MD simulations in that environment. The CRYSTAL module in CHARMM is useful for this purpose. From the MD simulation, which generally converges within 100 ps or less, the averages of the unit cell parameters, internal geometries, and nonbond interaction distances can be determined and compared directly with the experimental values. Such comparison are more valid for small molecule crystal simulations than for small-molecule-macromolecular complexes owing to the higher resolution of the former, although both are useful. Note that since crystal structures are obtained at finite temperatures it is appropriate to compare data from MD simulations at the same temperature rather than data from energy minimizations, as has been previously shown. If results from the crystal simulations are not satisfactory, then a careful analysis should be made to identify which parameters can be associated with the problem, and those parameters should be subjected to additional optimization as required. It should be noted that if discrepancies do occur, a careful examination of the experimental structure should also be made, as errors in the experimental structures are possible, especially for the lower resolution macromolecular structures.

CONCLUDING REMARKS

Improvements in empirical force fields continue, as evidenced by further improvements in the *AMBER* and *CHARMM* force fields and the new BMS force field of nucleic acids published during 1998 and 1999 based on the same from of energy function as that shown in Eqs. (1),(2), and (3). Thus, future efforts will continue to involve additional optimization of empirical force fields and their extension to new molecules. In certain cases, however, force fields based on Eqs. (1)-(3) are inadequate. One such case involves bond-breaking and-making events, where QM/MM-based approaches are relevant. Proper application of QM/MM methods will require special parametrization to ensure that the QM and MM regions and interactions between them are properly balanced. Other cases include biomolecules where local electronic or steric interactions dominate the biological phenomena being investigate. In these cases, extension of the force fields to include electronic polarizability or anisotropic (nonspherical) van der waals surfaces will be helpful.

Concerning anisotropic VDW parameter studies, examples include membrane bilayers that contain unsaturated lipids, for which the current spherical representations of atom are poor models for the carbons involved in double bonds. Efforts to introduce electronic polarizability and anisotropic VDW models have been undertaken but have not yet been applied to the study of biomolecular systems. Although it is to be expected that these advances will soon be used in biomolecular simulations, it should be emphasized that they should not be applied in all situations. As part of selecting the appropriate force field for a particular study, one should consider whether the additional computational costs associated with the use of a force field that includes electronic polarizability, anisotropic VDW surfaces, or both is worth the gains associated with those terms for the particular phenomenon under study. Furthermore, it should be emphasized that additional effects associated with going from a gas-phase environment to a condensed phase environment, such as changes in conformational energetics, may not necessarily be accurately modeled by inclusion of these terms alone, suggesting that additional extension of the potential energy function may be required.

Atomistic empirical force fields have been shown to be effective tools for the study of biomolecular systems, and it can be expected that their use will greatly expand in the future. However, these methods must be used with care, as the complexity of the systems involved and the number of simplifications employed in the mathematical treatment of these systems can yield results that may be misleading and consequently lead to the improper interpretation of data. To minimize the possibility of such an outcome, it is essential that the user of biomolecular force fields understand the assumptions implicit to them and the approaches used to derive and optimize them. That knowledge will allow users to the approaches used to derive and optimize them. That knowledge will allow users to better select the appropriate force field for their particular application as well as to judge if results from that application are significantly influenced by the simplifications used in the force fields. It is within this context that we can expect to gain the most from biomolecular simulations, there by increasing our understanding of a wide variety of biophysical and biochemical processes.

CHAPTER 3

BIO MOLECULAR MOTORS

Evolution has created class of proteins that have the ability to convert chemical energy into mechanical force. Some of these use the free energy of nucleotide hydrolysis as fuel, while others employ ion gradients. Some are "*walking motors,*" others rotating engines. Some are reversible; others are unidirectional. Could there are be any common principles among such among such diversity? The conversion of chemical energy into mechanical work is one of the main themes of modern biology.

Biochemists characterize energy transduction schemes by free energy diagrams. But thermodynamics tells us only whta *cannot* happen. Recent advances in laser trap and optical technology along with advances in molecular structure determination can augment traditional biochemical kinetic and thermodynamic analyses to make possible a more mechanistic view of how protein motors function. The result of these advances has been data that yield load-velocity curves and motion statistics for single molecular motors. This sort of data enables a more detailed, mechanistic level of modeling. At first, the mechanics of proteins may seem counterintuitive because their motions are dominated by Browning motion, the name given to the frequent changes in velocity of a macromolecule as it is buffeted about by random thermal motions of surrounding water molecules. In addition to " *smearing out*" deterministic trajectories, Brown an motion serves asan effective " *lubricant,*" allowing molecules to pass over high energy barriers that would arrest a deterministic system. More subtly, it makes possible "*uphill*" motions against an opposing force by "*capturing*" occasional large thermal fluctuations.

In this cahpter we will discuss protein motions on the molecular scale and derive a mathematical formalism to model such motions. To illustrate the formalism we will analyze (*i*) a "*switch*" controlling the direction of the bacterial flagellar motor,(*ii*) a polymerization ratchet, and(*iii*) a "*toy*" model related to the F_o motor of ATP synthase. These models are simple enough to yield analytical as well as numerical results, and to illustrate many of the principles involved in mechanochemical energy transduction. There is some ambiguity in what one calls a "*motor.*" here we take the narrow view that the principal-and proximate-function of a molecular motor is to convert chemical energy into mechanical force. This excludes, for example, ion pumps, which are surely protein machines that generate forces, but whose purpose is not force production.

Chemical energy comes in various forms, for example, in transmebrane ion gradients and in the covalent bonds of nucleotides such as ATP and GTP, and the designs of motor proteins are tailored to each energy form. Energy stored in one form frequently is converted into intermediate forms before being released as mechanical work. For example, a polymerizing actin filament or microtubule can generate a protrusive force capable of deforming a lipid vestcle or pushing out the leading edge of a cell.

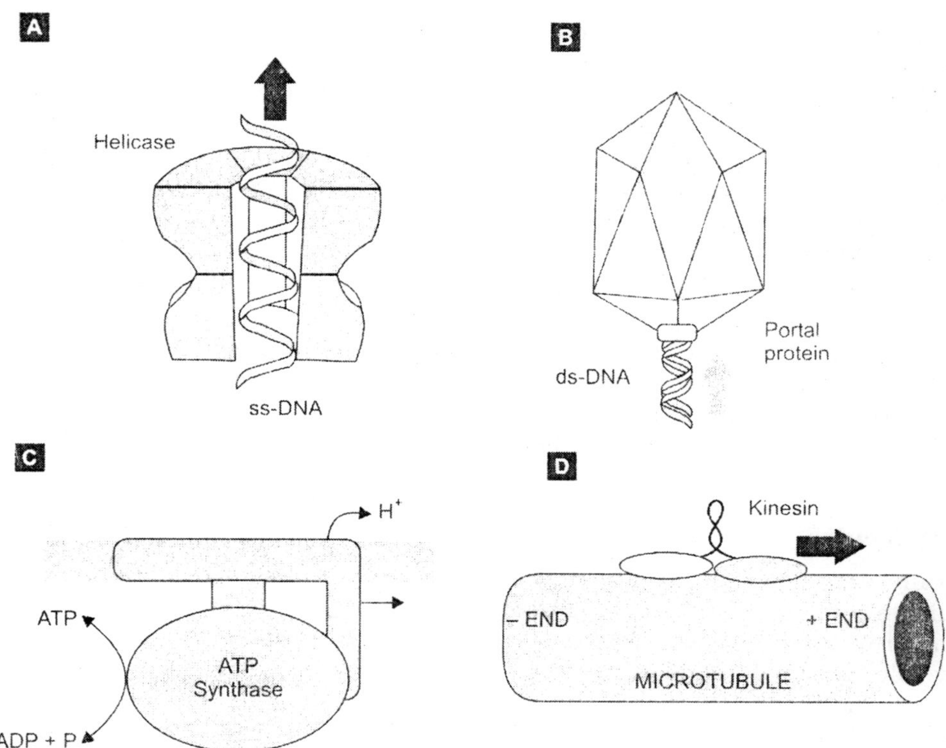

Fig. 3.1. Amazing variety of molecular motors : (A) Rotary motor DNA helicase translocates unidirectionally along the DNA strand using nucleotide hydrolysis as a "fuel."(B) Another rotary motor hydrolyzing ATP, bacteriophage portal protein, drives DNA in and out.(C) Reversible rotry motor ATP Synthase either produces ATP using ion gradient or pumps protons hydrolyzing ATP. (D) linear motor kinesin in a "walking enzyme," Utilizing chemical energy stored in ATP, it moves "head-over-head" toward the plus end of the microtubule "track." Some of these motors are discussed in this chapter.

The energy sources in this process is the free energy of binding monomers to the polymer tip. This energy is used to rectify the Browning motion of the load against which the polymer is pushing. Strictly speaking, the force is generated by thermal fluctuations of the load, and the binding free engrgy is used to rectify its thermal displacements. Energy conversion here is relatively direct. However, in the acrosomal process of the *lumulus* sperm, thermal fluctuations are first trapped as eleastic strain energy in the actin polymer by the binding of an auxiliary protein, scruin. Later, this strain energy is released to generate the force required to push the actin rod into the egg crotex. Many motors us nucleotied hydrolysis to generate mechanical forces, and it is frequently stated that the energy is stored in the yphosphate covalent bond. But releasing this energy to perform mechanical work can be quite indirect. The F_1 motor of ATP synthase uses uncleotide hydrolysis to generate a large rotary torque.

The actual force-generating step takes place during the binding of ATP to the catalytic site; the role of the hydrolysis step is to release the hydrolysis prodcuts, allowing the cycle to repeat. In some motors, not all of the nucieotide binding energy is used immediately for force production; some energy is stored in elastic deformation of the protein to be relased later as mechanical work. So energy transduction need not be a "pay as you go" process; deferred payments are permissible and common. The bacterial flagellar motor and the F_o motor of ATP synhthase both use transmembrance ion gradients to generate a rotary torque. Models of this process show how the chemical reaction of binding an ion onto a charge site creates an unbalanced electrostatic field tht rectifies the Browning motion of the motor and/or creates an electrostatic driving torque. Although the proximal energy transduction proces is a chemical binding event, the motion itself is productd by electrostatic forces and Brownian motion. Thus a common theme in energy transduction is that *chemical reactions power mechanical motion using free energy* released during binding events, but the final producttion of mechanical force may involve a number of intermediate energy transductions.

The most important quality of molecular motors that distinguishes them form macroscopic motors is the overwhelming importance of thermal fluctuations.For this reason, all protein motors must be regarded as "*brownian machines.*" This means that carelessly applying macroscopic physics, where Brownian motion is negligible, to microscopic situations inevitably leads to incorrect onclusions. Therefore, we must being our discussion by examining how to model molecular motions diminated by thermal fluctuations.

MOTIONS OF BIOMOLECULES

Protein Molecules

Generally, a stochastic process refers to a random variable that evolves in time. An example is a one-dimenisonal coordinate $x(t)$ locating protein diffusing in an aqueous solution. We will begin by approximating the coordinate $x(t)$ by a discrete random variable. The rational for this is twofold. Discreate random variables are conceptually simple than their continuous counterparts. The results for the discrete case are applicable in studying continuous random processes because continuous random variables represent limiting behavior of their discrete counterparts. Our discussion is restricted to Markov processes.

A *markov process* is a mathematical idealization in which the future state of a protein is affected by its current state but is independent of its past. That is, the system has no memory of how it arrived at its current state.To a very good approximation, all systems considered in this text satisfy the Markov property. The mathematics involved in studying stochastic processes that are non-Markovian is considerable more complicated.

In ths discrete model, a protein is initially started at $x = 0$. In each time interval Δt it takes one step of length Δx to the right with probability 1/2 or to the left with probability 1/2. Beacuse the length of the step that the protein takes is always the same, this example is referred to as a simple random walk. Let x_n denote the protein's positon at time $t = n\ \Delta t$ and define the set of random variables with z_m with $m = 1, 2.........,n$ to be independent and identically distributed with Prob$[z_m = -1] = 1/2$.

$$x_n = \Delta x(z_1 + z_2 + ... + z_n) \quad ...(3.1)$$

The collection of random variables $x = \{ x_o, x_1, x_2........\}$ represents a spatially and temporally discrete stochastic process. In Exercise 1,(3.1) is used to verify that $\langle x_n \rangle = 0$ and Var$[x_n]$ = $(\Delta x)^2 n = ((\Delta x)^2 / \Delta t)t$. Here we use the notation $\langle \cdot \rangle$ to to denote the average (expectation) and var [·] to denote the variance : verify: Var$[x] = (x)^2 - (x)$.Note that the variance in x grows linearly with time.This is a characteristic of diffusion; below we shoe in what sense the quanity D = $((\Delta x)^2 / (2\Delta t)$.can be interpreted as a diffusion coefficient.

To fully characterize x requires knowledge of the probability density for finding the particle at position x_n after k steps of size Δx: $p_k(n)$ = Prob[$x_n = k\Delta x$]. Note that $p_k(n) = 0$ if $n < |k|$. This comes form the fact that the protein can take only one step per time interval. At any time $n\Delta t$ the total number of steps taken by the protein is $n = R_n + L_n$, where R_n is the number of steps taken to the right and L_n is the number of steps taken to the right and L_n the number of steps taken to the left. Clearly, R_n is binomially distributed, like the number of "heads" in n flips of a coin:

$$\text{Prob} \qquad [R_n = m] = \binom{n}{m}\left(\frac{1}{2}\right)^n \frac{n!}{n!(n-m)!}\left(\frac{1}{2}\right)^n \qquad ...(3.2)$$

Using these definitions, x_n is written as

$$x_n = \Delta x(R_n - L_n) = \Delta x(2R_n - n), \qquad ...(3.3)$$

or equivalently

$$R_n = \frac{1}{2}\left(\frac{x_n}{\Delta x} + n\right). \qquad ...(3.4)$$

Thus $x_n/\Delta x = k$ if and only if $R_n = \frac{1}{2}(n + k)$. Furthermore, $x_n/\Delta x$ must be even if n is even adn odd if n is odd, since R_n must be an integer. Therefore, we immediately find that the distribution for x_n is

$$p_k(n) = \binom{n}{(k+n)/2}\left(\frac{1}{2}\right)^n \qquad ...(3.5)$$

for $n \geq |k|$ and k and n either both even or both odd.

Note that in (3.1) x_n is written as the sum of n independent and identically distributed random variables. Therefore, the central limit theorem of probability theory guarantees that as n gets large the distribution for x_n becomes progressively closer to normal with $\langle x_n \rangle = 0$ and Var[x_n] $=((\Delta x)^2/\Delta t)\, t = 2Dt$. That is,

$$\frac{pk(n)}{\Delta x} \approx p(x,t) = \frac{1}{\sqrt{4\pi D t}}\exp\left(-\frac{x^2}{4Dt}\right), x = k\Delta x, t = n\Delta t. \qquad ...(3.6)$$

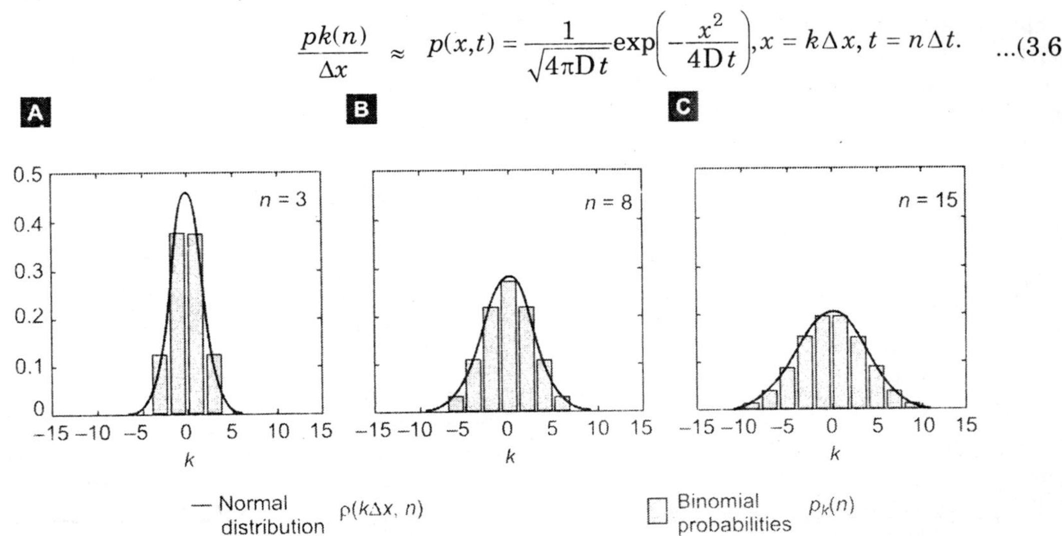

Fig. 3.2. In the limit of large n, the binomial distribution is well approximated by a normal distribution. In all three panels the bar graph represents the binomial probabilities and the solid line is the normal approximation.(A) $n = 3$, (B) $n = 8$, (C) $n = 15$.

Figure 3.2 shows the probability distribution for x_n and the normal approximation for various values of n. By the time $n = t/\Delta t = 15$, the distribution of x_n is close to normal, and the agreement gets better as n is increased. Physically, the normal approximation amounts to a "coarse graining" of the process in which only length scales much larger than Δx and time scales much large than Δt aie resolved. In this limit the random variable x_n, which is discrete in both space and time, is approximated by $x(t)$, a random variable that is continuos in space and time.The value of Δt can be approximated well by the "thermalization" time $\tau = 10^{-13}$. Thus , the continuous and discrete models of a protein's motion are equivalent at, alll time and distance scales of interest to us.

As an exercise, the reader is asked to verify that $p(x, t)$ satisfies the diffusion equation

$$\frac{\partial p(x,t)}{\partial t} = D\frac{\partial^2 p(x,t)}{\partial x^2},$$

justifying our association of the quanitity$(\Delta x)^2/(2\Delta t)$ with a diffusion coefficient.

Polymer Growth

Let us consider antother examples of a stochastic process: the number(n)(t) of monomers in a polymerizing biopolymer. There is an important distinction between the stochastic variables $x(t)$ and N(t) . in the first example, $x(t)$ is a continuous random variable, since it can take on any real value. On the other hand, the number of subunits in a growing polymer is restricted to the positive intergers, so that N(t) is a discrete random variable. Markov processes in which the random variable is discrete are oftaen referred to as Markov *chains* because they can be represented as a sequence of jumps between discrete states.

The simple random walk is an example of a spatially and temporally discrete Markov chain. As an example of a Markov chain in which time is contiuous we consider a polymerizing biological polymer(filament), *e.g.*, and actin flament or a microtuble. Figure 3.3. A depicts the type of proces we have in mind. In this example, two events change the length of the polymer by one monomer: polymerization and de-polymerization. Mathematically, the state of the system is specified by a single number N(t), the number of monomers in the filament at time t. The variable N(t) is random, because we have no way to predict when the next polymerization or depolymerization event will occur. There are two equivalent, but conceptually different, levels at which stochastic processes can be studied. The first is at the leavel of individual sample paths or realizations of the process.

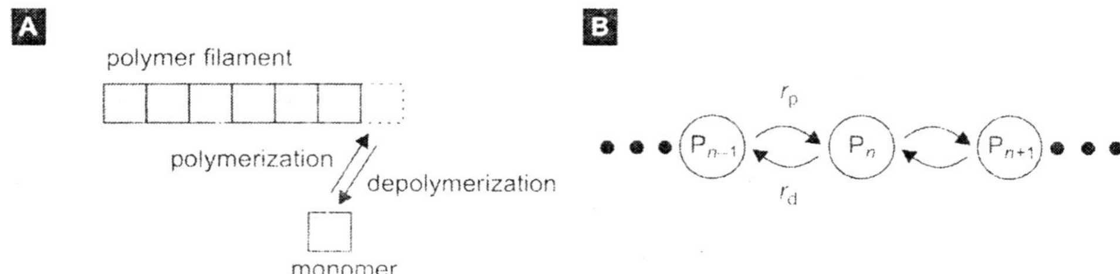

Fig. 3.3. A discrete Markov process(Markov chain).(A) A polymer filament grows by incorporating monomers from the solution onto its tip(polymerization). The process is stochastic. The monomer on the tip may dissociate from the filament into the solution(depolymerization). (B) A Markov chain model for the filament polymerization; p_n represents the state of the filament when its length is n monomers, r_p is the polymerization rate, r_d is the depolymerization rate r_d is the depolymerization rate. If $r_p > r_d$, the filament will grow over long times.

To understand what is meant by a sample path, suppose that at $t = 0$ we start with three filaments that are each exactly 5 monomers long. As time goes on, we observe that the number of monomers in each filament instantaneously changes , or "jumsp," by ± 1 at random times. Figure 3.4 graphically illustrates this behavior. Even though each sample path starts with $N(0) = 5$, they all evolve differently, illustrating the randomness of the process. The sample paths of a large ensemble of such filaments can be used to determine the statistics of N(t). The second approach is to ask how the probability $p_n(t)$ of having exacty in manomers in the filament at time + changes in time. If $p_n(t)$ can be determined for all t and n, then we have a complete characterization of the process. Both approaches are equally valid and are useful methods for studying stochastic processes.

The advantages of staying at the level of sample paths are that in general it is easy to unmerically generate single realizations of the process, and sample paths allow us to see the dyanmics of the system. The advantage of working directly with the probability distribution is that it fully characterizes the system without the need to average over many sample paths to compute the statistics. Of course, there is no free lunch: Obtaining all this information comes at a computaional price. Below we describe numerical techniques for treating both cases. To begin our discussion we derive an equation that governs the evolution of $p_n(t)$. Let us assume that we know $p_n(t)$ for a specific value of t. At a slightly later time $t + \Delta t$,we expect, $pn(t + \Delta t)$ to be equal to $p_n(t)$ plus a small correction.

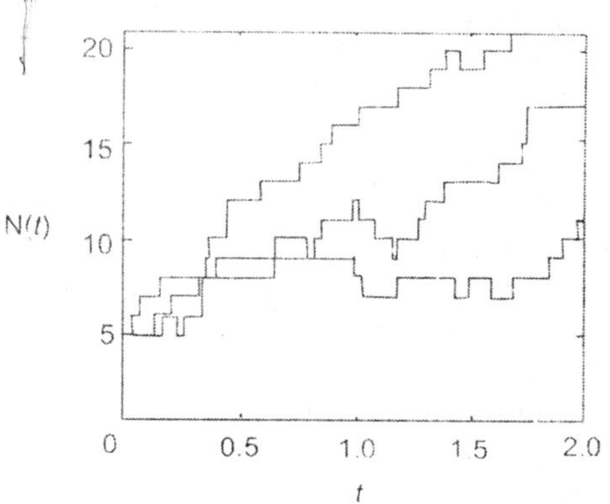

Fig. 3.4. Three sample trajectoris of the tip of a growing filament. The polymerization process is stochastic with occasional depolymerization events.

The key is to assume that Δt,is so small that the probability of two events in the interval $(t,t + \Delta t,)$ is very unlikely. Here an event means polymerization or depolymerization. Then we can write

$$\begin{aligned} p_n(t + \Delta t) &= \text{Prob}\,[\text{N}(t) = n \text{ and no event occurs in } (t,\ t + \Delta t,)] \\ &\quad + \text{Prob}[\text{N}(t) = n - 1 \text{ and polym. occurs in } (t,\ t + \Delta t,) \\ &\quad + \text{Prob}[\text{N}(t) = n + 1 \text{ and depol. occurs in } (t,\ t + \Delta t,), \end{aligned} \qquad \text{...(3.8)}$$

where the right-hand side follows from the fact that the three events described in the squares brackets are mutually exclusive. Next we make the reasonable assumption that the probability of polymerization or depolymerization is independent of the length of the filament. We also assume that these probabilities aee proportional to Δt, and let $r_p\,\Delta t$ and $r_d\,\Delta t$ be the probabilities of polymerization and depolymerization, respectively, in$(t,\ t + \Delta t,)$ Under these assumptions 3.8 can be written as

$$\begin{aligned} p_n(t + \Delta t) &= p_n(t)\,1 - (r_p + r_d)\,\Delta t) + p_{n-1}(t)\,r_p\,\Delta t + p_{n+1}(t)\,r_d\,\Delta t \\ &= p_n(t) + \Delta t[r_p\,p_n - 1)(t) + r_d\,p_{n+1}(t) - (\,r_p + r_d)\,p_n(t)], \end{aligned} \qquad \text{...(3.9)}$$

where$(1 - (\,r_p + r_d)\,\Delta t)$ is probability of no event in Δt . There are two important points to be drawn form(3.7). First, it is clear that if we know $p_n t)$ for all n at a given time, then we have a mechanism for updating the probabilities at all lateer times. This illustrates the Markov property. Secondly,(3.9) represents a numerical algorithm for updating $p_n(t)$. This is once a Δt is chodesn, we cn write a computer program to generate $p_n(t + k\,\Delta t)$, where k is positive interger. We now take the limit $\Delta t \to 0$ in (12.9):

$$\lim_{\Delta t \to 0} \frac{p_n(t+\Delta t) - p_n(t)}{\Delta t} = \frac{dp_n(t)}{dt} = -(r_p + r_d)\, p_n(t) + r_p\, p_{n-1}(t) + r_d\, p_{n+1}(t) \qquad ...(3.10)$$

Therefore, (3.9) is an algorithm for numerically solving the ordinary differential equation given By (3.10) . This algorithm is called the forward Euler method, and is a very useful numerical tool that works adequately for many situations. However, problems may arise in using this scheme, as discussed below. Also, note that (3.10) can be interpreted as a chemical rate equation, so that r_p and r_d are the rates of polymerization and depolymerization, respectively.

Sample Paths

The next question we address is how to numerically generate sample paths that are consistent with (3.10). To analyze this problem consider the following experiment. At time $t = 0$, we start with a filament containing exactly m monomers. That is, $p_m(0) = 1$. Next we watch the filament until the first event occurs (either polymerization or depolymerization). When this event occurs we record the time and start the experiment again. After doing this experiment many times, we find that the amount of time we must wait for the first event to occur is a random variable. Let us call it T. We are after the probability density $f_t(t)$ for T. Let $q(t)$ be the probability that no event has occurrred in $(0, t)$. Under the conditions of the experiment and from the derivation of (3.10), we have:

$$\frac{dq}{dt} = -(r_p + r_d)q \qquad ...(3.11)$$

Solving this equation, we obtain $q(t) = \exp(-(r_p + r_d)\, t)$. The probability $q(t)$ starts at 1 and decreases to 0 as time goes on. The probaility tht at least one event has occurred in(o, t) is $1 - q(t)$. Hence, we can use $1 - q(t)$ to define a probability density funtion $f_T(t)$ for the *waiting time* distribution:

$$1 - q(t) = \text{Prob [Waiting time T} < t] = 1 - q(t) = \int_o^t fT(t')dt'. \qquad ...(3.12)$$

By differentiating(12.12), we obtain the relationship

$$f_T(t) = -\frac{dq(t)}{dt} = (r_p + r_d)\exp(-(r_p + r_d)\, t) \qquad ...(3.13)$$

That is, the waiting time until the next event occurs has an exponential distribution with mean $1/(r_p + r_d)$. Thus, to produce realization of N(t), we need to be able to generate samples of an exponential random variable.

Most programming languages have built-in random number generators that produce number that are uniformly distributed between o and 1. If R is such random variable, its probability density function is $f_R(t) = 1$ in [0, 1]. The transformation that converts R to an exponential random variable with mean $1/(r_p + r_d)$ is

$$T(R) = -\frac{1}{(r_p + r_d)} \ln R. \qquad ...(3.14)$$

This can be verified mathematically as follows:

$$\int_0^t f_T(t)dt = \text{Prob } [T(R) < t] \qquad ...(3.15)$$

$$= \text{Prob } [R > \exp(-(r_p + r_d))]$$

$$= 1 - \exp(-(r_p + r_d)) = 1 - q(t) \qquad ...(3.16)$$

Given this way to computer when the next transition occurs, the next thing we need to determind is whether polymerization or depolymerization takes place. Remember that in any time interval of length Δt the probability of polymerization is $r_p \Delta t$. Likewise, the probability of depolymerization is $r_d \Delta t$. Therefore, given that an event has occurred, the probability that it was polymerization is

$$\text{P[polymerization | an even occurred at } t] = \frac{r_p}{r_p + r_d} \qquad ...(3.17)$$

We may now generate samle paths of the stochastic process as follows: Start N with a given value. Generate an exponentially distributed random number using(3.14) This determines when the next even takes place. To determine the type of event that occurred, generate a uniformly distributed random number R_2. If $R_2 < r_p/(r_p + r_d)$, then let $N \to N + 1$ otherwise $N \to N - 1$. Repeat the process and plot N as a funtioin of time. The trajectories shown in Figure 3.4 were generated in this way.

Before using this method to simulate protein motions, we briefly discuss the satistical behavior of polymer growth.

The Statistical Behaviour

Since the intervals of time between events of monomers assembly and or disassembly are random, one can measure only the statistical behaviour of polymer growth, such as the average velocity of the polymer's tip,(v) = L(n)/*t*, where L is the size of a monomer: Much useful information is buried in the statistical fluctuations about this mean velocity. One quantity that can be monitored as the polymer grows is the variance of the tip's displacement about the mean:

$$\text{Var}[x(t)] \equiv (x^2) = L^2(N^2) - (N^2).$$

It is easy to show (see Exercise 4) that the average velocity of the polymer tip and the variance of its displacement are

$$\langle V \rangle = L(r_p - r_d), \ \text{Var}[x(t)] = L^2(r_p + r_d)\, t.$$

Thus the variance grows linearly with time. In fact, a plot of Var[$x(t)$] vs t can be used to define an effective diffusion coefficient: $De_{ff} \equiv \text{Var}[x(t)]/2t$. The coefficient De_{ff} can be combined with the average velocity $\langle V \rangle$ to form a "randomnesss parameter"

$$r \equiv \frac{2D_{eff}}{L.\ (V)}. \qquad ...(3.18)$$

As an example of the utility of this randomness parameter, let us consider the case where there is no depolymerizaion: $r_d = 0$. Then, $\text{Var}[x(t)] = L^2 rp'$, $D_{eef} = L^2 rp/2$, $\langle V \rangle = Lrp$, and $r =$ 1. Now suppose that each polymerization event involves a sequence of reaction processes. Since chemical reactions are also stochastic processes, an additional variance is added to the spatial diffusion, so that the total varianc will grow faster, and the randomness parameter is greater than 1. In this case, $1/r$ give a lower bound on the number of reaction processes per step (*i.e.*, $1/r <$ number of reaction processes per step)

Similar agruments are applicable to some "*walking*" *e.g.*, kinesin that take a spatial step of constant size at random times. This time is determined by a sequence of hydrolysis reactions. If there is only one reaction, the walking motor is equivalent to the polymerizing filament, and is called a "Poirson stepper." Sucha stepper is characterized by randomness parameter

$\}r = 1$. In the fore going account we will show that the average velocity of a molecular motor is a function of the load force resisting the motor's advancement. The importance of considering the effective diffusion coefficient(or, equivalently, the randomness parameter) is that just as load-velocity data give information about the motor performacne, load-variance data can provide independent estimates of model parameters

MODELING MOLECULAR MOTIONS

The botanist Robert Brown first observed *brownian movement* in 1827. While studying a droplet of water number a microscope he noticed tiny specks of plant pollen dancing aruond. Brown first guessed, and later proved, that these were not motile, althought at the time he had no clue as to the meachanism of their motion. It was not until Einstein contemplated the phenomenon 75 years later that a quantitative explanation emerged. In order to develop an intuition about molecular dynamics we begin with some simple remarks on Brownian motion of proteins in aqueous solutions.

The Langevin Equation

The radius of a water molecules is about 0.1 nm, while proteins are two orders of magnitude larger, in the range 2-10 . This size difference suggests that we can view the fluid as a continuum. A protein moving through the fluid is acted on by frequent and uncorrelated momentum impulses arising from the thermal motions of the fluid. We model these fluctuations as a time-dependent random " Brownian force" $f_b(t)$ whose statistical properties can be mimicked by a random number generator in a computer in a fashion described below. At the same time, the fluid continuum exets on the movine protein a frictional drag force f_d proportional to the protein's velocity: $f_d = -\zeta v$, where ζ is the frictional drage coefficient. Thus we can write Newton's law for the motion, $x(t)$, of a protein moving in a one-dimensional domanin of length L:

$$\frac{dx}{dt} = v, m\frac{dv}{dt} = -\zeta v + f_B(t),\ 0 \le x(t) \le \mathrm{L}. \quad ...(3.19)$$

The mass m of a typical protein is about 10–^{21} kg, and the drag coefficient is about $10^7 \mathrm{p}^{\mathrm{N}}$. sec/nm.

If we multiply(3.19) by $x(t)$ and use the chain rule, we get

$$\frac{md^2(x^2)}{2\,dt^2} - mv^2 = mv^2 = -\frac{\zeta d(x^2)}{dt} + x \cdot f_{\mathrm{B}}(t). \quad ...(3.20)$$

In order to see the consquence of (3.20) for molecular motions we first must average (3.20) over a large number of proteins os that the peculiarities of any particular trajectory are averaged out. We use the notation $\langle\cdot\rangle$ to denote this *ensemble* average:

$$\frac{md^2(x^2)}{2dt^2} - mv^2 = -\frac{\zeta d(x^2)}{2\ dt} + \langle x \cdot f_{\mathrm{B}}(t)\rangle. \quad ...(3.21)$$

Next we take advantage of a central result from statistical mechanics called the *equipartition theorem*, which states that each *degree of freedom* of a Brownian particle carries an average energy

$$\langle \mathrm{E} \rangle = \frac{1}{2}k_{\mathrm{B}}\mathrm{T} \quad \text{[equipartition theorem]}, \quad ...(3.22)$$

where k_B *is* Boltzmann's constant and T the absolute temperature. Therefore, the second term in 3.21 is just twice the average kinetic energy of the protein: $(mv^2) = k_B T$. At room temperature, the quantity $k_B T. \simeq 4.1$ pN.nm is the "unit" of thermal energy.

Because the random impulses from the water molecules are uncorrelated with positon, $\langle x(t) \cdot f_B(t) \rangle = 0$. Introducing these two facts into (3.21) and intergrating twice between $t = (0, t)$ with $x(0) = 0$, we obtain

$$\frac{d(x^2)}{dt} = \frac{2k_B T}{\zeta}(1 - e^{-t/\tau}), \quad \langle x^2 \rangle = \frac{2k_B T}{\zeta}[t - \tau(1 - e^{-t/\tau})], \quad ...(3.23)$$

where we have introduced the time constant $\tau = m/\zeta$.

For very short times $t \ll \tau$ we can expand the exponential in (3.23) to second order to obtain

$$\langle x^2 \rangle = \frac{k_B T}{m} t^2 \ (t \ll \tau). \quad ...(3.24)$$

That is, at very short times the protein behaves like a ballistic particle moving with a velocity $v = \sqrt{k_B T/m}$. For a protein with $m \approx 10^{-21}$ kg [= 10^{-18} pN.sec^2/nm], $v \approx 2$ m/s. However, in a fluid the protein moves at this velocity only for a time $\tau \approx m/\zeta = 10^{-13}$ sec, much shorter than any motion of interest in a molecular motor. During this short time the protein travels a distance $v.\tau \sim 0.01$ *nm* before it collides with another molecule. This is only a fraction of a diameter of a water molecule, so the ballistic regime is verey short-lived indeed! Very quickly, the kinetic energy of the protein comes into thermal equilibrium(is "*thermalized*") with the fluid environment. Thus when $t \gg \tau$, the exponential term disappears, and (3.23) becomes

$$\langle x^2 \rangle = 2\frac{k_B T}{\zeta} t \ (t \gg \tau). \quad ...(3.25)$$

Einstein recognized that the frictional drag on a moving body is caused by random collisions with the fluid molecules, which is the same effect as the Brownian force $f_B(t)$ that gives rise to the diffusive motion of the body. Therefore, there must be a connection between the drage coefficient and diffusive motion. By comparing (3.25) to the relation we previously derived between the mean square displacement of diffusing particle and its diffusion coefficient

$$\langle x^2 \rangle = 2\,Dt, \quad ...(3.26)$$

we arrive at the famous relation dervied by Einstein in 1905:

$$D = k_B T/\zeta \quad \text{[Einstein relation]},$$

where D is the diffusion coefficient of the protein, typically $D \approx 10^7$ nm^2/sec. For diffusion in 2 and 3 dimensions the relations is

$$\langle x \cdot x \rangle = 2v \cdot Dt = 2v\frac{k_B T}{\zeta} t,$$

where $v = 2, 3$, respectively.

If an external force F acts on the protein, this can be added to (3.19), so that the equation of motion for a protein becomes $\zeta \cdot dx/dt = F(x, t) + f_B(t)$ (the inertial term has been neglected;In general, forces acting on proteins can be characterized by a potential $F(x, t) = -\partial\phi(x, t)/\partial x$, so the equation of motion for a protein moving through a fluid becomes

$$\zeta\frac{dx}{dt} = \frac{\partial\phi(x,t)}{\partial x} + f_B(t). \qquad ...(3.28)$$

Equation (3.28) is frequently referred to as a *langevin equation,* although this term more properly applies to the corresponding (3.19), which includes inertia.

Langevin Equation

Here we show how the stochastic algorithms developed above can be applied to a continuous Markov process describing a protein diffusing in water. We want a numerical algorithm for generating sample paths of equation (3.28). Let us intergrate both sides of this equation over the interval$(t,\ t + \Delta t)$:

$$x(t + \Delta t) = x = (t) - \frac{1}{\zeta}\int^{t+\Delta t}\frac{\partial\phi(x,t')}{\partial x}dt' + \frac{1}{\zeta}\int^{t+\Delta t} f_B(t')\,dt'$$

$$\approx x(t) - \frac{1}{\zeta}\frac{\partial\phi(x,t)}{\partial x}\Delta t + \frac{1}{\zeta}\int^{t+\Delta t} f_B(t')\,dt'. \qquad ...(3.29)$$

It has already been stated that the way to include the effect from the Brownian force, $f_b(t)$, is to use following numerical method for simulationg (3.29)

$$x(t + \Delta t) \approx x(t) - \frac{1}{\zeta}\frac{\partial\phi}{\partial x}\Delta t + \sqrt{2D\Delta t}z. \qquad ...(3.30)$$

where Z is a standard normal random variable, *i. e.,*with mean 0 and variance 1. Many numerical software packages have built-in random number generators that will generate samples of a standrad normal distribution. If one is not available, then a standard normal random variable Z can be generated from two independent uniform random variables R_1 and R_2:

$$Z = -\sqrt{-2\ln R_1}\cos(2\pi R_2). \qquad 3.31$$

A derivation of this result is similar to the one presented above for generating an exponential random variable.

Although simulating(3.30) on a computer is easy, it is also easy to generate erroneos results, *e.g.,* numerical in stabilities that look much like random displacements due to Brownian motion, or currents that do not vanish at equilibrium. in ordr to dervie a numeriacl method of simulating random motion that does not have these problems, we have to consider an alternative description of the molecular motion.

Drift Velocity

Consider a protein moving undr the influenc of a constant external force, for example an electric field. Because of Brownian motion, no two trajectories will look the same. Moreover, even a datailed examination of the path cannot distinguish whether a particular displacement "*Step*" was caused by a Brownian fluctuation or the effect of the field. Onlay by traking the particle for a long time and computing the average position vs. time can one detect that the diffusion of the particle exhibits a "*drift velocity*" in the direction of the force. Therefore, a better way to think about stoachastic motion is to imagine a large collection of independent particles moving together. Then we can define the *concentration* of particles at position x and time t as $c(x,t)$ [#/nm], and track the evolution of this *ensemble.*

As the could of particles diffuses and drifts, we can write an expression for the *flux* J [#/area/time] of particles passing through a unit area; in one dimension J_x has dimensions [#sec].

The diffusive motion of the particles is modeled well by Fick's law: $J_x = -D\,\partial x/\partial x$. The external field exerts a force $F = -\partial\phi/\partial x$ on each particle that, in the absence of any diffusive motion would impart a drift velocity proportional to the field: $v = F/\zeta$. Thus the motion of the body is the sum of the Brownian diffusion and the field-driven drift: $Jx = -D\,\partial c/\partial x + v.c$, which can be written in several ways

$$Jx = \underbrace{-D\frac{\partial c}{\partial x}}_{\text{Diffusion flux}} \underbrace{-\overbrace{\left(\frac{D}{k_B T}\cdot\frac{\partial \phi}{\partial x}\right)}^{\text{Drift velocity}} c}_{\text{Drift flux}} = -D\left(\frac{\partial c}{\partial x} + \frac{\partial(\phi/k_B T)}{\partial x}\cdot c\right)$$

$$= -\frac{1}{\zeta}\left(k_B T\frac{\partial c}{\partial x} + c\frac{\partial \phi}{\partial x}\right). \qquad \text{...(3.32)}$$

At equilibrium the flix vanishes: $(k_B T/c_{eq})\,(\partial c_{eq}/\partial x) + \partial\phi/\partial x = 0$. Intergrating with respect to x shows that the concentration of particles at equilibrium in an external field $\phi(x)$ is given by the *Boltzmann distribution:*

$$c_{eq} = c_0 e^{-\phi/k_B T} \qquad \text{[Boltzmann distribution]} \qquad \text{...(3.33)}$$

Since the number of particles in the swarm remains constant, $c(x, t)$ must obey a *conservation law.* This is simply a blance on a small volume element Δx:

$$\frac{\partial}{\partial t}(c\Delta x) = \text{net flux into } \Delta x\ J_x(\text{in}) - J_x(\text{out}) = J_x(x) - Jx(x + \Delta x),$$

or, taking the limit as $\Delta x \to 0$,

$$\frac{\partial c}{\partial t} = -\frac{\partial J_x}{\partial x} \qquad \text{[Conservation of paraticles].} \qquad \text{...(3.34)}$$

Rather than focusing our attention on the swarm of particles, we can rephrase our discussion in terms of the *probability* of finding a *single* particle at (x, t). To do this we normalize the concentration in (3.32) by dividing by the total population $p(x, t) \equiv c(x, t)/\left(\int_0^L c(x,t)dx\right)$. Inserting (3.32) expressed in terms of $p(x, t)$ into the conservation law (3.34) yields the *Smoluchowski equation*

$$\frac{\partial p}{\partial t} = D\left[\underbrace{\frac{\partial}{\partial x}\left(p\frac{\partial(\phi/k_B T)}{\partial x}\right)}_{\text{Drift}} + \underbrace{\frac{\partial^2 p}{\partial x^2}}_{\text{Diffusion}}\right] \qquad \text{[Smoluhowski equation].} \qquad \text{...(3.35)}$$

Comparing this with the Langevin equation (3.28) show that the Brownian force is replaced by the diffusion term, and the effect of the deterministic forcing is captured by the drift term.

We can nondimensionalize (3.35) by defining time and space scales. If the domanin is $0 \le x \le L$, the spartial variable can be normalized as x/L. A time scale can be defined by $\tau = L^2/D$. Introducing the space and time scales, (3.35) can be written in dimensionless form as

$$\frac{\partial p}{\partial t} = \frac{\partial}{\partial x}\left(p\frac{\partial \phi}{\partial x}\right) + \frac{\partial^2 p}{\partial x^2}, \qquad \text{...(3.36)}$$

where t and x are now dimensionless, and the potential ϕ is measured in units of $k_B T$.

Equation (3.36) must be augmented by appropriate boundary conditions specifying the value of $p(x = 0, t)$, $p(x = L, t)$, and $p(x, t = 0)$, where $p(x, t)$ is defined on the interval [0, L]. These will depend on the system being modeled.

First Passage Time

A very useful quantity in modeling protein motions is the average time it takes for a diffusing protein to first reach an absorbing boundary located at $x = L$, starting from postion $0 \le x \le L$. Denote the mean first passage time(MFPT) to position L starting from position x by T(x, L). The equation governing T(x, L). is derived as folows. A particle released at position x can diffuse either to the right or to the left. After a time τ it covers an average distance Δ so that it is locate at $x \pm \Delta$ with equal probability 1/2. The MFPT to L from the new positions are T(x, + ΔL). and T($x - \Delta$, L). The average value of T(x, L) is just $T(x, L) = \tau + (1/2)[T(x + \Delta, L) + T(x - \Delta L)]$. This equation can be rewriten in the form

$$\frac{1}{\Delta}\left(\frac{(T(x+\Delta, L)-T(x))}{\Delta} - \frac{(T(x)-(x-\Delta, L))}{\Delta}\right) + \frac{2\tau}{\Delta^2} = 0.$$

Taking the limit as $\Delta \to 0$ and $\tau \to 0$(so that $\Delta^2/2\tau$ = const) and recognizing that $\Delta^2/2\tau$ is just the diffusion coefficient D, the MFPT education becomes

$$D\frac{\partial^2 T}{\partial x^2} = -1 \qquad \text{[MFPT equation].} \qquad ...(3.37)$$

The boundary conditions for this equation are simple. At an absorbing boundary, T = 0 (it takes no time to get there). At a reflecting boundary, T is unchanged (*i.e.*,a constant), so $\partial T/\partial x = 0$. For example, releasing a particle at a position x with a reflecting boundary at x =L and absorbing boundary at $x = 0$, the MFPT is $T(x, L) = (2Lx - x^2)/2D$. The special case where x = L(releasing the particle at the reflecting boundary) is just $T = L^2/2D$. Note resemblance to the familiar equation $\langle x \rangle^2 = 2Dt$. This gives the mean squared distance diffused in tiem t, whereas the *MFPT* gives $\langle t \rangle = x^2/2D$, the mean time to diffuse a distanc x. This sugests that the MFPT equation might be related to the Smoluchowski diffusion equation; in fact, they are adjoints of one another.

BIO CHEMICAL REACTIONS

So far, we have paid attention exclusively to protein mechanics. To understand molecular motors, we have to consider chemical reactions, which supply the energy sources are nucleotide hydrolysis and transmembrane protonmotive force. The former uese the energy stored in the convalent bond that attaches the terminal phosphate(γ-phosphate) to the rest of the nucleotide. The latter uses the electrical and entropic energy arising from a difference in ion concentrations across a lipid bilayer. Hydrolysis is a complicated process, still incompletely understood. Therefore, we will intoduce the reaction model using the simple exmple of a positively charge ion (*e.g.*,H^+) binding to a negatively charged amono acid: $H^+ + A^- \to$ H.A. If we focus our attention on the amino acid, we seet that it exists in two states; charged(A^-) and neutral(H.A $\equiv A^0$), so that the neutralization reaction from the viewpoint of the amino acid is simply

$$A^- \underset{k^-}{\overset{k^+[H^+]}{\rightleftharpoons}} A^0. \qquad ...(3.38)$$

Here we use the chemist's convention of denoting concentrations in brackets: $k^{+} \cdot [H^{+}]$ and k^{-} are the forward and reverse rate constants; the forwared rate constant depends on the ion concentration $[H^{+}]$, which we will treat as a constant parameter (*i.e.*, we shorten our notation to $k^{+} \cdot [H^{+}] \equiv k^{*}$, where k^{*} is called a *pseudo*-first order(or pseudo-unimolecular) rate constant).

The rate constants in reaction(3.38) conceal a great deal of physics, for the process of even as simple a reaction as this is quite complex at the atomic level. To model this reaction at a more microscopic level involves introducing additional coordinates to describe the process by which an ionic chemical bond is made and broken.Thses coordinates have a spatial scale much smaller than the motion of the motor itself (*e.g.*,angstroms vs. nanometers), and a time scale much faster than any motion of the motor(picoseconds vs. microseconds). This is because all reactions involve a redistribution of electrons, being very small, move very rapidly. Moreover, in all but the simplest cases, their movements are governed by quantum mechanics rather than classical mechanics. Nevertheless, it is instructive to use the Smoluchowski model to dervie a more detailed expression for the rate constants.

The fundamental concept underlying the modeling of reactions is the notion of a "reaction coordinate," which we denote by ζ. In molecular dynamics simulations this is actually a 1-D path through a very high dimensional state space along which the system moves from reactants to products. For the reaction (3.38), $\zeta(t)$ is the distance between the ion (H^{+}) and the amino acid charage(A^{-}). The spatial scale of the motor's motion, but we can imagine a "super-microscopic" view of the process as shown in Figure 3.5A, where we have plotted the free energy change ΔG during a reaction as a function of the reaction coordinate ζ. The reason for using free energy is that there are many "hidden" degrees of freedom that must be handled statistically, as will become clear presently. Here the chemical states of the amino acid, A^{-} and A^{0}, are pictured as energy wells separated by barriers of heights ΔG_1 and ΔG_2, and whose difference in depth is ΔG. The "transition state"(TrSt) is located at top of the pass between the two wells.

For a fixed H^{+} is concentration, the forward chemical reaction $A^{-} \rightarrow A^{0}$ proceeds with a rate $k^{*} \cdot [A^{-}]$[#/sec]. However, this rate is a statistical average over many "*hidden*" events. For a particular reaction to take place, the proton must diffuse to within a few angstroms of the amino acid charge so that the electrostatic attraction between them is felt. Moreover, if the amino acid is located within a protein, there will be steric diffusion barriers that must be circumvented before the two ions "*see*" each other electrostatically.(Actually, protons insider proteins move by " *hopping*' along strings of water molecules, or "*water wires*"). As the concentration of H^{+} increases, there will be more "tries" at neutralization (*i.e.*, hops from the left well to the right well).

Similarly, the reverse reaction $A^{-} \longleftarrow A^{0}$ takes place when a thermal fluctuation confers enough kinetic energy on the proton to overcome the electrostatic attraction. Even then, the "free" proton will more often than not "jump" back and rebind to the amino acid, especially if the route between the solution and the amino acid is tortuous. Only when the proton manages a successful escape into solution (the left well) does it count in computing k^{-}

The *net* flux over the barriers is

$$J_{\zeta} = k^{*} \cdot [A^{-}] - k^{-} \cdot [A^{0}]. \quad ...(3.39)$$

After a long time the net flux between the two wells will vanish, $J_{\zeta} = 0$, so that the population of neutral and charged sites will distribute themselves between the wells in a fixed ratio, which we denote by K_{eq} (the equilibrium constant): $K_{eq} \equiv [A^{0}_{eq}]/[A^{-}_{eq}] = k^{*}/k^{-}$. If the transition state is high, then we can assume that population is apportioned between the two wells according to the Boltzmann distribution(3.33): k_{eq} = exp(G/k_BT), where Δ G is a free energy. The value of ΔG determines *how far the reaction goes,* but says nothing about the *rate* of the reaction. Now we know that ΔG = ΔH – TΔS. The enthalpy term ΔH is due to the electrostatic attraction between the proton and the charged site. The entropic term TΔS incorporates all the effects that influence the diffusion of the proton to the site and its escape from it, the "hidden coordinates." Thus we see that a thermodynamic equilibrium state ΔG = 0 is a compromise between energy(ΔH) and randomness(TΔS).

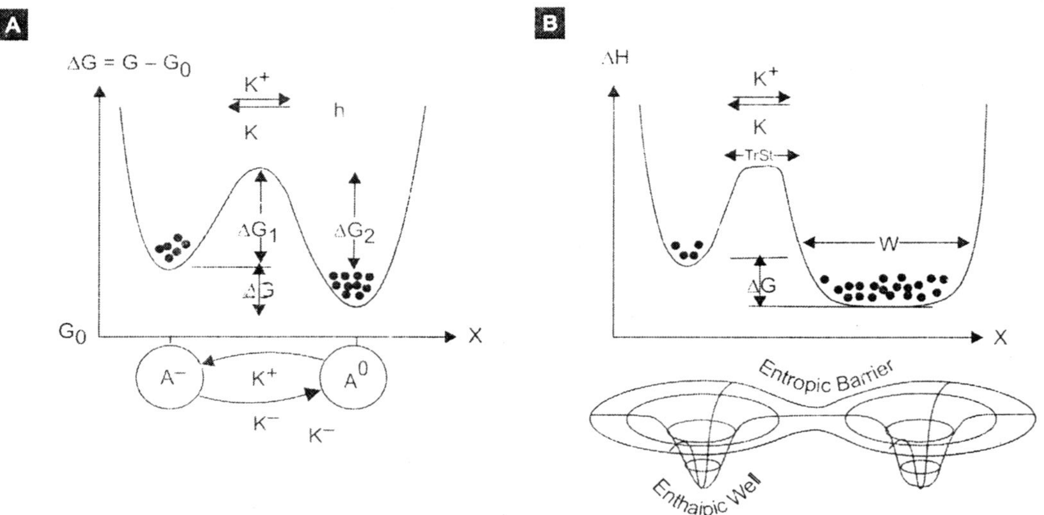

Fig. 3.5. (A) Free energy diagram illustrating the chemical reaction A ↔ B and the corresponding Markov model. The transition state TrSt ΔG_1^+ is above the left well and ΔG_2^+ above the right well. ΔG is the free energy difference between the well bottoms. The equilibrium distribution between the wells depends only on ΔG · (B) The effect of entropic factors on the reaction A ↔ B. potential, rather than free, energy is shown as a function of ξ, effective one dimensional reaction coordinate that involves concerted changes in both the chemical state and physical position along the path of the chemical reaction. The equilibrium populations in each well remain the same, but the transition rates between the wells are different due to the entropic effects of widening the transition state TrSt and the width W of the right well.

There is one very significant effect in biochemical reactions that illustrates the importance of entropic effects: hydration. Before a charged ion can bind to the amino acid, it must divest itself of several "*waters of hydration.*" This is because water, being a dipole, will tend to cluster about ions in solution, hindering them from binding to a charged site that is also insulated by its own hydration shell. Supoose for the sake of illustration that the energies binding the waters to the two reactants are just equal to the electrostatic energy of binding between the reactants. Binding seems unfavorable, since the ion will lose its translational and rotational degrees of freedom (≈ 3k_BT according to the equipartition theorem, The binding reaction can still proceed strongly because the liberation of the hydration waters is accompanied by a large entropy increase, since each water gains ≈ 3k_BT of rotational and translational energy, and so the term – TΔS is strongly negative.

All of this means that the rate constants summarize the statistical behavior of a large number of "hidden" coordinates that are very difficult to compute explicitly, but may be easy to measure phenomenologically. For our purposes, we shall adopt this phenomenological view of chemical reactions, and assume that the rate constants can be specified, so that the only entropic effect we need to deal with explicitly is the concentrations of the reactants, such as H^+ in (3.38). Therefore, we can treat reactions using Markov chain theory, as indicated by the 2-state model shown at the bottom of Figure 3.5A., whose equations of motion are $\frac{d[A^0]}{dt} = -\frac{d[A^-]}{dt}$ = net flow over the energy barrier = $J_\varsigma = k^*[A^-] - k^-[A^0]$, or in vector form

$$\frac{d}{dt}\mathbf{P} = \mathbf{J}_\varsigma = \mathbf{K}\cdot\mathbf{P},\ \mathbf{P} = \begin{pmatrix} p_- \\ p_0 \end{pmatrix},\ \mathbf{K} = \begin{pmatrix} k^* & -k^- \\ -k^* & k^- \end{pmatrix}. \quad \text{...(3.40)}$$

Here p_- and p_0 are the probabilities to have a negatively charged and neutral amino acide, respectively. In general, the reaction flux will have the form $\mathbf{J}_\varsigma = \mathbf{K(P)}\cdot\mathbf{P}$, where the matrix $\mathbf{K(P)}$ is the matrix of transition rates, *i.e.,* pseudo-first-order rate constants that may contain reactant concentrations that are held parametrically constant. Implicit in this formulation are the assumptions that (*i*) the actual reaction takes place instantaneously (electronic rearrangements are very fast), so that a substance remains in a chemical state for an exponentially distributed mean time before jumpeing (reacting) to anotheer state; (*ii*) the transition out of a state depends only on the state itself, and not on any previous history.

PHYSIOCHEMICAL MODEL

An important generalization is necessary to model molecular motors. We have spoken of the potential $\phi(x, t)$ that provides the deterministic forcing as an *external* force. However, a molecular motor $\phi(x, t)$ generally includes forces generated *internally* by the motor itself that drive the motor forward. Thus the potential term in (3.26) must be broken into two parts:

$$\phi(x, t) = \underbrace{\phi_1(x,t)}_{\text{internally generated forces}} + \phi_L \underbrace{(x,t)}_{\text{external load forces}}$$

where $\phi(x, t)$ is internally generated force and $\phi_L(x, t)$ is the external load force. A common situation is that of is a constant load force F_L, is which case $\phi_L = F_L \cdot x$, so that $\partial\phi/\partial x = -f_L$: *i.e.*, the load force acts to oppose the motor's forward progress. The internally generated force potential will generally depend on the chemical state of the system. That is, the mechanical evolution of the system's geometrical coordinates governed by (3.36) is coupled to the chemical reactions described by a Markov chain (3.40). Each *chemical state* is characterized by its own probability distribution $p_k(x, t)$, where k ranges over all the chemical states, and each chemical state is typically characterized by a separate driving potential $\phi(x, t)$ Thus there will be a Smoluchowski equation (3.36) for each chemical state, and these equations must be solved simultaneously to obtain the motor's motion.

For the neutralization reaction considered above, the total change in probability $p(x, \zeta\, t)$ is given by

$$\frac{\partial}{\partial t}\begin{pmatrix} p_1 \\ p_2 \end{pmatrix} = \text{net flow in space + net flow along reaction coordinates}$$

$$= -\overbrace{\begin{pmatrix} (\partial/\partial x_1)J_{x_1} \\ (\partial/\partial x_2)J_{x_2} \end{pmatrix}} + \overbrace{\begin{pmatrix} J\zeta_1 \\ J\zeta_2 \end{pmatrix}}$$

$$= -D\begin{pmatrix} -(\partial/\partial x_1)[p_1\partial(\phi_1/k_B T)/\partial x_1 + (\partial p_1/\partial x_1)] \\ -(\partial/\partial x_2)[p_2\partial(\phi_2/k_B T)/\partial x_2 + (\partial p_2/\partial x_2)] \end{pmatrix} + \begin{pmatrix} k^- p_2 - k^* p_1 \\ k^* p_1 - k_- p_2 \end{pmatrix} \quad \text{...(3.41)}$$

where the probability densities $p_i(x_i, t)$ $i = 1, 2$, now keep track of the motion along the spatial and reaction coordinates, and $J_{\zeta_1} = -J_{\zeta_2}$ keep track of flux along the reaction coordinate (since the reaction is of first order, *i.e.*, has only two states).

PROTEIN MOTION

We return to the problem of simulating the protein's motion numerically. Equation (3.30) is a very useful and easy to implement numerical scheme. However, one of its shortcomings is that it does not preserve the property of detailed balance. Detailed balance is the constraint placed on $c_{eq}(x)$ to ensure that systems in equilibrium do not experience a net drift. That is, when a system is in equilibrium, J in (3.32) is required to be identically zero. Detailed balance ensures that the equilibrium density has a Boltzmann distribution.

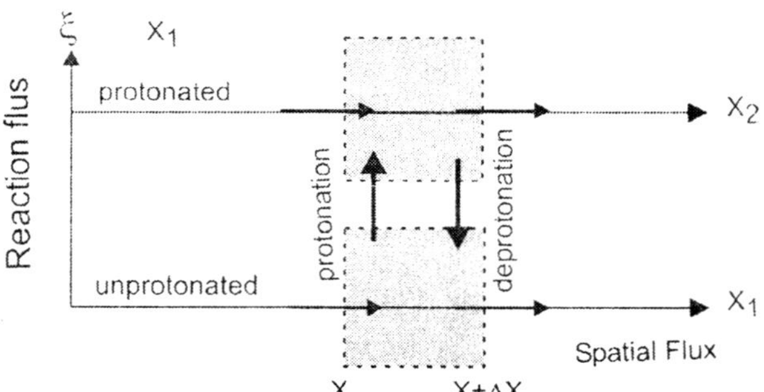

Fig. 3.6. The mechanochemical plase place. A point is defined by its spatial and reaction coordinates $(x(t), \zeta(t))$. The flow of probability in the spatial direction is given by the Smoluchowski model (3.35), and the flow in the reaction direction is given by the Markov model (3.40).

It is important to understnad the distinction between steady state and equilibrium. If we watch sample paths of $x(t)$ it is consistent for the trajectories to move with a mean velocity and for the system to have a steady state. In equilibrium the sample paths must not exhibit a mean velocity. It can be shown that (3.30) produces sample paths that show a net drift when the real system satisfies detailed balance. Clearly, it is desirable to have an algorithm that preserves detailed balance in equilibrium and can be used to simulate both equilibrium and nonequilibrium processes.

Numerical Algorithm

To obtain an algorithm that has detailed balance built in, we convert the problem into a Markov chain and use the procedurs described above to simulate it numerically. The numerical algorithm given by (3.30) is base on the discretization of time. To convert the problem into a Markov chain requires that we discretize space. Let $x_n = (n - 1/2)\, \Delta x$ for $n = 0, \pm 1, \pm 2, \pm 3, \ldots$ be the discrete sites on which the protein can reside. Site x_n is represented by the interval$[x_n - \Delta x/2, x_n + \Delta x/2]$. That is, when the protein is in the interval$[x_n - \Delta x/2, x_n + \Delta x/2]$. *we treat it as being at site* x_n. If the molecule is at site x_n,then it can jumps to either x_{n+1} or x_{n-1}. A diagram of this process is shown in

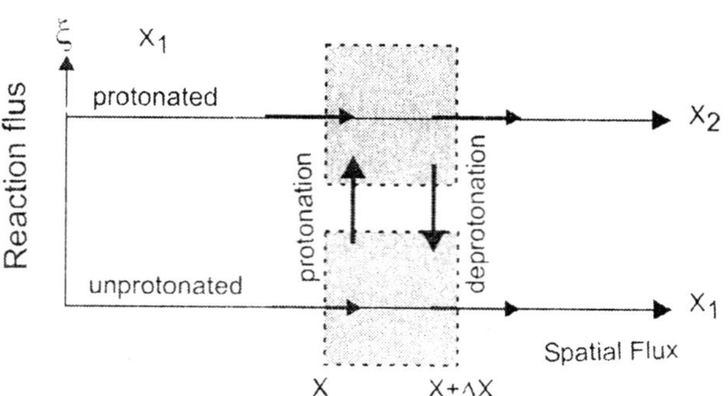

Fig. 3.7. The numerical discretization in spatial dimension. A continuous Markov process (the Langevin equation) is approximated to a set of discrete sites (x_n) and is allowed to jump only to the neighboring sites (x_{n-1}, X_{n+1}). The site x_n can be viewed as representing the interval $[x_n - \Delta - X/2, x_n + \Delta X/2]$.

Figure 3.7. The notation that we have adopted is that $F_{n+1/2}$ is the rate at which the protein jumps from x_n to x_{n+1} (F refers to a "*forward*" jump). Similarly, $B_{n+1/2}$ is the rate at which the protein jumps from x_{n+1} to x_n(B for "*backward*").

For small enough Δx, we have $p_n(t) \approx p(x, t)\,\Delta x$, where $p_n(t)$ is the probability that the protein is at x_n at time t. The governing equation for $p_n(t)$ is

$$\frac{dp_n}{dt} = -(B_{n-1/2} + F_{n+1/2})\,p_n + F_{n-1/2}\,p_{n-1} + B_{n+1/2}\,p_{n+1}$$

$$= (F_{n-1/2}\,p_{n-1} - B_{n-1/2}\,p_n) - (F_{n+1/2}\,p_n - B_{n+1/2}\,p_{n+1}) = J_{n-1/2} - J_{n+1/2}, \quad \text{...(3.42)}$$

where $J_{n+1/2}$ is the net flux between the points x_n and x_{n+1}.

In addition to preserving detailed balance, our numerical scheme must approximate the actual dynamics of the protein. In preserve Section we demonstrate that the following jump rates preserve the mean drift motion as well as detailed balance:

$$F_{n+1/2} = \frac{d}{(\Delta x)^2}\cdot\frac{\Delta\phi_{n+1/2/(k_B T)}}{\exp\left(\Delta\phi_{n+1/2/(k_B T)}\right)-1}, \quad \text{...(3.43)}$$

$$B_{n+1/2} = \frac{D}{(\Delta x)^2}\cdot\frac{\Delta\phi_{n+1/2/(k_B T)}}{\exp\left(\Delta\phi_{n+1/2/(k_B T)}\right)-1}, \quad \text{...(3.44)}$$

where

$$\Delta\phi_{n+1/2} = \phi(x_{n+1}) - \phi(x_n). \quad \text{...(3.45)}$$

Boundary Conditions

The algorithm described above must be complemented by boundary conditions. We discuss three types of boundary conditions: periodic, reflecting, and absorbing. In each case the total number of grid points with in the interval is M, and Δx = L/M, where L is the length of the spatial domain. The placement of the grid has been chosen such that $x_n = (n - 1/2)\,\Delta x$ for n = 1, 2 ..., M.

Periodic

Periodic boundary conditions require that $p_{M+1}(t) = p_1(t)$ and $p_0(t) = p_M(t)$. Using these two equalities in (3.42) for $p_1(t)$ and $p_M(t)$ Produces

$$\frac{dp_1}{dt} = -(B_{1/2} + F_{3/2})\,p_1 + F_{M+1/2}\,p_M + B_{3/2}\,p_2, \quad \text{...(3.46)}$$

$$\frac{dpM}{dt} = -(B_{M-1/2} + F_{M+1/2})\,pM + F_{m-1/2}\,p_{M-1} + B_{1/2}\,p_1. \quad \text{...(3.47)}$$

where we have also made use of the fact that $B_{1/2} = B_{M+1/2}$ and $F_{1/2} = F_{M+1/2}$.

Reflecting

Figure 3.8A shows a reflecting boundary condition locatded midway between the grid points M and M+ 1. A reflecting boundary requires that

$$J_{m+1/2(}t) = F_{m+1/2}\,p_M(t) - B_{M+1/2}P_{M+1}(t) = 0. \quad \text{...(3.48)}$$

we know that $pm + 1(t) = 0$, since it is located outside of the reflecting boundary and is inaccessible to the protein. Therefore, to enforce no flux through the boundary, we set $F^{\text{reflect}}_{M+1/2} = 0$. The equation for p_M is then

$$\frac{dp\text{M}}{dt} = -B_{M-1/2}p_M \; F_{M-1/2}p_{M-1} \quad ...(3.49)$$

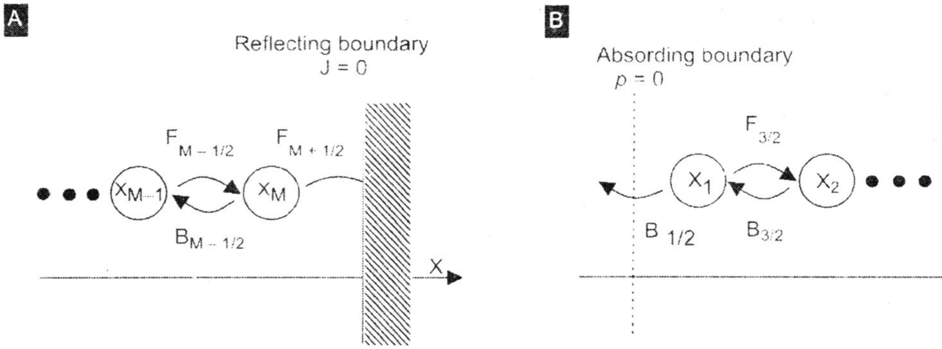

Fig. 3.8. Numerical treatments for two types of boundaries.(A) At a reflecting boundary, the particle is not allowed to jump through the boundary, and thus the flux through the boundary is zero.(B) At an absorbing bounday, the probability density is zero. Once the particle jumps out of the boundary, it should not be allowed to come back. However, in the numerical discretization, blocking the particle form coming back is not enoguh. The rate of the particle jumping out of the bounday has to be modified.

Absorbing

If the protein reaches an absorbing boundary it is instantaneously removed form the solution. Therefore, the probability of finding the protein at an absorbing bounday is zero. Figure 3.8B illustrates an absorbing boundary at x = 0. Thus, we must enforce the condition $p(0, t) = 0$. Thus derive the appropriate jump rate at this boundary:

$$B_{1/2}^{\text{absorb}} = \frac{\text{D}}{(\Delta x)^2}\cdot\frac{\alpha^2}{\exp(\alpha)-1-\alpha}, \quad \alpha\frac{\phi_0-\phi_1}{k_BT} \quad ...(3.50)$$

The equation for p_1 is then

$$\frac{dp_1}{dt} = -\left(\text{B}_{1/2}^{\text{absorb}}+\text{F}_{3/2}\right)P_1+B_{3/2}\,p_2. \quad ...(3.51)$$

It can be shown that this treatment of the absorbing boundary is accurate to second order in Δx and that it preserves the velocity of a perfect Brownian ratchet subject to any load force

Now we are ready to numerically intergrate p_n. However, before we turn to examples of implementing the algorithm, we must address the issue of numerical stability and introduce an implicit method for the time integration as an alternative to Euler's method.

Numerical Stability

For notational convenience let $p_n(k\Delta t) = p^k{}_n$. Then Euler's method has the form

$$p_n^{k+1} \approx p_n^k - \Delta t\left[\left(B_{n-1/2}+F_{n+1/2}\right)p_n^k + F_{n-1/2}\,p_n^{k+1} + B_{n+1/2}\,p_{n+1}^k\right] \quad ...(3.52)$$

Euler's method is called an explicit method because $p_n{}^{k+1}$ can be written explicitly in terms of $p^k{}_n$. Each time we use the above technique to update $p_n{}^{k+1}$, we introduce a small error due to the finite size of Δt and around-off error. In the absence of roudn-off error, we can achieve any desired accuracy by decreasing Δt. However, there are some problems with this approach. Usually, the biggest problem is the amount of computer time required when we choose a very samll Δt. However, the round-off error incurred in each step does not decrease with Δt. rather it accumulates. It is posible that if Δt is too small, the total error is dominated by the round-off error. In that situation, the more steps we take, the larger the accumulated error. So a careful choice of time step is important,

Numerical stability is another issue with which we have to contend. That is, we do not want our numerical solution to urn off to $\pm, \infty$ when the real solution is bounded for all time. It is possible to show that Euler's method is stable only if

$$\Delta t < \Delta t_c = \max_n \left(\frac{1}{F_{n+1/2} + B_{n+1/2}} \right) \qquad ...(3.53)$$

where max in the above equation means to use the value of n that produces the largest value of the quantity in the parentheses. Figure 3.9 illustrates this change in stability by using time steps slightly above and below Δt_c To get an intutive feel for this instability, let us consider one time step of the numerical scheme. At $t = 0$, we take $p_m^0 = 1$ and $p_n^0 = 0$ for $n \neq m$. From (3.52) we have

$$p_m^1 = \left[1 - \Delta t\left(B_{m-1/2} + F_{m+1/2}\right)\right] p_m^0. \qquad ...(3.54)$$

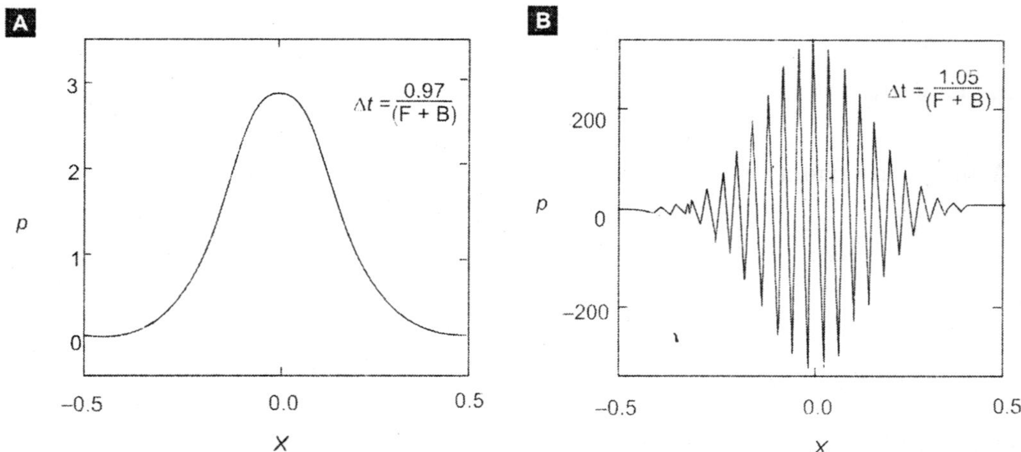

Fig. 3.9. Numerical stability/ instability.(A) When the time step is slightly below the critical step size, the numerical solution is stable. (B) When the time step is sligthly above the critical step size, the numerical solution is unstable.

It is clear that if $\Delta t > \Delta t_c$, then p_m^1 will be negative. Since p_m^1 is a probability, negative values clearly do not make sense. The condition on Δt for stability is rather restrictive. Using the jump rates given in(3.84) and(3.85), it is possible to show that

$$\Delta t_c < \frac{(\Delta x)^2}{2D}. \qquad ...(3.55)$$

This implies that in order to reduce the spatial step by a factor of 10(which could be necessary to model accurately spatial fluctuations of the force, for example), the time step must be reduced by a factior of 100.

Implicit Discretization

We now improve upon Euler's method in two ways. First, we use second-order algorithm that improves the accuracy of the soluttion for fixed Δt. Second-order algorithm we employ is called the Crank-Nicolson method. For a simple one-dimensional differential equation $dx/dt = h(x)$, the Crank-Nicolson method has the form

$$\frac{x^{k+1}-x^k}{\Delta t} = \frac{h\left(x^{k+1}\right) h\left(x^k\right)}{2} \quad ...(3.56)$$

For (3.42) this scheme becomes:

$$\frac{p_n^{k+1}-p_n^k}{\Delta t} = -\left(B_{n-1/2}+F_{n+1/2}\right)\cdot\frac{p_n^{k+1}-p_n^k}{2} + F_{n-1/2}\cdot\frac{p_{n-1}^{k+1}-p_{n-1}^k}{2} + B_{n+1/2}\cdot\frac{p_{n+1}^{k+1}-p_{n+1}^k}{2} \quad ...(3.57)$$

If we now bring all the p^{k+1} terms to the left-hand side and use the vector notation

$$\begin{pmatrix} p_1^k \\ p_2^k \\ \vdots \\ p_M^k \end{pmatrix} \quad ...(3.58)$$

(3.57) can be written in matrix form as

$$\mathbf{A}\,\mathbf{p}^{k+1} = \mathbf{C}\,\mathbf{p}^k, \quad ...(3.59)$$

where **A** and **C** are tridiagonal matrices with elements

$$A_{nn} = 1+\frac{\Delta t}{2}\left(B_{n-1/2}+F_{n+1/2}\right),\ A_{n,n-1} = -\frac{\Delta t}{2}F_{n-1/2},\ A_{n,n+1} = \frac{\Delta t}{2}B_{n+1/2}. \quad ...(3.60)$$

and

$$C_{nn} = 1-\frac{\Delta t}{2}\left(B_{n-1/2}+F_{n+1/2}\right),\ C_{n,n-1} = \frac{\Delta t}{2}F_{n-1/2},\ C_{n,n+1} = \frac{\Delta t}{2}B_{n+1/2}. \quad ...(3.61)$$

Equation (3.59) reveals why this method is called implicit. At each time step we must solve a linear set of coupled equations. Luckily, **A** is a sparse matrix, and it is therefore computationally fast to solve(3.59) for $\mathbf{p}^{k+1}$.

THE DRAG COEFFICIENT

The natural units if distance and force on the molecular scale are nanometers($1nm = 10^{-9}$ m) and piconewtons (1 pN $= 10^{-12}$ N), respectively. In these units, the viscosity of water at room temperature is $\eta \approx 10^{-9}$ pN.sec/nm^2 Then, a typical value for the hydrodynamic drage coefficient of a sphere of radius R = 10 nm is $\zeta = 6\pi\eta R \approx 10^{-7}$ $pN.sec/nm$ A dimensionless number that measures the ratio of inertial to viscous forces is the Reynolds number: $Re = \rho v R/\eta$, where ρ is the density of water (10^3 kg/m^3 = 10^{-21} pN.sec^2/nm^4). Typical velocities of molecular motors are $v < 10^3$ nm/sec, so on the molecular scale, the Reynolds number is very small indeed : R $\approx 10^{-8}$. This confirms our conjecture that we can safely ignore the inertial term in (3.19). If the fluid can truly be viewed as continuum, then ζ can be computed from hydrodynamics.The frictional drage coefficient ζ depends on the particle shape and size as well as the fluid viscosity: ζ = (dimensionless geometric drag coefficient) × (size factor) ×(size factor) × (shape factor). For a sphere, $\zeta = 6\pi\eta$ R; drag coefficients for other shapes are given in Berg (1993), a good source of intuition on Brownian motion.

The Equipartition Theorem

Let us consider a collision of two particles of masses m_1 and m_2 with velocities v_1 and v_2 before the collision, and with velocities v'_1 and v'_1 after collision, respectively. Conservation of energy and momentum guarantee conservation of the velocity of the center of mass after the collision, as well as of the absolute value of the relative velocity. One of the central assumptions of statistical mechanics is that the velocities of the scattered particles are uncorrelated.

From this one can show that $\langle m_1 v'^2_1 \rangle = \langle m_2 v'^2_2 \rangle$. A more general result that can be dervied from statistical mechanics is that each quadratic degree of freedom(*e.g*, linear or angular momentum) of a particle carries an average amount of energy $\langle E \rangle = k_B T/2$. Thus the mean kinetic energy of a point particle moving in the x direction is $\langle mv_x^2/2 \rangle = k_B$ T/2, or $v_x^2 = k_B$T/M. If the particle is moving in a harmonic potential well (*i.e.*, on a spring), its mean potential energy is $k\langle x^2 \rangle/2 = k_B$T/2. Thus the mean total energy is $\langle E \rangle = \langle E_{kin} \rangle + \langle E_{pot} \rangle = k_B T$.

Langevin Equation

For (3.29) to be useful, we must specify the statistical properties of $f_B(t)$ models the net effect of many protein-water interactions, from the central limit the orem it seems reasonable to assume that $f_B(t)$ is normally distrubuted. physically, we also want $(f_B)(t) = 0$, since a protein that is undergoing pure diffusion does not experience a net force. All that is left to fully characterize $f_B(t)$, is to specify its covariance cov $[f_B(t)\ f_B(s)]$. Remember that a necessary condition for two random variables to be independent is that their covariance is zero. Since the motion of the water molecules is very fast compared with the motion of the diffusing protein, we take $f_B(t)$ and $f_B(s)$ to be statistically independent whenever $t \neq$ s. When $\phi = 0$, we shoud recover diffusive motion. That is, the variance of a Brownian particles started at $x(0) = 0$ is

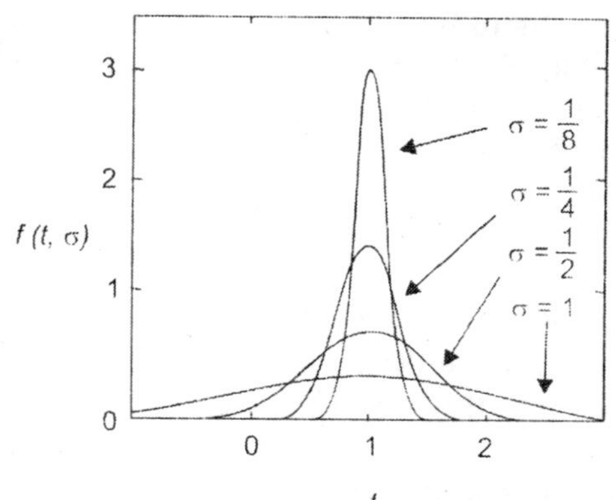

Fig. 3.10. The Dirac delta function can be viewed as the limit a sequence of normal probability density functions as the standard deviation goes to zero.

$$\text{VAR}[x(t)] = \langle x(t)^2 \rangle = 2\text{Dt}. \quad ...(3.62)$$

we claim that an appropriate choice for the covariance is

$$\text{cov}[f_B(t)\, f_B(s)] = f_B(t)\, f_B(s) = 2k_B\, \text{T}\, \zeta\delta(t{-}s). \quad ...(3.63)$$

The Dirac delta function $\delta(t - s)$ is(3.63) is a mathematical concept that is best understood as the limit of a normal distribution centered at s as the variance goes to zero:

$$\delta(t-s) = \lim_{\sigma \to 0} \frac{1}{\sqrt{2\pi\sigma^2}} \exp\left(-\frac{(t-s)^2}{2\sigma^2}\right). \quad ...(3.64)$$

This is illustrated in Figure 3.10. The only property of the Dirac delta function that we need is $\int_{-\infty}^{\infty} g(t)\delta(t-s)\,dt = g(s)$, which is easily understood when the Dirac delta function is interpreted as a sharply peaked probability density. We now have a complete description of $f_B(t)$. A random variable described as such is referred to as *gaussian white noise.*

Next we illustrate that our choice of $f_B(t)$ markes sense both mathematically and physically. Note that form (3.28) with $\phi = 0$ and $x(0) = 0$, we have

$$x(t) = \frac{1}{\zeta}\int_0^t f_B(t')dt', \qquad \text{...(3.65)}$$

from which it is clear that $\langle x(t)\rangle = 0$. Assume without loss of generality that $t > s$. One can show(from the theory of Dirac's delta function) that

$$\cos[x(t)\,x(s)] = \frac{1}{\zeta^2}\left\langle\left(\int_0^t f_B(t')\,dt'\right)\left(\int_0^s f_B(t'')\,dt''\right)\right\rangle$$

$$= 2D\int_0^t \int_0^s \delta(t'-t'')\,dt'dt'' = 2DS, \qquad \text{...(3.66)}$$

which is consistent with (3.62) when $s = t$. Furthermore, since $f_B(t)$ is a normal stochastic variable, so is $x(t)$.

If we define the new random variable $W(t) = x(t) = x(t)\sqrt{2D}$, then $W(t)$ is a normal random variable characterized by

$$\langle W(t)\rangle = 0, \text{cov}[W(t)W(s)] = \min(t, s), \qquad (3.67)$$

where the min in (3.67) means to use the value of t or s that is smaller. The random variable variable $W(t)$ is referred to as a *wiener process* and possesses some interesting mathematical properties, which we will not go into here. Note that form (3.29) what we need for our numerical algorithm is actually the incremental Wiener process defined as

$$\Delta W(t) = W(t + \Delta t) - W(t) = \frac{1}{\sqrt{2k_B T\zeta}}\int_t^{t+\Delta t} f_B(t')\,dt' \qquad \text{...(3.68)}$$

Following a procedure similar to that used in (3.66), it is straightforward to show that $\Delta W(t)$ is a normal random variable with mean zero and standard deviation $\sqrt{\Delta t}$. It is also possible to show that all $\Delta W(t)$ and $\Delta W(s)$ are statistically independent for $t \neq s$. This gives us a way to generalize Elur's method to include Gaussian white noise. That is, a numerical method for simulating (3.28) is (3.30).

Thermodynamics Relation

Note that the flux (3.32) can also be written as $J_x = -(c/\zeta)\partial/\partial x(k_B T \text{ in } c + \phi)$. There can be many *steady states* characterized by a constant flux J_x = const; one of these is the special case of *eduilibrium:* $J_x = 0$. At equilibrium distribution, one can define the quanitity $\mu = (k_B T \text{ In } c_{eq} + \phi)$ called the *chemical potential.* The equilibrium distribution of $c_{eq}(x)$ can be computed by setting the gradient in chemical potential to zero, so that μ =const; this is exactly equivalent to enforcing a Boltzmann distribution (3.33).

The chemical potential is also the free energy *per mole,* $G = \mu N$, where N is the mole number: A mole is an Avogadro's number N_a of objects(*e.g*, molecules), where $N_a = 6.02.\ 10^{23}$[#/mol]. At *equilibrium* we can define the *entropy* $S \equiv -k_B N \text{ In } c_{eq}$ and the *enthlapy* $H = \phi N$.Then we arrive at the definition of the free energy $G = H - TS$. These definitions will prove useful when we discuss chemical reactions. Here we note simply that diffusion smoothese out the concentration, leading to an increase in entropy. Thus entropic increase accompanying the motion of the ensemble is handled by the Fickian diffusion term in the flux (3.32).

When the particles are charged (*e.g.*, protons, H^+), then the chemical potential difference between two states,or across a membrane is written as $\Delta\mu = \mu_2 - \mu_1 = (\phi_2 - \phi_1) + k_B T$ (In c_2 – In c_1) = $\Delta\phi$ – 2.3 kgT ΔpH. Here pH = – $log_{10}c_{H^+}$, where c_{H^+}, is the proton concentration. The protonmotive force is defined as p.m.f. = $\Delta\mu/e = \Delta\psi - 2.3(k_B T/e)\,\Delta$pH, were e is the electronic charge, and $\Delta\psi = \Delta\phi$/e(mV) is the transmembrane electric potential.

Consider the simple case where a protein motor is propelled by an internally generated motor force f_M and opoosed by a constant load force f_L. Forexample, $f_M = \Delta\, G/l$, where l is the length of the power stroke and ΔG is the free energy drop accompanying one cycle of the chemical reaction that is supplying the energy to the motor.(This would be an ideal motor: It used *all* of the chemical energy to produce a constant-force power stroke!) Then the Langevin equation(3.19) becomes

$$\frac{dx}{dt} = v,\ m\frac{dv}{dt} = -\zeta v + f_M - f_L + f_B(t). \qquad ...(3.69)$$

The diffusion equation associated with(3.69) for the probability density $p(x, v, t)$ is called the *kramers equations:*

$$\frac{\partial p}{\partial t} = -\frac{\partial}{\partial x}(vp) - \frac{1}{m}\frac{\partial}{\partial v}\left[(f_M - f_L) - \zeta v)p - \left(\zeta\frac{k_B T}{m}\right)\frac{\partial p}{\partial v}\right]. \qquad ...(3.70)$$

The Smoluchowski diffusion equation (3.35) is a special case of the Kramers equation; both are generically referred to as *Fokker&Planck equations.* However, deriving(3.35 form 3.70) is not trivial: it requires a *singular perturbation treatment* that is beyond the scope of this chapter, but is discussd briefly in Appendix A.

We set $\partial p/\partial t$ = 0(3.70) to look for the steady state. Multiplying by v^2 and taking the average by intergrating over x and v, and using the equipartition theorem, we obtain

$$0 = \underbrace{-\zeta\langle v^2\rangle}_{1} + \underbrace{f_{M\langle v\rangle}}_{2} - \underbrace{f_L\langle v\rangle}_{3} + \underbrace{\frac{k_B T}{m}\zeta}_{4}. \qquad ...(3.71)$$

At constant temperature, the terms in (3.71) have the following interpretation:

(*i*) The rate at which the motor dissipates energy via frictional drag with the fluid.

(*ii*) The rate at which energy is being absorbed by the motor from the chemical reaction.

(*iii*) The rate of work done by the load force against the motor.

(*iv*) The rate at which the motor absorbes energy from the thermal fluctuations of the fluid.

Thus we see that when the chemical reaction is turned off, ΔG = 0, the heat absorbed by the motor from the thermal environment(term 4) is

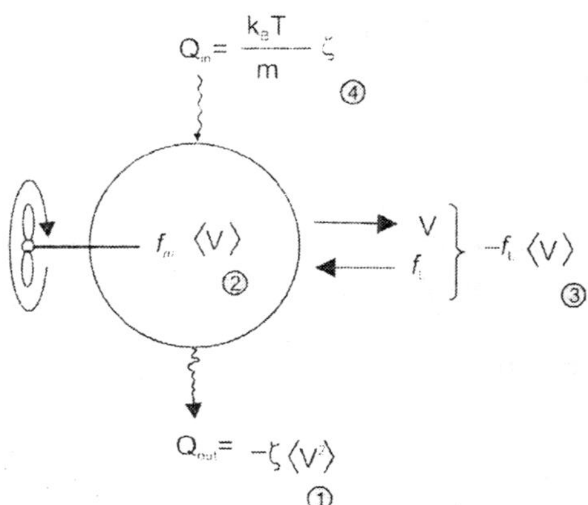

Fig. 3.11. Energy blance on a protein motor. Q_{out} is the heat dissipated by the motion of the motor(term 1). Q_{in} is the heat supplied to the motor by thermal fluctuations of the fluid(term4). The rate of work done by the load force is $-f_L$.(v)(term 3). The rate of work done by the chemical reaction driving the motor is f_M.(v)(term 2).

just equal to the heat returned to the environment by frictional drag(term 1) in the absence of the load force. If the reaction driving the motor were *endothermic,* then it is possible for the motor to move by taking heat form the environment without violating the second law of thermodynamics.

Jumping Beans and Entropy

An analogy may make the role of entropic factors clearer. Imagine that the left enthalpic well in Figure 3.5B is filled with jumping beans whowe hops are random in height and angle. We can vary the equilibrium populations of beans in each well without altering the height of the enthalpic barrier by simply increasing the width of the transition state or of one of the wells.This is shown in Figure 3.5B. Now a bean in the right well may execute many more futile jumps before hurdling the barrier: If it jumps from the right side of the well, it will fall back into the well even if its jump is high enough, or if it reaches the transition state it must diffuse(hop) along the plateau randomly with a high probability of hopping back in to the right well. Both of these effect make it more difficult to eascape from the right well, and so the equilibrium population there will increase, as will k_{eq}, the equilibrium population ratio.

The rate at which beans can pass the barrier from right to left will have the form

$$k^{-} = v \cdot \exp\left(\Delta G_2^{\dagger}/k_B T\right)$$

$$= \underbrace{v}_{1} \cdot \underbrace{\exp\left(\Delta S/k_B\right)}_{2} \cdot \underbrace{\exp\left(-\Delta H/k_B T\right)}_{3} \cdot \underbrace{\exp\left(-\Delta F_L\, L/k_B\, T\right)}_{4},$$

Where(1) v is frequency factor (number of jumps/unit time.). For reactions that involves an atomic vibration, this is approximately $k_B T/h$, where h is Planck's constant. For diffusion controlled reactions this is of order D/L^2, where D is the diffusion coefficient and L a characteristic dimension.(2) The entropic term $e^{\Delta / kB}$ accounts for geometric and " *hidden variables*" effects.(3) The enthalpic term $e^{-\Delta H/kBT}$ is free energy "payoff" for a successful jimps;it accounts for the electrostatic and/ or hydrophoaic interactions.(4) If the reaction involves a meachanical step that is opposed by a load force, F_L, then fourth term accounts for the penalty exacted by performing work against the load. All of these effects are contained in the kinetic rate constants and can estimated from more detailed models.

Note that the height $\Delta G_2^{\dagger}$ of the free energy barrier determines *how fast the reaction goes.* The exponent exp $\left(\Delta G_2/k_B T\right)$ is called the *Arrhenius factor.* Because of this factor the reaction rate depends dramatically on the height of the free energy barrier.

Jump Rates

Here we determine the appropriate values of $F_{n+1/2}$ and $B_{n+1/2}$ in(3.42) Suppose the system, with proper boundary restrictions, attains an equilibrium as time goes to infinity: $p_n^{(eq)} = \lim_{t \to \infty} p_n(t)$. Then the rates should preserve the property of detailed balance. That is,

$$J_{n+1/2} = F_{n+1/2}\, p_n^{(eq)} - B_{n+1/2}\, P_{n+1}^{(eq)} = 0. \quad ...(3.72)$$

Making use of the equilibrium distribution given by (3.33) we have

$$\frac{F_{n+1/2}}{B_{n+1/2}} = \frac{p_{n+1}^{(eq)}}{p_n^{(eq)}} = \frac{P_{eq}(x_{n+1})}{P_{eq}(xn)} \exp\left(\frac{-\Delta\phi_{n+1/2}}{k_B T}\right) \quad ...(3.73)$$

where

$$\Delta\phi_{n+1/2} = \phi(x_{n+1}) - \phi(x_n). \qquad ...(3.74)$$

equation (3.73) is our first constraint on the jump rates.

Besides preserving detailed balance, our numerical scheme must of course approximate the actual dynamics of the protein. Let us consider the two simplest statistical properties of the random variable $x(t)$, namely, the mean and the variance. To simplify the presentation, we make the assumption that $\phi(x) = -fx$. That is, our Brownian particle feels a constant force of strength f. For this problem(3.28) reduces to

$$\frac{dx}{dt} = \frac{f}{\zeta} + \frac{f_B(t)}{\zeta} \qquad ...(3.75)$$

Assuming $x(0) = 0$, the above equation can be integrated to produce

$$x(t) - \frac{f}{\zeta}t + \frac{1}{\zeta}\int_0^t f_B(s)\,ds. \qquad ...(3.76)$$

Using(3.76) the mean and variance of $x(t)$ are found to be

$$\langle x(t)\rangle = \frac{f}{\zeta}t, \qquad ...(3.77)$$

$$\text{Var}[x(t)] = 2Dt. \qquad ...(3.78)$$

Remember that in the discrete version, $x(t) = \Delta x n(t)$. Since the force acting on the protein is constant, the forward and backward rates are independent of n. Therefore, we drop the subscripts and use F and B. Using(12.42), it is straightforward to show that

$$\langle x(t)\rangle = \Delta x\langle n(t)\rangle = (\text{F} - \text{B})t, \qquad ...(3.79)$$

$$\text{Var}[x(t)] = (\Delta x)^2\,\text{Var}[n(t)] = (\Delta x)^2(\text{F}+\text{B})t. \qquad ...(3.80)$$

Equating the mean and the variance given in (3.77) and (3.78) with those of (3.79) and (3.80) gives us two more constraints on the jump rates. To summarize, we would like F and B to satisfy the following three equations:

$$\frac{\text{F}}{\text{B}} = \exp\left(\frac{f\Delta x}{k_B\text{T}}\right) \quad \text{[detailed balance]} \qquad (3.81)$$

$$(\text{F}-\text{B})\,\Delta x = \frac{f}{\zeta} \quad \text{[mean]}, \qquad (3.82)$$

$$(\text{F}+\text{B})(\Delta x)^2 = 2\text{D} \quad \text{[variance]}. \qquad (3.83)$$

Generally, its is impossible to satisfy three equations with two unkonwns. Let us ignore the constraint on the variance for the time being. The rates that satisfy(3.81) and(3.82) are

$$\text{F} = \frac{\text{D}}{(\Delta x)^2}\cdot\frac{-f\Delta x/(k_\text{B}\text{T})}{\exp(-f\Delta x/(k_\text{B}\text{T}))-1}, \qquad ...(3.84)$$

$$\text{B} = \frac{\text{D}}{(\Delta x)^2}\cdot\frac{f\Delta x/(k_\text{B}\text{T})}{\exp(f\Delta x/(k_\text{B}\text{T}))-1}, \qquad ...(3.85)$$

Additionally, this set of jump rates satisfies(3.83) with an error of $O((\Delta x)^2)$. We point out that this choice of F and B is an improvement over the rates used by Elston and Doering, since the mean is exactly preserved and F and B have finite values as $k_B T \to 0$.

In general, the force f will not be constant, but will depend on x. In this case the jumps rates will depend on n, and are given by (3.43) and (3.44).

Absorbing Boundary

To derive an appropriate jump rate at this boundary, we approximate(3.35) near $x = 0$ by

$$\frac{\partial p}{\partial t} = D\frac{\partial}{\partial x}\left(-\frac{f}{k_B T^p} + \frac{\partial p}{dx}\right) \quad ...(3.86)$$

where $f = -(\phi_1 - \phi_0)/\Delta x$ is an approximation for $-\partial\phi/\partial x$ in $(0, \Delta x)$. To derive a second-order treatment of the boundary, we need only a first-order approximation for this derivative.

Next we assume that at any given time $p(x, t)$ in the interval $(0, \Delta x)$ is approximately at steady state. This assumption is valid because at small length scales diffusion is the dominant effect. The scale for a particle with diffusion coefficient D to diffuse a distance Δx is $\Delta t_{dif} = (\Delta x)^2/2D$, which is proportional to $(\Delta x)^2$. The time scale for a flow with velocity v to travel a distance of Δx is $\Delta t_{flow} = \Delta x/v$, which is proportional to Δx. For small Δx, $\Delta t_{diff} \ll t_{flow}$. At small length scales, diffusion relaxes the system to steady state immediately after it is disturbed by the flow. Thus, at any given time, the local structure of the solution is given approximately by the steady-state solution. At an absorbing boundary, the steady-state assumption in $(0, \Delta x)$ can also be justified mathematically by examining (3.86) at $x = 0$, the left-hand side of is exactly zero at $x = 0$, Therefore, we set the left-hand of (3.86) to zero in the interval $(0, \Delta x)$ and solve the resulting ordinary differential equation subject to the following two conditons:

$$p(0) = 0, \int_0^{\Delta x} p(x)dx = p_1 \quad ...(3.87)$$

The solution is

$$p(x) = p_1 \frac{\exp(fx/(k_B T)) - 1}{\frac{k_B T}{f}\left[\exp(f\Delta x/(k_B T)) - \right] - \Delta x} \quad ...(3.88)$$

Using the above expression for $p(x)$, the flux is found to be

$$J = D\frac{f}{k_B T}p - D\frac{\partial p}{\partial x} = -p_1 \frac{D}{(\Delta x)^2}\cdot\frac{\alpha^2}{\exp(\alpha) - 1 - \alpha'} \quad \alpha\frac{f\Delta x}{K_B T} \quad ...(3.89)$$

In the numerical scheme,the flux at the boundary is

$$J_{1/2} = -p_1 B_{1/2} \quad ...(3.90)$$

This equation reflects the facts that once the protein is absorbed, it does not return to the fluid. Comparing (3.89) with (3.90), we get (3.50).

CHAPTER

4 IONIC CHANNEL GATING

The time course of voltage changes in a whole cell is the result of the average behavior of many individual channels. Our understanding of individual channel gating comes largely from experiments using the patch clamp technique. For example, typical measurements from a so-called on-cell patch of T-type calcium currents in guinea pig cardiac ventricular myocytes are the shown in the middel panel of Figure 4.2 The small current deviations in the negative direction indicate the opening of individual T-type calcium channels gating in response to a command membrance voltage stepped from –70 m V to –20 m V (top panel). Notice that two conductance

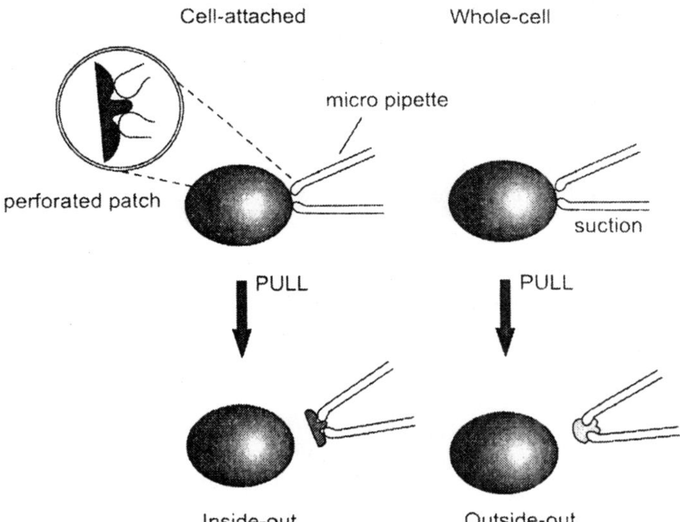

Fig. 4.1. Four methods of measuring electrical responses in cells with the patch clamp technique. In the patch technique, a pipette with an opening of $\approx 1\mu m$ is used to make a high-resistance ("gigaohm" = 10^9 ohm) seal onto a cellular membrane. In the on-cell patch configuration all the current into the pipette flows directly through the patch, which can contain as few as one or two ion channels. in the whole-cell configuration, the patch is broken and more accurate whole cell recording can be made as compared with a relatively leaky sharpe electrode puncture. In a perforated patch configuation, an ionophore such as nystatin is introduced into the pipette in order to allow whole-cell-like access while minimizing exchange of the cell contents with the contents of the pipette. Alternatively, patches of membrane can be torn off, leading to inside out and outside-patches that can be studied in isolation. Adapted from Hille (2001).

states of the channel are visable: a closed state with no current flowing and an open state with unitary current of $\approx 10^{-12}$ amperes (1 picoampere or 1 pA). While transitions between these two conductance states are random in time, the mean of several hundred records (bottom panel of Figure 4.2). smoothes out these current fluctuations and demonstrates that *on average* the stochastically gating channel activates and subsequently inactivates with time constants of $\approx$ 5 ms and 50 ms, respectively. Interestingly, the average dynamics of individual T-type calcium channels is strikingly similar to " whole cell" measurements of the activation and inactivation of T-type calcium currents.

MARKOV PROCESS

The stochastic gating of a single ion channel can be modeled as continuous-time Markov process. Consider the simple transition-state diagram the kinetic scheme for ion channel with two states, one closed (C) and the other open (O),

$$\text{C (closed)} \underset{k^-}{\overset{k^+}{\rightleftharpoons}} \text{O(open)}. \qquad ...(4.1)$$

Define s to be a random variable taking values s $\in$ {C, O} corresponding to these two states, and write Prob $\{s = i, t\}$ (or for short, $P_i(t)$ to represent the probability that $s(t) = i$; that is, the molecule is in state i at time t. Because the molecule must always be in one of the two states, total probability must be conserved and we have

$$Pc\ (t) + Po\ (t) = 1.$$

Now consider the possibility that the two-state ion channel is in state C at time t. If this is the case, then the rate k^+ (*e.g.*, with units of ms^{-1}) is related to the probability that in a short interval of time (Δt) the two-state ion channel will open. The relationship is given by

$$k^+\Delta t = \text{Prob}\ \{s = O,\ t + \Delta t / s = C,\ t\}, \qquad ...(4.2)$$

where $k^+ \Delta t$ is dimensionless (a pure number) and Prob $\{s = O, t + \Delta t/s = C, t\}$, is a shorthand notation for the probability, given that the channel is closed at time t, of c C $\rightarrow$ O transition occurring in the interval $[t, t + \Delta t]$. Multiplying by $Pc(t)$, the probability that the ion channel is indeed in state C, we find that $k^+ Pc(t) \Delta t$ is the probability that the transition C $\rightarrow$ O actually occurs.

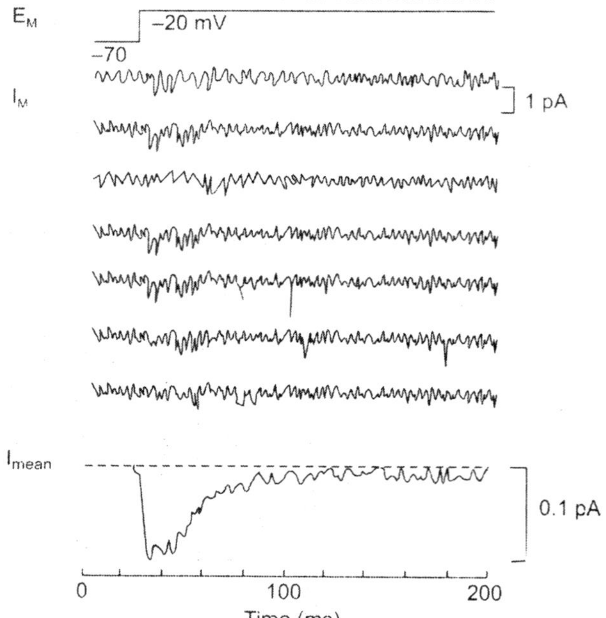

Fig. 4.2. On-cell patch clamp measurements of T-type calcium currents in guinea pig cardiac ventricular myocytes. The upper recordings show currents due to one (or a few) stochastically gating single channels when the command voltage is stepped from -70 mV to –20 mV. The lower plot is an average of severl hundred such records that shows rapid activation followed by slow inactivation, proportional to macroscopic T-type calcium currents measured in whole-cell configuration.

The transition-state diagram (4.1) indicates two possible ways for the ion channel to enter or leave the closed state. Accounting for both of these, we have

$$\text{Pc}(t + \Delta t) = \text{Pc}(t) - k^{+} \text{Pc}(t) \Delta t + k^{-} \text{Po}(t) \Delta t.$$

Writing a similar equation relating $\text{Po}(t + \Delta t)$ and $\text{Po}(t)$ and taking the limit $\Delta t \to 0$ gives the system of ODEs

$$\frac{d\text{Pc}}{dt} = -k^{+} \text{Pc} + k^{-} \text{Po}, \quad \text{...(4.3)}$$

$$\frac{d\text{Po}}{dt} = +k^{+} \text{Pc} - k^{-} \text{Po}. \quad \text{...(4.4)}$$

Because conservation of probability ensures that $\text{Pc}(t) = 1 - \text{Po}(t)$, (4.3) can be eliminated to give

$$\frac{d\text{Po}}{dt} = k^{+}(1 - \text{Po}) - k^{-} \text{Po}.$$

Note that the similarity of this euation to the kinetic equation is not accidental. The equation governing changes in probabilities for a single molecule always has the same form as the rate equation for a large number of molecules.

The Transition Probability Matrix

From our analysis of the two-state ion channel above, we know that for a channel closed at time t, $k^{+} \Delta t$ is the probability that it undergoes a transition and opens in the time interval $[t, t + \Delta t]$, provided that Δt is small. By conservation, we also know that the probability that the channel remains closed during the same interval is $1 - k^{+} \Delta t$. Because a similar argument applies when the channel is open at time t, we can write the *transition probability matrix*

$$Q = \begin{bmatrix} \text{Prob}\{C,t+\Delta,t|C,t\} & \text{Prob}\{C,t+\Delta|O,t\} \\ \text{Prob}\{C,t+\Delta,t|O,t\} & \text{Prob}\{C,t+\Delta|O,t\} \end{bmatrix} = \begin{bmatrix} 1-k^{+}\Delta t & k^{-}\Delta t \\ k^{+}\Delta t & 1-k^{-}\Delta t \end{bmatrix}, \quad \text{...(4.5)}$$

where the elements of Q_{ij} (row i, column i) correspond to the transition probability from state j to state i, and conservation of probability ensures that all the columns sum to one, that

The is, for each column j,

$$\sum_i Q_{ij} = 1. \quad \text{...(4.6)}$$

The Transition probability matrix is especially useful when we write the current state of the channel as the vector

$$\vec{\text{P}}(t) = \begin{bmatrix} \text{Prob}\{C,t\} \\ \text{Prob}\{O,t\} \end{bmatrix}. \quad \text{...(4.7)}$$

Using this natation, the state of the channel at $t + \Delta t$ is given by the matrix multiplication

$$\vec{\text{P}}(t+\Delta t) = Q\vec{\text{P}}(t) \quad \text{...(4.8)}$$

For example, if the channel is known to be close at time t, then

$$\vec{\text{P}}(t) = \begin{bmatrix} 1 \\ 0 \end{bmatrix},$$

and the distribution of probability after one time step is

$$\vec{\text{P}}(t+\Delta t) = \begin{bmatrix} 1-k^{+}\Delta t & k^{-}\Delta t \\ k^{+}\Delta t & 1-k^{-}\Delta t \end{bmatrix} \begin{bmatrix} 1 \\ 0 \end{bmatrix} = \begin{bmatrix} 1-k^{+}\Delta t \\ k^{+}\Delta t \end{bmatrix},$$

Applying (4.8) iteratively, we see that if the channel is closed at time t, the probability that it is closed or open at time $t + 2\Delta t$ is given by

$$\vec{P}(t+2\Delta t) = Q\left[Q\vec{P}(t)\right], = Q^2\vec{P}(t)$$

or more generally,

$$\vec{P}(t+n\Delta t) = Q^n\vec{P}(t). \quad ...(4.9)$$

This iterative procedure can be used to calculate the evolution of the probability that the two-state channel is in an open or closed state. It amounts to using Euler's method to intergrate (4.3) and (4.4)

Dwell Times

Using the transition probability matrix, it is possible to derive an expression for the average anount of time that the channel remains in the open or closed state, *i.e,.* the open and closed *dwell times.* We have already seen that if a channel is closed at tiem t, the probability that it reamins closed at time $t + \Delta t$ is $1- k^+ \Delta t$. The probability that the channel remains closed for the following time step as weel is thus $(1- k^+ \Delta t.)^n$ general, we can write

$$\text{Prob}\ \{C, [t, t + n\Delta t]\ |C, t\} = (1 - k^+ \Delta t)^n. \quad ...(4.10)$$

This expression is actually much simpler than (4.9) because here we are insisting that the channel remain closed for the entire interval $[t, t + n\Delta t\ ]$, while (4.9) accounts for the possibility that the channel changes states multiple times. If we define $\tau = n\Delta t$, we can rewrite (4.10) as

$$\text{Prob}\ \{C, [t, t + \tau\ ]|C, t\} = \left(1-\frac{k^+\tau}{n}\right)^n,$$

which is an approximate expression that becomes more accurate (for fixed τ) as $\Delta t \to 0$ and $n \to \infty$ simultaneously. Taking this limit and using

$$\lim_{n\to\infty}\left(1-\frac{\alpha}{n}\right)^n = e^{-\alpha},$$

we obtain

$$\text{Prob}\{C, [t, t + \tau]\ |C, t\ \} = e^{-k^+\tau}. \quad ...(4.11)$$

Thus, the probability that a channel closed at time t remains closed until $t + \tau$ is a decreasing exponential function of τ.

In order to complete our calculation of the closed dwell time for the two-state channel, we must consider the probability that a channel closed at time t stays closed during the interval $\{t, t + \tau]$ and then opens for the first time in the interval $[t + \tau, t + \tau + \Delta t]$. This probnability is given by

$$\text{Prob}\{C, [t, t + \tau]\ |C, t\}\ \text{Prob}\ \{O, t + \tau + \Delta t|C, t + \tau\} = e^{-k^+\tau} k^+ \Delta t.$$

Thus, the average closed time will be given by

$$\langle \tau c\rangle = \int_0^\infty \tau e^{-k^+\tau} k^+\, d\tau = \frac{1}{k^+},$$

where we have used

$$\int_0^\infty \tau e^{-t}\, dt = 1.$$

Similarly, the average open time of the two-state channel model is

$$\langle \tau o \rangle = \int_0^\infty \tau e^{-k^- \tau} k^{-d\tau} = \frac{1}{k^-}.$$

Monte Carlo Simulation

The elements Q_{ij} of the transition probability matrix represent the probability of making a transition from state j to state i in a time step of duration Δt. A simple method for simulating the transitions of a two-state channel is base on (4.6) Because conservation of probability ensures that each column of Q will sum to unity, we can divide the interval into regions, each corresponding to a possible change of state (or lack of change of state). Next, we chosse a random number Y uniformly distributed on the interval, and make a transition (or not) based upon the subinterval in which Y falls. For example, let us return to the transition probability matrix for the two-state channel given by (4.5) . If the current state is O (open), then transition to the closed state occurs if $0 \leq Y \leq k$- Δt, while the channel remains open if $k^- \Delta t \leq Y \leq 1$, an interval of lenght $1 - k^-\Delta t$. Similarly, if the channel is closed, it remains closed if $0 \leq Y \leq 1 - k^+ \Delta t$, and a transition to the open state occurs if $1 - k^+ \Delta t \leq Y \leq 1$.

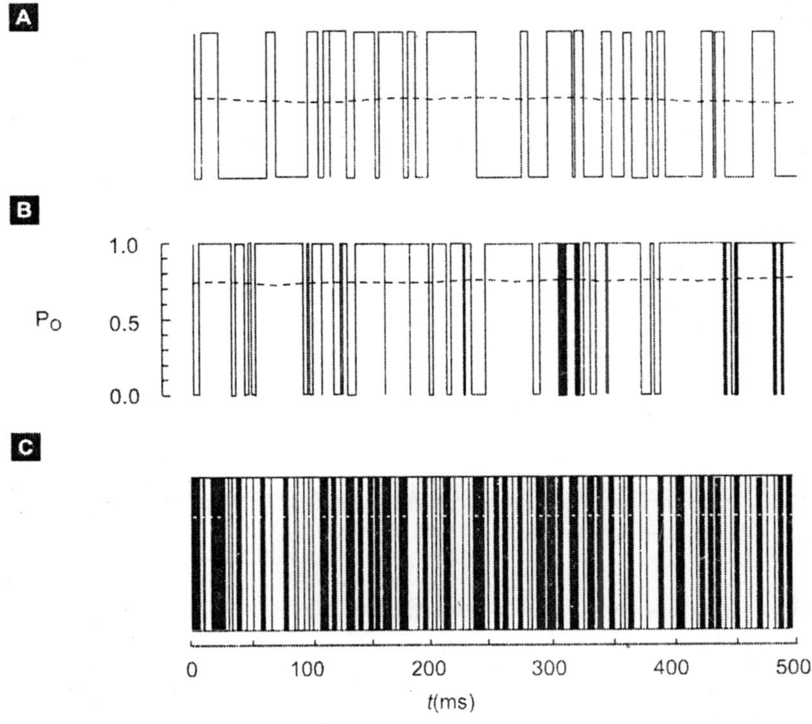

Fig. 4.3. Monte Carlo simulation of the two-state ion channel. (A) $k^+ = 0.1/ms$, $k^- = 0.1$/ms, given an equilibruim open probability (dotted lines) of 0.5 (B) k^+ changed to $0.3/ms$, and now the equilibrium open probability is 0.75. (C) Transition probabilities increased by factor of 5 ($k^+ = 1.5/ms$, $k^- = 0.5/ms$). Note that average open time and average close time are shorter in this case, as evidenced by many more transitions between states.

Several example simulations of stochastic gating of a two-state channel model jusing the Monte Carlo method are shown in Figure 4.3. By comparing open probabilities and dwell times in the three simulations shown, one can see how the transition probabilities k^+ and k^- lead to distinct channel kinetics.

Channels

The gating of multiple independent channels can be simulated in one of several ways. An obvious possible method for simulating a small number of independent two-state ion channels is to implement N Markov variables with identical transition probability matrices given by (4.5)

Under the assumption of identical and independent channels, an alternative method is to simulate a single Markov process that accurately tracks the number of open chanels. Note that for an ensemble of N Two-state ion channels there are N + 1 possibilities for the number of open channels (*i. e.,* {0, 1, 2, ... N – 2, N – 1, N} and thus N + 1 distinguishable states for the ensemble. If we label these states S_o through S_n, we can write the transition-state diagram

$$S_0 \underset{k^-}{\overset{Nk^+}{\rightleftharpoons}} S_1 \underset{2k^-}{\overset{(N-1k^+}{\rightleftharpoons}} S_2 \underset{3k^-}{\overset{(N-2)k^+}{\rightleftharpoons}} \cdots \underset{(N-2)k^-}{\overset{3k^+}{\rightleftharpoons}} S_{N-2} \underset{(N-1)k^-}{\overset{2k^+}{\rightleftharpoons}} S_{N-1} \underset{Nk^-}{\overset{k^+}{\rightleftharpoons}} SN, \quad ...(4.12)$$

Where the factors modifying the rate constants k^+ and k^- account for combinatorics. For example, the transition probability Nk^+ leading out of state S_o accounts for the fact that any one of N closed channels can open (at rate k^+), resulting in one open channel and ensemble configurations S_1.

The transition probability matrix for the Markov process diagrammed in (4.12) is tridiagonal

$$Q = \begin{bmatrix} D_0 & k^-\Delta t & & & \\ Nk^+\Delta t & D_1 & & & \\ & (N-k)k^+\Delta t & \ddots & & \\ & & (N-1)\,k^-\Delta t & & \\ & & & D_{N-1} & Nk^-\Delta t \\ & & & k^+\Delta t & D_N \end{bmatrix} \quad ...(4.13)$$

where the diagonal terms are such that probability is conserved and each column sums to 1:

$$D_0 = 1 - Nk^+ t,$$
$$D_1 = 1 - k^- \Delta t - (N-1)\,k^+ \Delta t,$$
$$D_{N-1} = 1 - (N-1)\,\Delta k^- t - k^+ \Delta t,$$
$$D_N = 1 - Nk- \Delta t.$$

The reader is encouraged to implement a simulation of a small number (*e. g.,* N = 4) of identical and independent two-state channels.

Gillespie's Method

Both of the simulation methods described above involve iterating a transition probability matrix and can be quit cumbersome when the number of ion channels being considered is large. Forthuratelу, an alternative that works well for large N has been devised

Consider a single two-state ion channel obeying the transition-state diagram (4.1) and recall that the probability that a channel closed at time t remains closed until $t + \tau$ is an exponentially decreasing function of τ given by (4.11). Thus the colsed dwell time (τc) of the two-state channel is an exponentially distributed random variable; that is, the probability distribution function of τc is

$$\text{Prob}\{\tau < \tau c \leq \tau + d\tau\} = k^{+} e^{-k+\tau} d\tau.$$

Similarly, the open dwell time (τo) of the channel is an exponentially distributed random variable with probability distribution function

$$\text{Prob}\ \{\tau < \tau o \leq \tau + d\tau = k^{-} e^{-k-\tau} d\tau.$$

Thus, we can simulate a two-state ion channel by alternately chossing open and closed dwell times consistent with these distributions. If one has no subroutine for simulating an exponentially distributed random variable, simply choose a uniformly distributed random variable U on the interval and use the relations.

$$\tau c = -\frac{1}{k_{+}} \text{In U},$$

$$\tau c = -\frac{1}{k_{-}} \text{In U}.$$

Gillespies method is much faster computationally than the Monte Carlo methods described above. Furthermore,because there is no time step involved, the method is exact.

Gillespie's method becomes more involved when the transition-state diagram indicates that more than one possible transition contributes to the dwell time for a given state. This possibility is handled by first choosing an exponentially distributed random number for the dwell time that accounts for all of the posible transitions out of the current state (*i.e.,* using the sum of the transition probabilities.) After the lenght of the dwell time in the current state is thus determined, the destination state is selected by choosing a uniformly distributed random vaiable on a appropriately partitioned interva, a process similar to the selection of transition during Monte Carlo simulation

TWO-STATE CHANNLES

In the previous section we claimed that the equation governing changes in probabilities for a single molecule has the same for as the equation for a large number of molecules. This connection can be made more rigorous by specifying the number of molecules we are considering in advance. To smiplify calculations we will again consider the two-state ion channel diagrammed in 4.1.

Finding N Channels

Let us write N as the number of molecules, and let $Po(n, t)$ and $Pc(n, t)$ be the probabilies of having n molecules in states O and C respectively. Because we will ultimately be interested in the statistics of current fluctuations, we will focus our attention on $Po(n, t)$. In any case,the presence of n onen channels implies $N - n$ closed channels, *i. e.,*

$$Pc(n, t) = Po(N - n, t) \qquad (o \leq n \leq N).$$

Assume that all N molecules are independent and consider a time interval $[t, t + \Delta t]$ short enough that only one molecules has appreciable probability of making a C $\rightarrow$ O or O $\rightarrow$ C transtion. During this short time interval, there are four events that can influence $Po(n, t)$, the probability that there are n open channels. For example, it is possible that there are currently n open channels, and during the time interval $\{t, t + \Delta t]$ one of these channels closes. This probability is given by

$$\text{loss}_- = k^- n \, \text{Po}(n, t) \, \Delta t,$$

where the parameter k^- is the transition probability for O $\rightarrow$ C, Po (n, t) is the probability that there were n open channels to begin with, and the n scales this probability to account for the fact that any one of the n independent open chanels can close with equivalent result. Similar reasoning leads to the expression

$$\text{Po}(n, t + \Delta t) + \text{Po}(n, t) \, \text{gain}_+ - \text{loss}_+ + \text{gain}_- - \text{loss}_-,$$

where

$$\text{gain}_- = k^- (n + 1) \, \text{Po}(n + 1, t) \, \Delta t$$
$$\text{loss}_+ = k^+ (\text{N} - n) \, \text{Po}(n, t) \, \Delta t,$$
$$\text{gain}_+ = k^+ (\text{N} - n + 1) \, \text{Po}(n - 1, t) \, \Delta t \, .$$

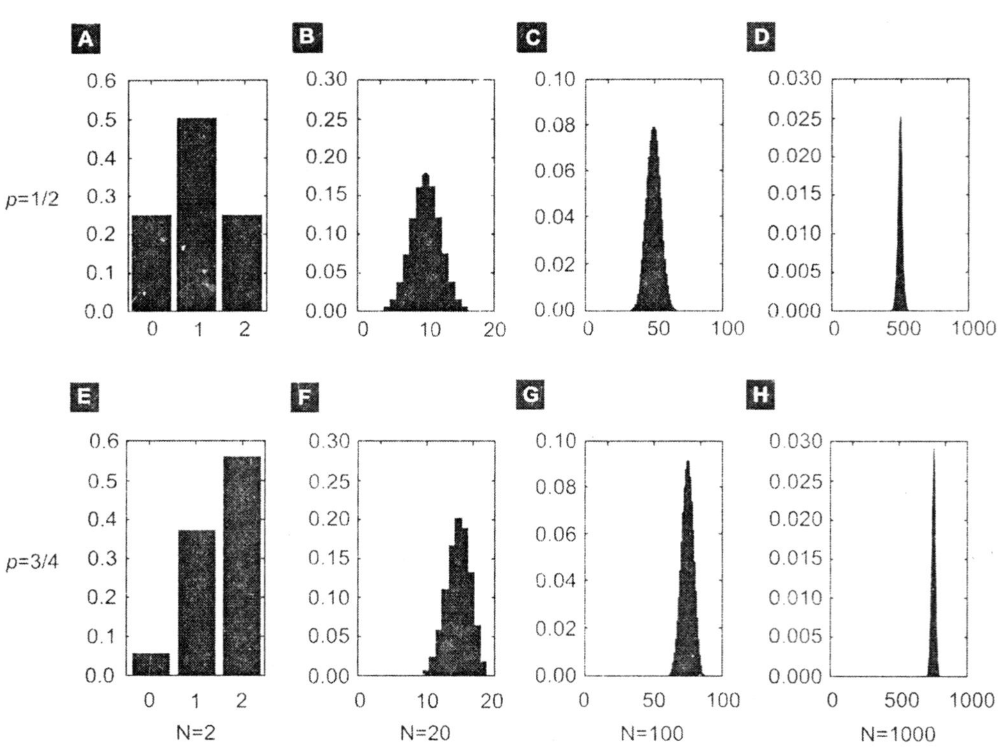

Fig. 4.4. For an equilibrium ensemble of N two-state channels with open probability p, the likelihood of observing n open channels is given by the binomial probability distribution (4.16) with parameters N and p. The binomial probability distribution has mean Np, variance Np (1 – p) and coefficient of variation $[(1 - p) / \text{Np}]^{1/2}$. Note that as the equilibrium open probability, p, icreasses the mean number of open channels shifts rightward. The $N^{-1/2}$ factor in the coefficient of variation is reflected in the narrowing of the distributions (from left to right).

To give one more example, the gain_+ term in this equation represents a probability flux due to the possibility that there are n –1 open chanels and one of the closed channels opens.

This transition probability is given by k^+ (N – n +1) Po(n – 1, t) Δt, because any one of the N – (n – 1) = N – n + 1 closed channels can open with equivalent result.

Taking the limit $\Delta t \to 0$ of (4.14) gives the ordinary differential equation

$$\frac{d}{dt}\mathrm{Po}(n,t) = k^+ (\mathrm{N} - n + 1)\,\mathrm{Po}(n-1, t) - k^+ (\mathrm{N} - n)\,\mathrm{Po}(n, t) + k^- (n+1)\,\mathrm{Po}(n+1, t) - k^- n\,\mathrm{Po}(n, t). \quad ...(4.15)$$

This rather complicated expression is called a *master equation*. It actually represents N + 1 coupled ordinary differential equations, one for each Po (n, t) for $0 \le n \le \mathrm{N}$ (all possible values for the number of open channels).

The equilibrium solution to the master equation is N +1 time-independent probabilits $\mathrm{P}_o^{\infty}(n)$, given by the binomial distribution

$$\mathrm{P}_o^{\infty}(n) = \binom{\mathrm{N}}{n} p^n (1-p)^{\mathrm{N}-n}, \quad ...(4.16)$$

where

$$p = k^+/(k^+ + k^-) \text{ and}$$

$$\binom{N}{n} = \frac{\mathrm{N}!}{n!(\mathrm{N}-n)!}.$$

Although this may not be obvious at first, the mathematically inclined reader can use the method of substitution to confirm that the binomial distribution satisfies a time-independent version of (4.15).

Figure 4.4 shows several binomial probability distributions with parameters N and p varied. Given an ensemble of n tow-state channels, these distribution represent the equilibrium probability of finding n channels in the open state. in the top row, the equilibrium open probability of p = 0.5 result in a centered distribution:The likelihood of observing n open channels is equal to the likelihood of observing N – n open channels. In the bootom row p = 0.75, and the enhanced likelihood that channels are open is evident in the rightward shift of the distributions.

The Equilibrium Solution

The equilibrium solution to the master equation for the two- state channel given by (4.16) is the binomial distribution, and thus the average number of open channels at equilibrium is $\langle \mathrm{No} \rangle_{\infty} = \mathrm{N}p$. But what about the time-dependence of the average number of open channels? Because the average number of open channels is given by

$$\langle \mathrm{No} \rangle = \sum_{n=o}^{\mathrm{N}} n\,\mathrm{Po}(n, t), \quad ...(4.17)$$

we can find an equation for $d\langle \mathrm{No} \rangle/dt$ by multiplying (4.15) by and summing. This gives

$$\frac{d\langle \mathrm{No} \rangle}{dt} = k^+ \sum_{n=o}^{\mathrm{N}} n(\mathrm{N}-n+1)\,\mathrm{Po}(n-1,t) - k^+ \sum_{n=0}^{\mathrm{N}} n(\mathrm{N}-n)\,\mathrm{Po}(n,t)$$

$$+k^- \sum_{n=0}^{\mathrm{N}} n(n+1)\,\mathrm{Po}(n+1,t) - k^- \sum_{n=o}^{\mathrm{N}} n^2\,\mathrm{Po}(n,t). \quad ...(4.18)$$

In Exercise 6 the reader can show that this equation can be reduced to

$$\frac{d\langle \mathrm{No} \rangle}{dt} = k^+(\mathrm{N} - \langle \mathrm{No} \rangle) - k^- \langle \mathrm{No} \rangle, \quad ...(4.19)$$

where

$$N-\langle No\rangle = \langle Nc\rangle. \qquad ...(4.20)$$

Note that 4.17 is identical to the rate equation for a population of two-state channels dervied by other means. For the duration of this chapter we will refer to such an equation as an *average* rate equation. Also note that the equilibrium average number of open $(\langle N\rangle_{\infty})$ and closed $(\langle Nc\rangle_{\infty})$ channels can be found by setting the left-hand side of (4.19) to zero that is,

$$\langle Nc\rangle_{\infty} = N\frac{k^+}{k^+ + k^-} = Np, \qquad ...(4.21)$$

$$\langle Nc\rangle_{\infty} = N\frac{k^-}{k^+ + k^-} = N(1-p), \qquad ...(4.22)$$

in agreement with our knowledge of the mean of a binomial distribution.

If we divide (4.19) by the total number of channles N, we find the average rate equation for the fraction of channels

$$\frac{d\langle fo\rangle}{dt} = k^+(1-\langle fo\rangle) - k^-\langle fo\rangle, \qquad ...(4.23)$$

where $\langle fo\rangle = \langle No\rangle / N, \langle fc\rangle = \langle Nc\rangle / N$, and (4.20) implies $\langle fo\rangle + \langle fc\rangle = 1$. The equilibrium fractions of open and closed channels are $\langle fo\rangle_{\infty} = k^+/(k^+ k^-) = p$ and $\langle fc\rangle_{\infty} = k^-/(k^+ k^-) = 1-p$. We thus see explicitly for a two-state channel that the master equation implies an average rate equation of the soft introduced which is true in general.

Variability

One advantage of beginning with a master equation is that in addition to the average rate equation, an evolution equation for the variance in the number of open channels can be derived. The variance in the number of open channels is defined as

$$\sigma^2_{No} = \left\langle (No-\langle No\rangle)^2 \right\rangle = \sum_{n=0}^{N} (n-\langle No\rangle)^2 \, Po\langle n,t\rangle. \qquad ...(4.24)$$

Similarly, the variance in the number of closed channels is

$$\sigma^2_{N_c} = \left\langle (Nc-\langle Nc\rangle)^2 \right\rangle = \sum_{n=0}^{N} (n-\langle Nc\rangle)^2 \, Pc\langle n,t\rangle. \qquad ...(4.25)$$

Again, we are ultimately interested in the statistics of current fluctuations, so we focus on σ^2_{No}. For the two-state channel under consideration, it is shown in Exercise 8 that these quantities are equal.

Beginning with (4.24) and the master equation (4.15), it can be shown that the variance σ^2_{No} satisfies the ODE

$$\frac{d\sigma^2_{No}}{dt} = -2(k^+ + k^-)\sigma^2_{No} + k^+(N-\langle No\rangle + k^-\langle No\rangle. \qquad ...(4.26)$$

The equilibrium variance $\left(\sigma^2_{No}\right)_\infty$ is thus given by steady steady states of this equation. Setting the left-hand side of this expression to zero, we obtain

$$\left(\sigma^2_{No}\right)_\infty = N \frac{k^+k^-}{\left(k^+ + k^-\right)^2} Np(1-p). \qquad ...(4.27)$$

From this equation it is clear that the equilibrium variance is proportional to N, the total number of channels. However, a relative measure of the variance known as the *coefficient of variation* is more meaningful. The coefficient of variation of the number of open chanels, CV_{No}, is given by the ratio of the standard deviation σ_{No} (the square root of the variance) to the mean $\langle No \rangle$. At equilibrium, we have

$$(CV_{No})_\infty = \frac{(\sigma_{No})_\infty}{\langle No \rangle_\infty} = \frac{1}{\sqrt{N}}\sqrt{\frac{k^-}{k^+}} = \sqrt{\frac{1-p}{Np}},$$

where the last equality is in agreement with the mean and variance of a binomially distributed random variable being Np and $Np(1-p)$, respectively. From this expression it is cxlear that the equilibrium coefficeint of variation for the number of open channels is inversely proportional to the square root of the number of channels N. Thus, in order to decrease this relative measure of channel noise by a factro of two, the number of channels must be increased bya factor of 4.

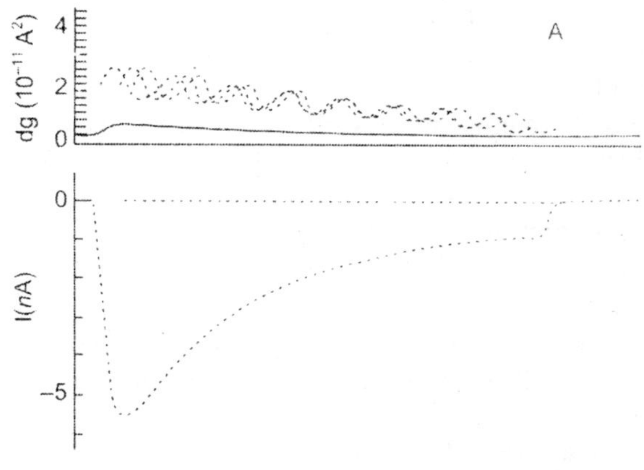

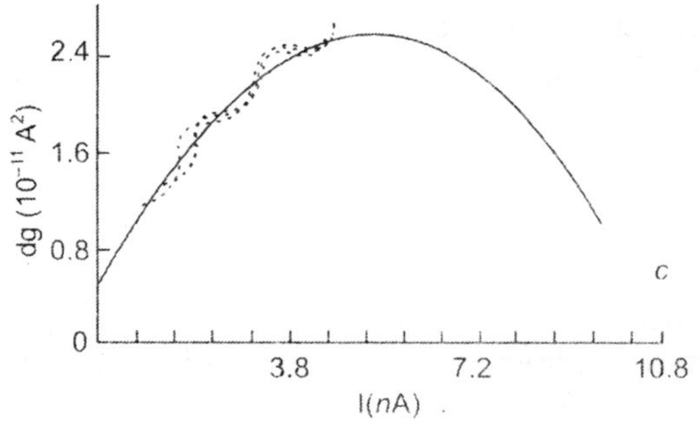

Fig. 4.5. Variance and mean sodium current measured from voltage-clamped single myelianted nerve fibers from *rana pipiens* depolatized to –15 mV after 50 ms Prepulses to –105 mm (A) Variance arising from the stochastic gating of sodium channels (dots) and thermal noise (solidline). (B) Mean current. (C) After carefully accounting for contributions to the variance due to thermal noise, the parabolic relationship between variance and mean current suggests N = 20,400 sodium channels at this node of Ranvier each with single channel conductance of i_{unit} = 0.55 pA.

MACROSCOPIC FLUCTUATIONS

When the volatate clamp techinque is applied to isolated membrane patches,openings and closings of single ion channels can be observed. Recall the single-channel recordings of T-type Ca^{2+} currents shown in the top panel of Figure (4.2) importantly, the bottom panel of Figure (4.2) shows that when several hundred single-channle recordings are summed, the kinetics of rapid activation and slower inactivation of the T-type Ca^{2+} current are evident. In this summed trace the relative size of the fluctuations in the macroscopic current is much smaller than those observed inthe single-channle the macrosopic current is much smaller than those observed in the single-channle recordings; however, the fluctuations in ionic current are still noticeable.

During voltage clamp recordings of large numbers of ion channels, stochastic gating leads to current fluctuations. For example, Figure 4.5 shows the time evolution of the mean sodium current measured form voltage-clamped single myelinated nerve fibers of *Rana pipiens* (frog) that were depolarized to –15 mV after 50 ms prepulsed to –150 mV. After a careful accounting ofr contributions to the variance of the sodium current due to thermal noise, the variance arising from the stochastic gating of sodium channels remains. Figure 4.6 shows macroscopic current fluctuations induced by the intophoretic application of acetylcholine (ACh) to voltage-clamped end-plates of a *Rana pipiens* nerve-muscle preparation. Interestingly, iontophoretic application of ACh increased the variance of the end-plate current as well as the mean. While the first trace in Figure 4.6 shows a spontaneous miniature end-plate current (sharp peak), the phenomenon of interest is the 10-fold increase in variance observed throughout the duration of the second trace.

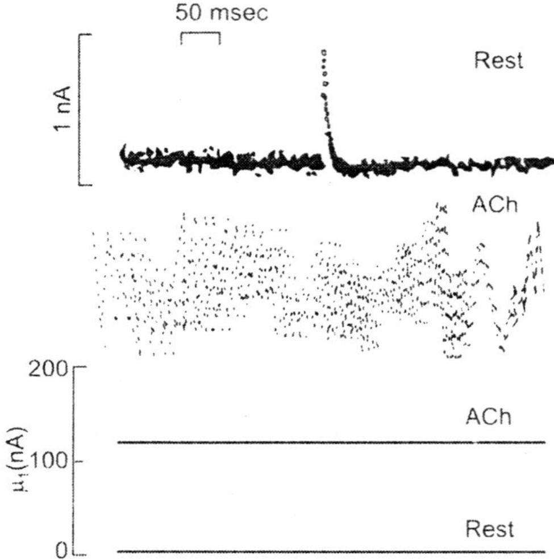

Fig. 4.6. Acetylcholine-produced current noise due to fluctuations in ionic conductance of voltage clamped end-plates of *Rana pipiens* nerve-muscle perparation. Iontophoretic application of ACh resulted in an increase in mean current as well as variance. The second trace, labeled "Rest", also shows a spontaneous miniature end-plate current.

In order to understand the relationship between fluctuations in macrosopic currents and the underlying single-channel kinetics, consider the statistics of ionic current implied by the two-state channel model presented in the previous section. In the simplest case,the unitary current of each two-state channel will be a random varibale taking the value zero when the channel is closed or a fixed value i_{unit} when the channel is open. That is, the unitary current will be a random variable I_{unit} given by

$$I_{unit} = \begin{cases} i_{unit} = g_{unit}(V - V_{rev}) & \text{when open,} \\ 0 & \text{when closed,} \end{cases} \qquad ...(4.28)$$

where V is a fixed command volage, V_{rev} is the reversal potential for the single-channel conductance g_{unit}, and i_{unit} is directly proportional to the conductance of the open channel. With these assumptions, it is straightforward to apply the results of Section 4.2 and derive the statistics of a fluctuating current that will result form N two-state channels with unitary current given by (4.28). The fluctuating macroscopic current will be a random variable defined by

$$I_{macro} = No\, i_{unit} \qquad (0 \le No \le N),$$

where N*o* is the number of open channels (also a random variable). Because the marcoscopic current it direclty proportional to No, we can use (4.17) to find the equilibrium average macroscopic current

$$\langle I_{macro} \rangle = i_{unit} \langle No \rangle_{\infty}.$$

Similarly, the equilibrium variance in the number of open channels, $\left(\sigma^2_{No}\right)_{\infty}$ given by (4.27) determines the equilibrium variance of the macroscopic current

$$\left(\sigma^2_{I_{macro}}\right)_{\infty} = i^2_{unit} \left(\sigma^2_{No}\right)_{\infty}.$$

Recall that if we write $p = k^+/(k^+ + k^-)$, the equilibrium mean and variance for the number of open channels are given by $\langle No\rangle_\infty = Np$ and $\left(\sigma^2_{No}\right)_\infty = Np =(1-p)$. Thus, the equilibrium mean and variance for the macroscopic current are given by $\langle I_{macro}\rangle_\infty = i_{unit}\ Np$ and $\left(\sigma^2_{I_{macro}}\right)_\infty = i^2_{unit}\ Np(1-p)$. Combining these expressions and eliminating p gives

$$\left(\sigma^2_{I_{macro}}\right)_\infty = i_{unit}\langle I_{macro}\rangle_\infty - \langle I_{macro}\rangle^2_\infty/N, \qquad \text{...(4.29)}$$

where both $\langle I_{macro}\rangle_\infty$ and $\left(\sigma^2_{I_{macro}}\right)_\infty$ are parameterized by p.

Equation (4.29) is the basis of a standard technique of membrane noise analysis whereby current fluctuations can be used to estimate the number of ion channels in a membrane patlich. By repeatedly manipulating the fraction of open channels p an estimate of $\left(\sigma^2_{I_{macro}}\right)_\infty$ as a function of $\langle I_{macro}\rangle_\infty$ is obtained. According to (4.29), the relationship will be parabolic with zero variance at $\langle I_{macro}\rangle_\infty = 0$ and $\langle I_{macro}\rangle_\infty = i_{unit}$ N and a maximum variance of $\left(\sigma^2_{I_{macro}}\right)_\infty = Ni^2_{unit}/4$ at $\langle I_{macro}\rangle_\infty\ i_{unit}\ N/2$. Figure 11.5C this techique was applied to voltage-clamped single myelinated nerve fibers from *Rana pipiens*. Equations (4.29) and a visual fit resulting in a maximum of $\left\langle\sigma^2_{I_{macro}}\right\rangle_\infty = 2.5 \times 10^{-21}\ A^2$ at $\langle I_{macro}\rangle_\infty = 5$ nA suggests that $N = \langle I_{macro}\rangle^2_\infty/\left(\sigma^2_{I_{macro}}\right)_\infty = 10{,}000$ and $i_{unit} = 1$ pA. However, after carefully accounting for contributions to the variance due to thermal noise, and adjusted fit gives N = 20, 400 sodium channels at this node of Ranvier and a unitary current of $i_{unit} = 0.55$ pA.

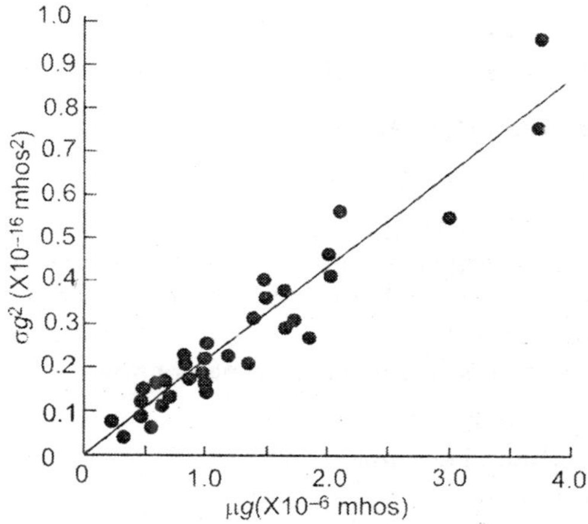

Fig. 4.7. Variance of conducatance fluctuations as a function of mean end-plate conductance of *Rana pipiens* never-muscle preparation. Because the unitary conducatance of end-plate channels is small, the relationship is linear and the slope of 0.19×10^{-10} mho= 19 pS gives the single channel conductance.

In Figure 4.7 this technique was applied to end-plate conductance fluctuations of *Rana pipiens* never-muscle preparation. Here, the equilibrium variance of the macroscopic conductance $\left\langle\sigma^2_{g_{macro}}\right\rangle_\infty$ is plotted against the mean conductance $\langle g_{macro}\rangle_\infty$, where the macroscopic conducatance is related to the unitary conductance through $g_{macro} = Ng_{unit}$. Using (4.29) and the relations

$$\langle g_{macro}\rangle_\infty = \frac{\langle I_{macro}\rangle_\infty}{V - V_{rev}},\quad \left(\sigma^2_{g_{macro}}\right)_\infty = \frac{\left(\sigma^2_{I_{macro}}\right)_\infty}{(V - V_{rev})^2} \text{ and } i_{unit} = g_{unit}(V - V_{rev}),$$

the reader can confirm that this relationship is also expected to be parabolic, that is,

$$\left(\sigma^2_{g_{macro}}\right)_\infty = g_{unit}(g_{macro})_\infty - (g_{macro})^2_\infty/N$$

However, because the unitary end-plate channel conductance of *Rana pipiens* nervemuscle preparation is very small (quadratic term negligible), The relationship is nearly linear:

$$\left(\sigma^2_{g_{macro}}\right)_\infty = g_{unit}\langle g_{macro}\rangle_\infty.$$

Indeed, the slope of the line in Figure 4.7 gives a single channel conductance of 19 pS for the open end-plate channel.

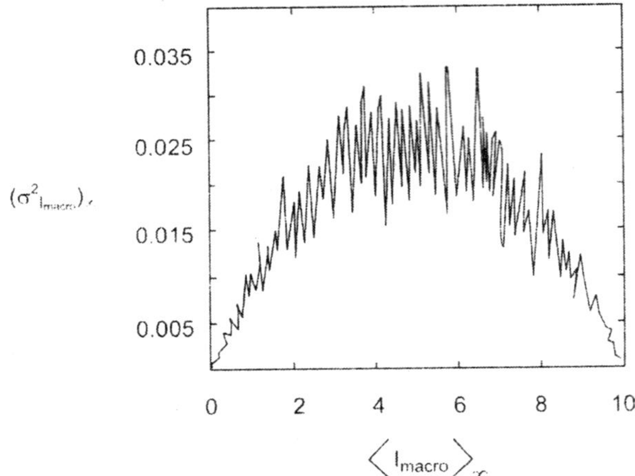

Fig. 4.8. The parabolic relationship between the variance and mean of ionic current through N = 10, 000 two-state channels each with single channel conductance of i_{unit} = 0.01 pA. The parabolic relationship betwen the variance $\left(\sigma^2_{I_{macro}}\right)_\infty$ and mean $\langle I_{macro}\rangle_\infty$ of current fluctuations is calculated form 100 simultaneously intergrtated Langevin equations.

STOCHASTIC ODES

Figure 4.8 shows a simulation reproducing the parabolic relationship between the variance $\left(\sigma^2_{I_{macro}}\right)_\infty$ and mean, $\langle I_{macro}\rangle_\infty$, of current fluctuations due to the stochastic gating of ion channels. This simulation includes N = 10, 000 identical two-state channels with uitary conductance of i_{unit} = 0.01 pA. A hundred trials were simultaneously performed and averaged how this simulation was performed.

Indeed, in simulating the stochastic gating of large numbers of ion chanels, Monte Carlo methods becomes impractical. However, when N is large, fluctuations on macroscopic currents can instead be described using a stochastic ordinary differential equation, called Langevin equation, that takes the form

$$\frac{df}{dt} = g(f) + \xi, \qquad \text{....(4.30)}$$

In this equation,the familiar deterministic dynamics given by $g(f)$ are supplemented with a rapidly varying random forcing term $\xi(t)$. Because ξ is a random function of time, solving (4.30) often means finding a solution $f(t)$ that satisfies the equation for a particular instantiation of ξ. Alternatively, if the statistic of ζ are given, we may be interested in derving the statistics of the new random variable $f(t)$ that is formally defined by (4.30).

The most common fluctuating force to consider are the increments of a Wiener process, Similar to the unbiased random walk a Winear process B(t) is a “Gaussian” random process that has zero mean

$$\langle B\rangle = 0 \qquad \text{...(4.31)}$$

and variance directly proportional to time,

$$\sigma^2_B = \left\langle (B - \langle B\rangle)^2 \right\rangle = \left\langle B^2 \right\rangle = t. \qquad \text{...(4.32)}$$

Indeed, the instantiations of a Wiener process B_1 and B_2 shown in Figure 4.9 are similar to the random walks presented already. Just as the increments of an unbiased random walk are ± Δx with equal probability, resulting in an increment with mean zero $\langle \Delta X \rangle = 0$. the increments of the numerical approximation to a Wiener process shown in Figure 4.9 are normally distributed with mean zero,

$$\langle \Delta B \rangle = 0.$$

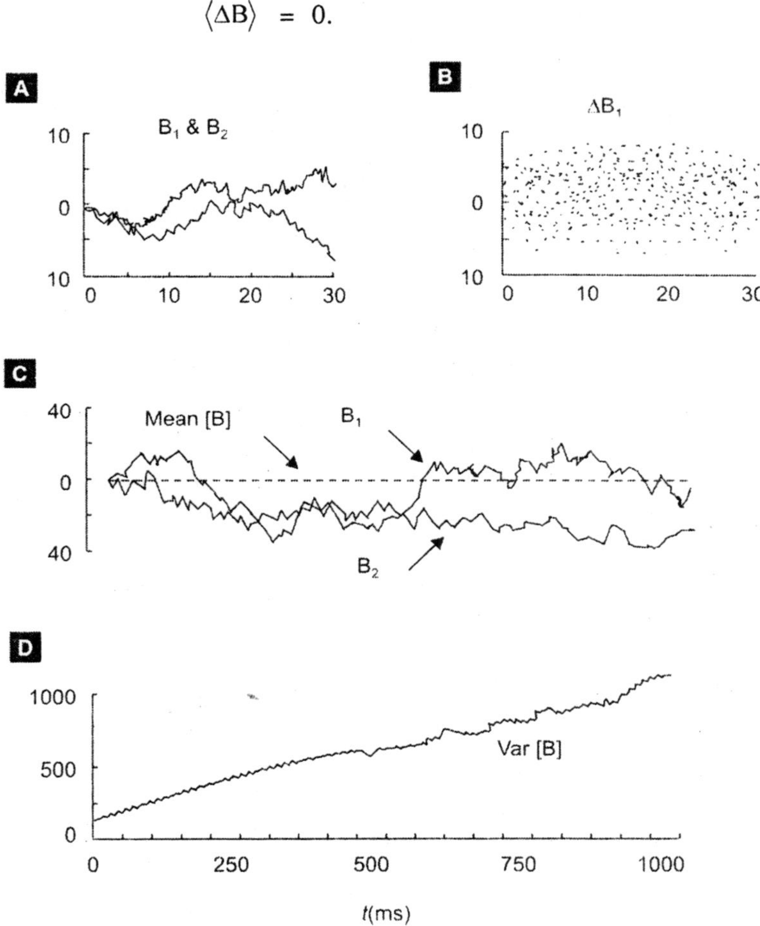

Fig. 4.9. Two instantiations of a Winear Process, B_1 and B_2, have trajectories similr to the random walks. The mean of 100 trails is near zero, while the variance of 100 trials increases linearly with time. In these simulations the increments (ΔB) are normally distributed with mean zero and variance $1/\Delta t$, where Δt is the integration time step.

In order to understand the variance of the increments of this simulated Wiener process, we must remember that unlike an unbiased random walk, a Winear process is a continuous function of time, B(t) A relevant statistic ofr the increments of a Wiener process is the two-time covariance or *autocorrelation function* $\langle \Delta B(t) \Delta B(t') \rangle$. Because nonoverlapping increments of a Wiener process are statistically independent and uncorrelated, $\langle \Delta B(t) \Delta B(t') \rangle = 0$ for t' $t + t\Delta$, in Figure 4.9. In the limit as $\Delta t \to 0$ (*i.e.*, for a "real" Wiener process), we might even write $dB(t)/dt = \xi(t)$, where $\langle \xi(t) \rangle = 0$ and

$$\langle \xi(t)\,\xi(t') \rangle = \delta(t-t') \qquad ...(4.33)$$

although technically this derivative does not exist. Equation (4.33) may appear unusual, especially if the reader is unfamiliar with the Dirac delta function, defined by $\delta(t) = 0$ for $t \neq 0$ and

$$\int_{-\infty}^{\infty} \delta(t)dt = 1.$$

A Wiener process B (t) can be simulated by numerically integrating a piecewise constant approximation to the Winear increment B Δ(t) . The Winear increment is a normally distributed random variable with zero mean that is held fixed during the time interval $[t, t + \Delta t]$ and updated after the intergration time step is complete. Intergrating this erratic function of time results in the Winear trajectories of Figure 4.9 and Figure 4.9. If we rewrite 4.33 to account for this piecewise constant approximation to the Wiener increments, we obtain

$$\langle \Delta B(t)\Delta B(t') \rangle = \begin{cases} 1/\Delta t, & t' \in [t, t+\Delta t] \\ 0 & \text{otherwise.} \end{cases}$$

Like the unbiased random walk, the variance of this simulated Wiener process is proportional to time. The reader can confirm through simulations that in order to achieve this macroscopic behaviour, the variance of the Wiener increments B$\Delta(t)$ must be adjusted according to the integration time step (*i. e.,* Var [ΔB) = $1/\Delta$ gives Var [B] = t.

Langevin Equation

In order to use a Langevin equation of the form of (4.30) to simulate a large number of ion channels, we must make an appropriate choice for both the deterministic function $g(f)$ as well as the statistics of the random variable ξ. Recalling the average rate equation for the dynamics of the open fraction of channels 4.23 we write

$$\frac{dfo}{dt} = k^+(1-fo) - k^- fo + \xi \qquad ...(4.34)$$

$$= \frac{fo - \langle fo \rangle_\infty}{\tau_f} + \xi, \qquad ...(4.35)$$

where fo = No/N is random variable, the fluctuating fraction of open channels, $\langle fo \rangle_\infty = k^+/(k^+ + k^-)$, and $\tau_f = 1/(k^+ + k^-)$. For 4.35 to be meaningful, we must specify the statistics of ξ. An appropriate choice for ξ is a fluctuating function of time that has zero mean.

$$\langle \xi(t) \rangle = 0,$$

and an autocorrelation function given by

$$\langle \xi(t)\xi(t') \rangle = \gamma\delta(t-t').$$

By methods of statistical physics beyond the scope of this book, γ can be shown to be inversely proportional to M and proportional to the sum of the rates of both the O $\rightarrow$ C and C $\rightarrow$ O transitions, that is,

$$\gamma(fo) = \frac{k^+(1-fo) + k^- fo}{N}. \qquad ...(4.36)$$

An appropriate choice for ξ is thus $\xi = \sqrt{\gamma}\,\Delta B$, where the ΔB are the increments of a Winear process.

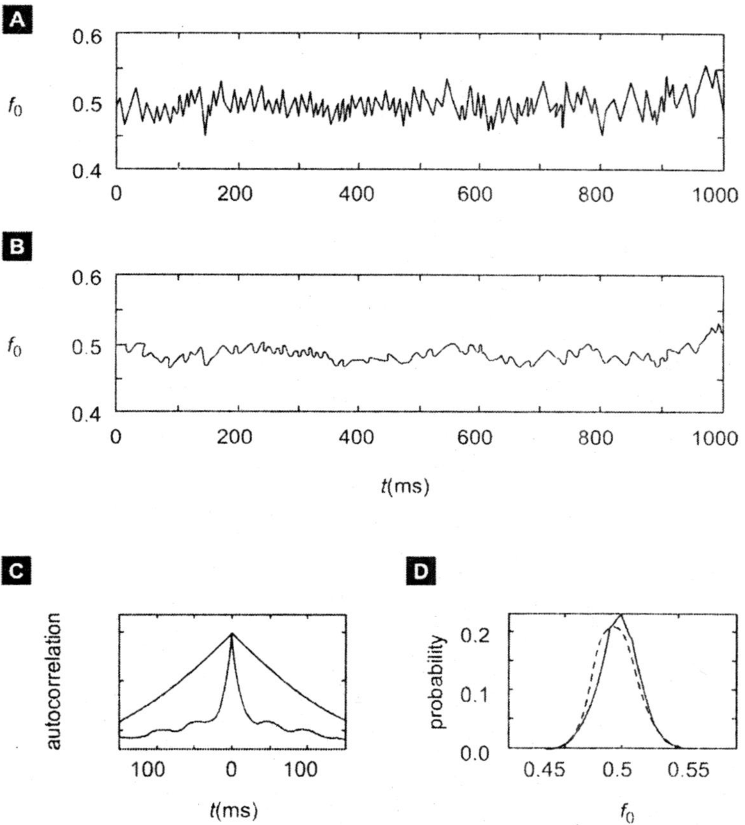

Fig. 4.10. (A,B) The open fraction *fo* of 1000 two-state ion channels simulated using a Langevin equation. Transition rates are ten times faster in (A) that (B) so than the time constant τ_f is 10 and 100 ms, respectively. (C) Numerically calculated autocorrelation function of *fo*. (D) The equilibrium mean and variance of *fo* are approximately equal in the two cases.

Although we haven't fully justified this choice for ξ, we can check that this random variable and (4.35) define the random variable *fo* in a manner consistent with the work in previous sections. To do this we use the fluctuation-dissipation theorem from statistical physics that relates γ, which occurs in the correlation function ξ, of to the equilibrium variance of *fo*. The relationship depends on the relaxation time constant τ_f and is given by

$$\gamma_\infty = \frac{2\left(\sigma_{fo}^2\right)_\infty}{\tau_f}. \qquad ...(4.37)$$

Using (4.27), and remembering that $\left(\sigma_{fo}^2\right)_\infty = \left(\sigma_{No}^2\right)_\infty / N^2$, the reader can confirm that the last equality holds at equilibrium.

Stochastic simulaitons of the open fraction, *fo*, of 1000 two-state ion channels calculated by intergrating (4.35) are shown in Figure 4.10. The transition rates used were $k^+ = k^- = 0.05/$

in panel (B) giving time constants (tf) of 10 and 100 ms and equilibrium open fraction $\langle fo \rangle \infty$ = p = 0.5 in both cases. This difference in relaxation time constants is evident in the (normalized) autocorrelation functions compared in C. It can be shown that for an infinitely long simulation the autocorrelation functions for *fo* is

$$\langle fo(t)\, fo(t') \rangle = \left(\sigma^2_{fo}\right)_\infty e^{-|t-t'|/\tau_f}$$

The narrower autocorrelation function in Figure 4.10c thus corresponds to the case with small time constant τ_f. Note that although the time constant for reaxation to $\langle fo \rangle_\infty$ = 0.5 is faster in A than in B, the equilibrium variance $\left(\sigma^2_{fo}\right)_\infty$ shown in Figure 4.10D is approximately equal in the two cases, as expected according to (4.27)

Fokker-Planck Equation

Rather than calculating trajectories for the fraction of open channels *fo* using a Langevin equation, an alternative is to calcuate the evolution of the probability distribution function (PDF) for *fo*. While the binomial distribution is an example of a discrete probability distribution (No takes on N + 1 discrete values), the Langevinequation for *fo* (4.35) implies that *fo* can take on any value on the interval. Thus, the PDF for *fo* is continuous and defined as

$$P(f, t)\, df = \text{Prob}\,\{f(t) < fo < f(t) + df\},$$

where conservation of probability gives

$$\int_0^1 P(f,t)\, df = 1. \qquad ...(4.38)$$

We can write an evolution euation for P (f,t), known as Fokker-Planck equation, that corresponds to the Langevin description given by (4.35)

$$\frac{\partial P(f,t)}{\partial t} = -\frac{\partial}{\partial f}[J_{adv}(f,t) + J_{dif}(f,t)]. \qquad ...(4.39)$$

In this equation, $J_{adv}(f, t)$ is a probability flux due to advection (that is, transport) governed by the deterministic terms in (4.35):

$$J_{adv}(f, t) = -\frac{f - \langle fo \rangle_\infty}{\tau_f} P(f, t). \qquad ...(4.40)$$

In contrast, $J_{dif}(f, t)$ is a diffusive flux that accounts for the spread of porbability induced by the random variable ξ. This diffusive probability flux is given by

$$J_{dif}(f, t) = -\frac{1}{2}\frac{\partial}{\partial f}\left[\gamma(f)\, P(f, t)\right], \qquad ...(4.41)$$

where γ is given by (4.36). Rewriting (4.40) in terms of the total prtobability flux $J_{tot} = J_{adv} + J_{dif}$, we have

$$\frac{\partial P(f,t)}{\partial t} = \frac{\partial J_{tot}(f,t)}{\partial f}, \qquad (4.42)$$

with associated boundary conditions

$$J_{tot}(0, t)\, J_{tot}(1, t) = 0$$

that imply no flux of probability out of the physiological range for *fo*. An appropraiate choice of initial conditions would be $P(f, 0) = \delta(f - \langle fo \rangle_\infty)$, implying that the system is known to be in equilibruim at $t = 0$.

Setting the left-hand side of (4.39) equal to zero, we see that equilibrium probability distribution $P_\infty(f)$ solve $J^{\infty}_{tot} = 0$. That is,

$$-\frac{f-\langle fo\rangle_\infty}{\tau_f}P_\infty(f)-\frac{1}{2}\frac{d}{df}\left[\gamma(f)P_\infty(f)\right] = 0. \qquad \text{...(4.43)}$$

This differential equation can be solve numerically. However, we obtain more insight by approximating $\gamma(f)$ by

$$\gamma_\infty = \frac{k^+\left(1-\langle fo\rangle_\infty\right)+k^-\langle fo\rangle_\infty}{N}.$$

This procedure is valid when fluctuations of *fo* away form equilibrium, $\langle fo\rangle_\infty$ are small (that is, when N is large). If we maek this approximation, it can be shown that the probability distribution

$$P_\infty = A\exp\left[\frac{f\left(2\langle fo\rangle_\infty - f\right)}{\gamma_\infty \ddot{A}_f}\right]. \qquad \text{...(4.44)}$$

satisfies (4.43), where the normalization constant A is chosen to satisfy conservation of probability (4.38). While this expression may not look familiar, when γ_∞ is sufficiently small (N is sufficiently large), $P_\infty(f)$ is well pproximated by the Gaussian

At equilibrium *fo* will be a normally distriuted random variable with mean $\langle fo\rangle_\infty$ and variance $\left(\sigma^2_{fo}\right)_\infty = \gamma_\infty \tau_f/2$, in agreeement with (4.37). The distribution $P_\infty(f)$ is approximately Gaussian for N = 1000 as shown in Figure 4.10D.

MEMBRANE VOLTAGE FLUCTUATION

In Section 4.3 we discussed macroscopic current fluctuations experimentally observed in voltage clamp recordings and a memberane noise analysis. in this section we will simulate electrical recordings in which the memberane potential is not clamped, but rather fluctuates under the influence of two-state in channels. Although a misnomer, such measurements are referred to as *curen clamp* recordings. For now we asume that ion channel gating is voltage-independent.

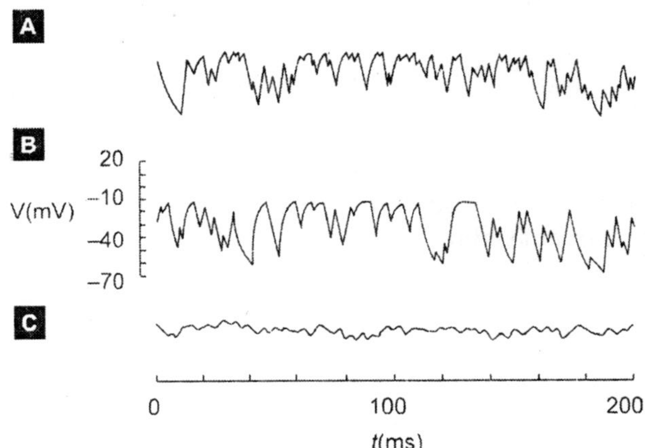

Fig. 4.11. Membrane voltage fluctuations due to the stochastic gating of one or more sodium channels. (A,B) Single channel simulations. Transition probabilities are a factor of two slower in (B), leading to longeer dwell times and fewer transitions as evidenced by 'kinks' in graph, (C) Twenty channels are simulated. As the number of sodium channels increases, the variance in membrane voltage decrease.

Simulations of membrane voltage fluctuations due to the stochastic gating of a single sodium channel obeying the transition-state diagram(4.1) are shown in Figure 4.11A and Figure 4.11B. The gating of the two-state channel is simulated using MonteCarlo methods, while memberane voltage is simultaneously calculated using the current blance equation

$$C\frac{dV}{dt} = \text{I}g_L(V - V_L) - g_{Na}(V - V_{Na}), \quad ...(4.45)$$

where g_L is leakage conductance with reversal potential $V_L = -70$ mV, and gNa (a random variable taking values of 0 or g_{Na}^{max} depending on channle state) is a sodium conductance with reversal potential $V_{Na} = 60$ mV. Because transition probabilities are a factror of two slower in Figure 4.11B longer dwell times and fewer transitions are observed. For comparison, Figure 4.11C shows the result when twenty channels are simulated. Here g_{Na} takes on 21 possible values between 0 and g_{Na}^{max} and as condequence the variance in the fluctuating membrane voltage decreases.

Membrane voltage fluctuations such as those shown in figure 4.11 can also be modeled by tracking the evolution of probability distribution functions (PDFs) for membrane voltage conditioned on the state of the ion channel. The governing equations are coupled partial differential equations each of which is similar to the advective component of the Fokker-Planck equations described above. We will refer to this method as the *ensemble density approach.*

If membrane voltage fluctuations are due to a single two-state sodium channel, there are two relevant conditional PDFs,

$$Pc(v, t)dv = \text{Prob } \{v < V < v + dv|C, t\},$$

$$Po(v, t)dv = \text{Prob } \{v < V < v + dv|o, t\},$$

where $Pc(v, t)$ and $Po(v, t)$ are conditioned on the channel being closed or open, respectively. Conservation of probability implies

$$\int_{-\infty}^{\infty} Pc(v, t)dv = 1, \qquad \int_{-\infty}^{\infty} Po(v, t)dv = 1,$$

The equations for the evolution of these conditional PDFs are

$$\frac{\partial}{\partial t}Pc(v,t) = -\frac{\partial}{\partial v}Jc(v, t) - k^+ Pc(v, t) + k^- Po(v, t), \quad ...(4.46)$$

$$\frac{\partial}{\partial t}Po(v, t) = -\frac{\partial}{\partial v}Jo(v, t) + k^+ Pc(v, t) - k^- Po(v, t), \quad ...(4.47)$$

where $Jc(v, t)$ and $Jo(v, t)$ are advective probability fluxes due to membrane voltage obeying the current balance equation (4.45):

$$Jc(v, t) = -\frac{1}{C}\left[g_L(v - V_L)\right] Pc(v, t)$$

$$Jo(v,t) = -\frac{1}{C}\left[g_L(v - V_L) + g_{Na}^{max}(v - V_{Na})\right] Po(v, t)$$

Notice that the sodium current term occurs only in $Jo(v, t)$ because if the channel is colsed, $g_{Na} = 0$. The reaction terms that appear in (4.46) and (4.47) account for probability flux due to the stochastic gating of the sodium channel. Regradless of membrane voltage, the conditional probability $Pc(v, t)$ can decrease due to channel opening at a rate of $k^+ Pc(v, t)$ and increase due to closing of open channels at a rate of $k^- Fo(v, t)$, The reactionterms occur with opposite sign because any increase or descrease inthe conditional probability $Pc(v, t)$ due to ta channel gating implies commensurate change in $Po(v, t)$.

The equilibrium conditional probability distribution functions for the memberane voltage, $P_c^{\infty}(v)$ and $P_o^{\infty}(v)$, calculated numerically form (4.46) and (4.47), are shown in Figure 4.12A.

The simulation ran for 1 second, corresponding to approximately 1000 changes in channel state. As expected, $P_o^{\infty}(v)$ is shifted toward the right (depolarized V) relative to $P_c^{\infty}(v)$. The astute reader will note that the PDFs are not symmetric, indicating that when the channel is open probability advects toward V_{Na} faster than it advects toward V_L when the channle is closed; *i.e.*, an open sodium channel leads to a smaller membrane time constant. In Figure 4.12B, the rate constants K^+ and K^- are a factor of the slower, In this case more probability accumulates near both V_L and V_{Na}.

Density

The ensemble density formulation described above can be extended to the case where membrane voltage fluctuations are due to an ensemble of N two-state channels. If we write Po (n, v, t) for the conditional probability density for membrane voltage given n open sodium channels, we have

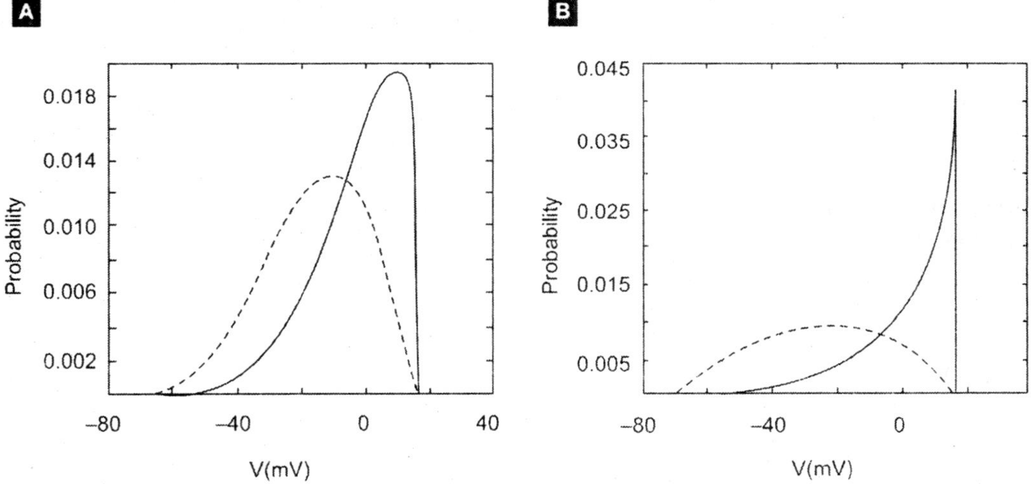

Fig. 4.12. Equilibrium conditional probability distribution functions (PDFs), $P_c^{\infty}(v)$ (dotted lines) and $P_o^{\infty}(v)$ (solid lines) for the membrane voltage conditioned on the state of a single two-state sodium channel. In (A), $k^+ = k^- = 1$/ms, while for (B) the transition probabilities are a factor of two slower.

$$\begin{aligned}\frac{\partial}{\partial t}\mathrm{Po}(n,v,t) &= -\frac{\partial}{\partial v}\mathrm{Jo}(n,v,t)\\ &\quad + k^+(\mathrm{N}-n+1)\,\mathrm{Po}(n-1,v,t) - k^+(\mathrm{N}-n)\,\mathrm{Po}(n,v,t)\\ &\quad + k^-(n+1)\,\mathrm{Po}(n+1,v,t) - k^- n\mathrm{Po}(n,v,t), \qquad \text{...(4.48)}\end{aligned}$$

where the reaction terms are based on the master equation formulation presented in Section 11.2.1, Jo(n, v, t) is given by

$$\mathrm{Jo}(n,v,t) = -\frac{1}{\mathrm{C}}\left[g_{\mathrm{L}}(v-\mathrm{V_L}) + g_{\mathrm{Na}}^{\max}\frac{n}{\mathrm{N}}(v-\mathrm{V_{Na}})\,\mathrm{Po}(n,v,t),\right]$$

and $g_{Na}^{\max}$ is the sodium conductance when all N channels are open. Note that (4.48) represents N + 1 coupled partial differential equations, one for each Po(n, v, t), where $0 \le n \le N$.

Figure 4.13 shows equilibrium conditional PDFs $P_o^{\infty}(n, v)$ for membrane voltage fluctuations induced by 20 two-state sodium channels. These PDFs are calculated by numerically solving (4.48) until a steady state is achieved. Careful inspection of the figure shows that in the case of high n (more open channels) the equilibrium distribution of membrane voltage is shifted toward V_{Na}. Note that these PDFs appear to be consistent with a binomial distribution for the total equilibrium probability for a given value of n. That is,

$$P_o^{\infty}(n) = \int_{-\infty}^{\infty} P_o^{\infty}(n, v)\,dv$$

is in agreement with (4.16).

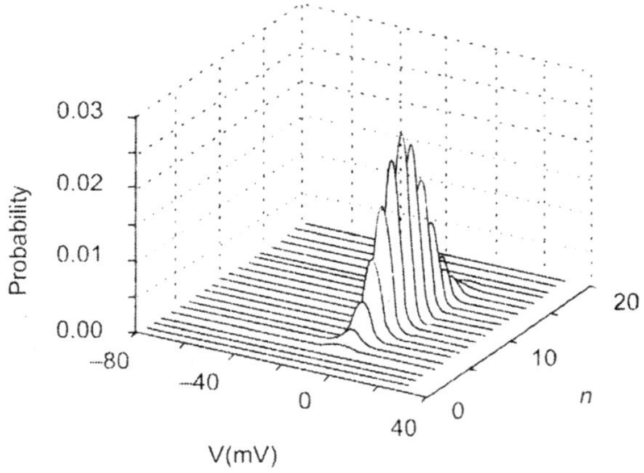

Fig. 4.13. Conditional probability distribution functions for membrane voltage fluctuations due to stochastic gating of 20 two-state sodium channels.

STOCHASTICITY AND DISCRETENESS

Using the results of previous sections, we are prepared to explore the consequences of stochasticity and discreateness in an excitable membrane model. The deterministic Morris-Lecar model is

$$C\frac{dV}{dt} = I_{app} - g_L(V - V_L) - g_K w(V - V_K) - g_{Ca}\, m_{\infty}(V - V_{Ca}) \qquad ...(4.49)$$

$$\frac{dw}{dt} = \frac{w_{\infty} - w}{\tau}, \qquad (4.50)$$

where the activation function for the Ca^{2+} current, $m_{\infty}(V)$; the activation function for the K^+ current, $w_{\infty}(V)$; and voltage-dependent time scale for activation of K^+ current, $\tau(V)$ have already been discuased.

In (4.50), w is usually though to represent the fraction of open K^+ channels. However, we now understand that this differentail equation is actually an average rate equation similar to (4.23). To be clear, let us write this deterministic average rate equations as

$$\frac{d\langle w\rangle}{dt} = \frac{w_{\infty} - \langle w\rangle}{\tau},$$

where w (random variable) represent the fraction of open K^+ channels. The reader can easily verify that this average rate equation corresponds to the two-state kinetic scheme

$$\text{C(closed)} \underset{\beta(v)}{\overset{\alpha(v)}{\rightleftharpoons}} \text{O(open)},$$

where C and O indicate closed and open states of the K+ channel, and the voltage-dependent transtions rates $\alpha(V)$ and $\beta(V)$ are given by

$$\alpha(V) = \frac{w_{\infty}}{\tau}$$

$$\beta(V) = \frac{1 - w_\infty}{\tau},$$

This, in turn, implies that equilibrium fraction of open K^+ channel is

$$\langle w \rangle_\infty = w_\infty(V)\frac{\alpha}{\alpha + \beta},$$

and the time constant $\tau(V)$ is

$$\tau(V) = \frac{1}{\alpha + \beta}.$$

With these preliminaries, we can see that a Morris-Lecar simulation that includes channel noise due to a small number of K^+ channles could be performed with several Markov variables obeying the voltage-dependent transition probability matrix

$$Q = \begin{bmatrix} 1 - \alpha\Delta t & \beta\Delta t \\ \alpha\Delta t & 1 - \beta\Delta t \end{bmatrix}$$

for each channel. Alternatively, a larger collection of N channels can be simulated by tracking only a single Markov variable, the number of open K^+ channels. In this case, the following tridiagonal transition probability matrix would be used:

$$Q = \begin{bmatrix} D_0 & \beta\Delta t & & & \\ N\alpha\Delta t & D_1 & & & \\ & (N-1)\alpha\Delta t & \ddots & & \\ & & & (N-1)\beta\Delta t & \\ & & & D_{N-1} & N\beta\Delta t \\ & & & \alpha\Delta t & D_N \end{bmatrix}$$

with diagonal terms (D_0, D_1, ... D_{N-1}, D_N) such that each colomn sums to unity.

Basic Mechanism

Figure 4.14 and Figure 4.15 show Morris-Lecar simulations that include stochastic voltage-dependent gating of 100 K^+ channels. Spontaneous action potentials are induced by this simulated channel noise (Figure 4.14A). We will refer to this phenomenon as "stochastic excitability," because it is understood as a sampling of the excitable phase space of the deterministic model made possible by membrane potential fluctuations due to the stochastic gating of K^+ channles. The (V, w) phase plane trajectories for Figure 4.14 are shown in Figure 4.15. The discretensess and stochasticity of the K^+ gating variabel w allows trajectories to fluctuate around the fixed point of the deterministic model seen in the lower left of Figure 4.15. Occasioally, K^+ channels spontaneously inactivate (w fluctuates toward 0) and a regenerative Ca^{2+} current leads to an action potential. This type of spontaneous activity ahs been observed in stochastic versions of the Hodgkin-Huxley equations and is though to influence subthreshold membrane potential oscillations and excitability of stellate neurons of the medial entorhinal cortex of the hippocampal region.

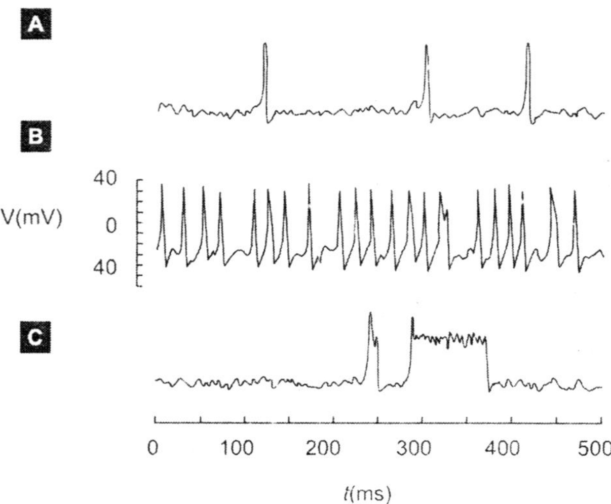

Fig. 4.14. Morris-Lecar simulations including stochastic gating of 100 K^+ channels. (A) Spontaneous excitability driven by channel noise is observed when I_{app} = 10 and the deterministic model is excitable. (B) Stochastic oscillations are observed when I_{app} = 12 and the deterministic model is oscillatory. (C) Stochastic bistability is observed when the deterministic model is bistable (I_{app} = 12 and V_3 = 15 mV rather than standard value of 10 MV).

In Figure 4.14B parameters are such that the deterministic model (as N $\rightarrow \infty$) is oscillatory. However, when N = 100 channel noise results in irregular oscillations. In Figure 4.14B stochastic bistability is observed. When parameters are chosen so that the deterministic model is bistable, channel noise allows the alternate sampling of two stable fixed points in the (V, w) phase plane, a phenomenon known as basin hopping.

Ensembly Density Approach

The ensemble density approach described in Section 4.42 can be applied to the stochastic Morris-Lecar model described above. The evolution equations for the conditional PDFs take the form

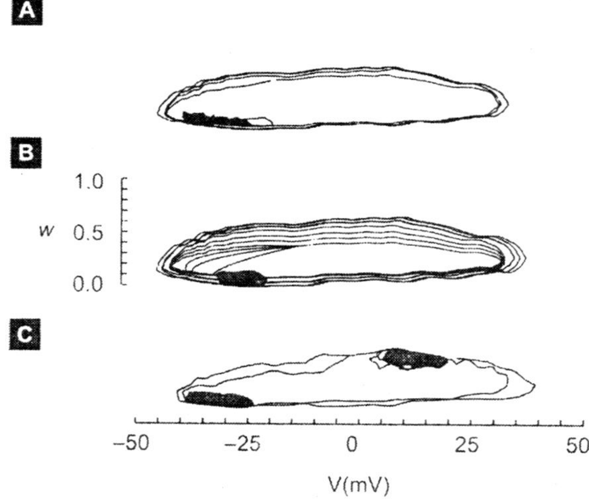

Fig. 4.15. Morris-Lecar simulations with stochastic gating of 100 k+ channels shown in the (V, w) phase plane. Panels correspond to (A) stochastic excitability, (B) oscillations, and (C) bistablity shown in Figure 4.14.

$$\frac{\partial}{\partial t} \text{Po}(n, v, t) = -\frac{\partial}{\partial v} \text{Jo}(n, v, t)$$

$$+ \alpha(v)(\text{N} - n + 1)\,\text{Po}(n, v, t) \qquad ...(4.51)$$

$$- \alpha(v)(\text{N} - n)\,\text{Po}(n, v, t)$$

$$+ \beta(v)(\text{N} + 1)\,\text{Po}(n, + 1\ v, t) - \beta(v)\, n\, \text{Po}(n, v, t).$$

This form is similar to (4.48) except that the transition probabilities are now voltage dependent, and the probability fluxes Jo(n, v, t) are given by the Morris-Lecar current balance equation (11.49). That is,

$$\text{Jo}(n, v, t) = -\frac{1}{\text{C}}\left[\text{I}_{app} - g_{\text{L}}(v - \text{V}_{\text{L}}) - g_{\text{K}}^{\max}\frac{n}{\text{N}}(v - \text{V}_{\text{k}}) - g_{Ca}\, m_{\infty}(v - \text{V}_{Ca})\right] Po\,(n, v, t).$$

Figure 4.16 shows equilibrium PDFs for the membrane voltage of the stochastic Morris-Lecar model conditioned on the number of open K+ channles. These equilibrium PDFs are steady-state solutions to (4.51) and correspond to the three types of trajectories shown in Figure 4.15. The amount of time that trajectories spend in different regions of the (V, w) phase plane is reflected in these distributions. It is clear form Figure 4.15A that the Morris-Lecar model exhibiting stochastic excitablity spends a large proportion of time near the threshold for excitation.

Langevin Formulation

To consider the behaviour of the Morris-lecar model under the influnce of channel noise from a large number of K^+ channels, it is most convenient to use the langevin formulation presented in Section 4.4.1 We do this by supplementing the rate equation for the average fraction of open K^+ channels, (4.50), with a rapidly varying forcing term

$$\frac{dw}{dt} = \frac{w_{\infty} - w}{\tau} + \xi,$$

when w is a random variable, $\langle \xi \rangle = 0$, and the autocorrelation function of ξ is given by

$$\langle \xi(t)\xi(t') \rangle = \gamma(w, \text{V})\,\delta(t - t')$$

Following (4.36), $\gamma\,(w\ \text{V})$ is chosen to be

$$\gamma(w, \text{V}) = \frac{\alpha(1 - w) + \beta w}{\text{N}} = \frac{1}{\text{N}}\frac{(1 - 2w_{\infty})w + w_{\infty}}{\tau}. \qquad (4.52)$$

Thus, ξ is a random variable defined by $\xi = \sqrt{\gamma(w, \text{V})}\,\Delta\text{B}$, where the ΔB are the increments of a Wiener process.

Figure 4.17 Presents stochastic Morris-Lecar model simulations implemented using the Langevin formulation described above. Intersetingly, the existence of stochastic excitability depedns on the number of K^+ channels included. When N is relatively small (N = 25, 50 100) membrane potential fluctuations are large, and spontaneous action potentials are frequent. However,when more K^+ channels are included (N = 500, 1000), the model becomes essentially deterministic. Although the model is still excitable, as $\text{N} \to \infty$ fluctuations in w become smaller, and spontaneous action potentials are no longer observed.

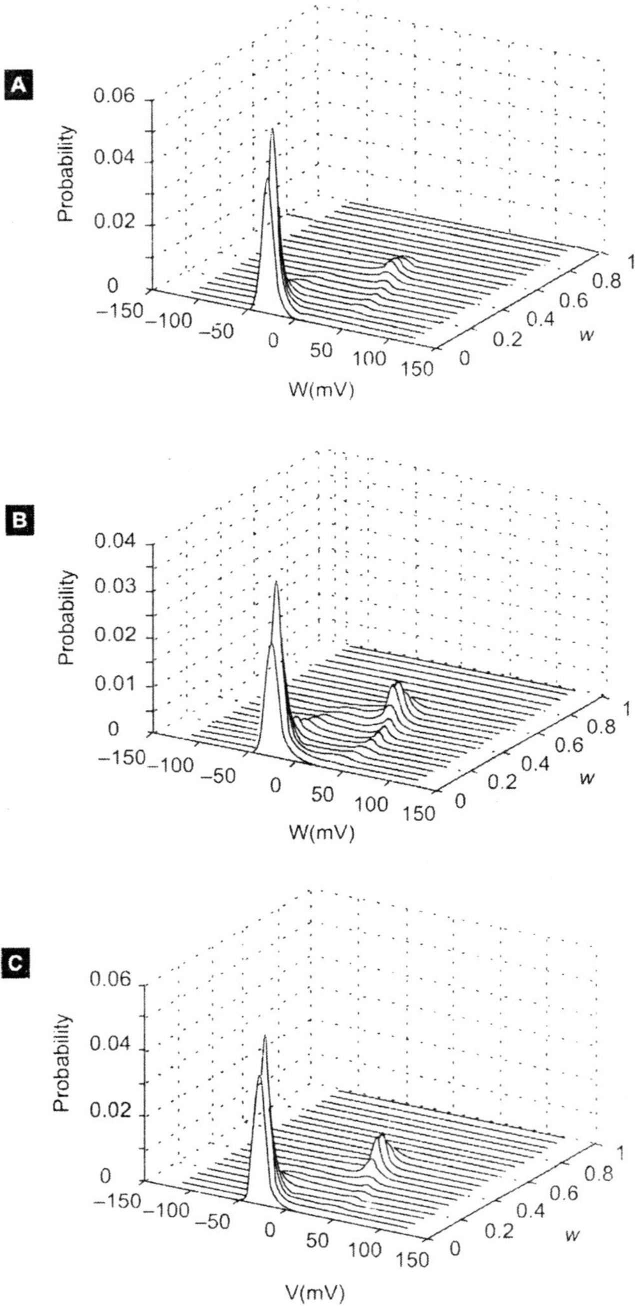

Fig. 4.16. Probability distribution functions for the membrane voltage of the stochastic Morris Lecar model conditioned on the number of open k+ channels. The equilibrium PDFs show evidence of stochastic excitablity (A), oscillations (B), and bistabillity (C), corresponding to the trajectories shown in Figure 4.15.

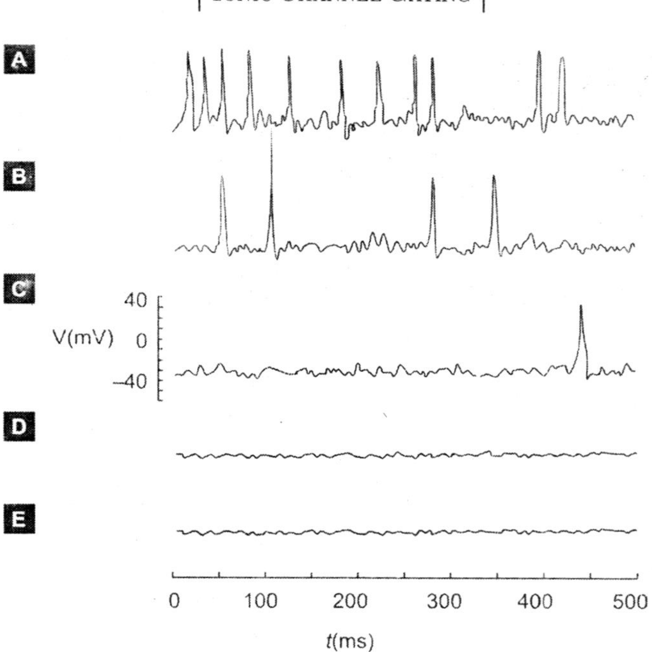

Fig. 4.17. The stochastic Morris-Lecar model simulated using a langevin equation for W, the fraction of open K+ channels. As the number of K+ channels is increased (n = 25, 50, 100, 500 or 1000) Spontaneious action potential induced by stochastic gating are eliminated. For large N, the model is excitable, but essentially deterministic; i.e., fluctuations in w are small and spontaneous action potentials are no longer observed without applied current.

CHAPTER 5

BIOMOLECULAR ANALYSIS

All living organisms are composed of the same types of substances, namely water, inorganic ions, and organic copounds. The organic compounds including carbohydrates, lipids, proteins, and nucleic acids are often called biomolecules. Biochemical studies of biomolecules start with isolation and purification, the chromatographic techniques that are the most commonly employed. The most informative method to investigate structures of biomolecules is spectroscopic techiniques. The computational adjuncts to these techniques are presented. The Internet search for biomolecular structures and information is described.

SURVEY OF BIOMOLECULES

The bulk of all cells is water, in which relatively small amounts of inorganic ions and several organic compounds are dissolved. The number of distinct organic chemical species in a cell is large, but most may be classified as carbohydrates, lipids, proteins, nucleic acids, or derivatives thereof. Those few compounds not so classified account for only a small fraction of the mass of the cell and are metabolically derived from substances in one of the four major classes.The four major classes of organic compounds have markedly different structures, properties, and biological functions, but each class contains two types of compounds: monomeric species with molecular weights of about 10^2 to 10^3 and polymeric biomolecules (biomacromolecules), with molecular weights of about 10^2 to 10^{10}. The Web site of International Union of Biochemistry and Molecular Biology is an excellent starting point for the classififcation and nomenclature of biochemical compounds.

Carbohydrates

The simple sugars (monosaccharides, glycoses), their oligomers (oligosaccharides), and their polymers (polysaccharides, glycans) are collectively known as carbohydrates. They constitute a major class of cell components. Monosaccharides, commonly named sugars, are aldoses (polyhydroxyaldehydes) or ketose (polygydroxyketones) with a general formula, $[C(H_2O]n$. Sugars contain several chiral centers at secondary hydroxy carbons resulting in numerous diastereomers. These various diastereomers are given different names; for example, ribose (Rib) and xylose (Xyl) are two of the four distereomeric aldopentoses; galactose (Gal), glucose (Glc, and mannose (Man) are three of the eight diastereomeric aldohexoses; fructose (Fru) is the most common one of the four diastereomeric ketohexoses. Each of the sugars exists as enantiomeric pairs, D and L forms, according to the configuration at the chiral center farthest from the carbonyl group.

The carbonyl functional group forms a cyclic hemiacetal with one of the hydroxy groups to yield a five-member *furanose* ring (for pentoses) or a six-member *pyranose* rng (for hexoses). The cyclization of sugars yields a new chiral center at the carbonyl carbon atom (anomeric carbon atom), giving rise to α- and β-anomers that mutarotate in solution via an open-chain structure. The common pyranoses occur in the chair conformation. Many sugar derivatives occur in nature. Among these are aldonic acids such as 6-phosphogluconic acid, uronic acids such as glucuronic acid, and galacturonic acid. The hydroxyl group at the 2 position of glucose and galactose may be replaced by an amino group to form glucosamine (GlcN) and galactosamine (GalN), respectively. The amino group of these amino sugars is often acetylated to yield N-acetylglucosamine and N-acetylgalactosamine.

Oligosaccharides (sugar units ≤ 10) and polysaccharides are formed from joining the monosaccharide units by glycosidic linkages. Several disaccharides are important to the metabolism of plants and animals. Examples are maltose [αD-Glc-(1→4)-αD-Glc], lactose [βD-Gal-(1→4)-αD-Glc], sucrose [αD-Glc-(1→2)-βD-Fru], and trehalose [αD-Glc-(1→1)-D-αGlc]. polysaccharides are present in all cells and serve in various functions. Despite the variety of different monomer units and types of linkage present in carbohydrate chains. the conformational possibilities of oligo-and polysaccharides are limited. The sugar ring is a rigid unit and the connection of two torsion angles, ϕ and ψ, which are conveniently taken as 0º when the two midplanes as defined by the linked glycosidic atoms of the sugar rings are coplanar.

A systematic examination of the possible value for ϕ and ψ for β-1, 4- linked glucose units in cellulose (β-glucan) shows that these angles are constrained to an extremely narrow range around 180º, placing the monomer units in an almost completely extended confromation. Each glucose unit is flipped over 180° from the previous one so that the plane of the rings extends in a zigazg manner. In glycogen and starch (α-glucan), glucose units are joined in α-1,4-linkages, and the chain tends to undergo helical coiling. The helix contains 6 residues per turn with a diameter of nearly 14 nm.

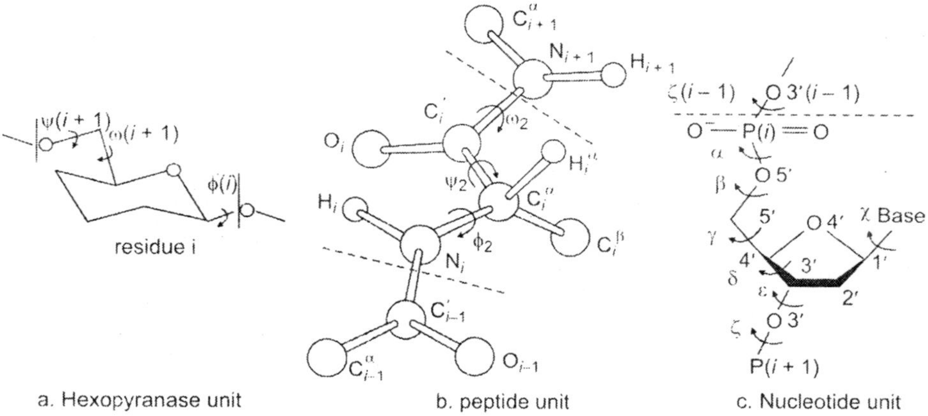

a. Hexopyranase unit b. peptide unit c. Nucleotide unit

Fig. 5.1. Notation for torsion angles of biopolymer chains. Torsion angles (ϕ and ψ) that afect the main chain confromations of biopolyers are shown for polysaccharide (*a*), polypeptide (*b*), and polynucleotide (*c*) chains according to the IUBMB notation. The two torsion angles, ϕ and ψ, specified around the phosphodiesteric bonds of nucleic acids correspond to α and ξ, respectively.

Many oligosaccharides and polysaccharides are often attached to cell surface or secreted in the froms of glycoproteins. This may be done via *O*- glycosidic linkages to hydroxy groups of serine, threonine, and hydroxylysine or in N-glycosidic linkages to the amide nitrogen of asparagine. Because sugar chains can occur between gydroxyls of different positions or configurations, the oligo-and polysaccharide molecules are branched with varied linkage (unlike linear nucleic acids with 3′, 5′-diphosphoesteric linkages or linear proteins with amide linkages). *O*-Linked glycans contain an N-acetylagalactosamine residue at their reducing termini conferring particular physicochemical properties to the proteins. N-Linked glycans act as recognition signals of cell surface proteins. They contain the triasaccharide core, Manβ1→4GlcNACβ1→4GlcNAc, from which various mono- bi- tri-, tetra-, or penta-antennary sugar chains are attached.

Lipids

Lipids refer to a large, heterogeneous group of substances classified together by their hydrophobic properties. Most lipids are derived from linking together small molecules such as fatty acids, or they are derived from acetate units by complex condensations reactions. Fatty acids such as palimtic (C_{16}), Palamitoleic (9E-C_{16}), stearic (C_{18}), oleic (9E-C_{18}), linoleic (9E,12E-C_{18}), linolenic (9E,12E, 15E-C_{18}), and arachidonic (5E,8E,11E,14E-C_{20}) acids are commonly found in lipids. They have even carbon number and may contain double bonds in *cis* (E) configuration. The hydrocarbon chains of the fatty acids tend to assume the extended conformation, but double bonds induce kinks and bends. Triglycerides, which are commonly found in adipose fats and vegetable oils, are formed from esterification of glycerol by three fatty acids. The phosphatides are fatty acid derivatives of *sn*-glycerol-3-phosphate which form phosphoesteric linkagges with choline, serine, or inositol ot yield phospholipids. A second group of phospholipids, the sphingomyelins, contain sphingosine as the central unit.

Phospholipids are amphipatic (molecules containing both hydrophobic ahydrophilic ends) and areimportant constituents of biomembranes. Lipids are the only class of four major biomolecules which do not undergopolymerization, but they may associate together to form large aggregate such as liposomes or membranes. Biomembranes are made up of protein and lipid wilth the ratio (by weight) varying from 0.25 to 3.0 and having a typical value of 1:1. According to lipid bilayer model for membrane structure, hydrophobic bonding holds the fuly extended hydrocarbon chains together while the polar groups of the phospholipid molecules interact with proteins that line both sides of the lipid bilayer. Phospholipids, which make up from 40% to over 90% of the membrane lipid, are predominated by five types: phosphatidylcholine, phosphatidylethanolamine, phosphatidylserine, diphosphatidylglycerol, and sphingomyelin.

Polyprenyl compounds are formed by condensation of isoprene units that are derived from acetate by reductive trimerization and decarboxylation. These compounds include terpenes, carotenoids, sterols, and steroids,

Proteins

The complete hydrolysis of proteins produces 20 α-amino acids that also occur as free metabolic intermediates. As free acids, they exist mostly as dipolar ions (zwitterions). Except for glycine, they contain chiral α-carbons and therefore exist in the D- and L- enantiomeric pair, of which the L-isomers are the monomeric units of proteins. They are differentiated strucaturally bly their side-chain groups with varying chamical reactivities that determine many of the chamical and physical properties of proteins. These side-chain groups include:

(*i*) **Hydrogen**—that is glycine (G).

(*ii*) **Alkyl groups**—for example, alanine (A), valine (V), leucine (L), and isoleucine (I), which participate in hydrophobic interactions and tend to cluster inside protein molecules.

(*iii*) **Aromatic rings**—for example, phenylalanine (F), tyrosine (Y), and tryptophane (W), which give proteins characteristic UV absorption at 280 nm and form effective hydrophobic bonds especially in bonding to flat moecules.

(*iv*) **Hydroxyl functionality**—for example, serine (S) and threonine (T), which provide glycosylation and phosphorylation sites and are found at the active center of some enzymes.

(*v*) **Carboxylic group**—for example aspartic acid (D) and glutamic acid (E), which provide anionic charges on the surface of proteins and constitute catalytic residues of glycosidases.

(*vi*) **Amide derivative**—for example, asparagine (N) and glutamine (Q), which may participate in hydrogen bonding and provide glysosylation site.

(*vii*) **Sulphydryl functionality**—for example, cysteine (C) and methionine (M), of whilch cysteine is noted for its easy formation of disulfide linkages and its involvement in the activities of oxidoreducatases.

(*viii*) **Amino group**—for example, lysine (K) provides cationic charge on the surface of proteins and is the site of glycation by blood glucose.

(*ix*) **Imidazole ring**—for example, histidine (H), which may participate in general acid and general base catalyses. It is found at the catalytic site of many enzymes.

(*x*) **Guanidino group**—for example, arginine (R), which may involve in the binding of phosphate groups of nucleotide coenzymes.

(*xi*) **Imino structure**—for exmple, proline (P), whose amino group forms imino ring with the rigid conformation therefore its presence disrupts the formation of regular secondary structures.

Proteins are the cornerstones of cell structure and the agents of biological function. They function as enzymes that catalyze virtually all cellular reactions, serve as regulators of these reactions, act as transporters, function as transducers of energy/electrical pulses/mechanincal motion/light, form an essentialbiological defense system, and provide structure and storage materials of the cells.

Proteins are formed by joining α-amino acids together *via* peptide linkages (polypeptide chains) with monomer units in the chain known as amino acid residues. A polypeptide chain usually has one free terminal amino group (N-terminal) and a terminal carboxyl group (C-terminal) at the other end, though sometimes they are derivatized. Polypeptide chains formed by polymerization of α-amino acids in specific sequences provide the *primary structure* of proteins.

Knowledge of amino acid sequences is of major importance in understanding ht behaviour of specific proteins. A polypeptide (protein) can be thought of as a chain of flat peptide units for which each peptide unit is connected by the α-carbon of an amino acid. This carbon provides two single bonds to the chain, and rotation can occur about both of them (except proline). To specify the conformation of an amino acid unit in a protein chain, it is neccessary to specify torsion angles about both of these single bonds. These torsion angles are indicated by the symbols ϕ (around C_α-N bond) and ψ (around C_α-C_0 bond), and they are assigned the value 0º for the fully extended chain. Both ϕ and ψ can vary by rotation, resulting in a large number of conformations. Because no two atoms may approach one another more closely than is allowed by their van der Waals radii, the restricted rotation around the two single bonds at C_α atom imposes steric constraints on the torsion angles, ϕ and ψ, of a polypeptide backbone that limits the number of permissible conformations.

The whole range of possible combination of ϕ and ψ plotted (ϕ versus ψ) in the conformational map (Ramachandran plot) indicating the allowable combinations of the two angles within the blocked areas. A polypeptide chain twisted by the same amount about each of its C_α atoms assumes a helical conformation, the most important of which is α helical structure. The α helix (3.6_{13} helix) is found when a stretch of consecutive amino acid residues all have $\phi = -57°$ and $\psi = -47°$ twisting right-handed 3.6 residues per turn with hydrogen bonds between C = O of residue n and NH of residue $n + 4$. The β strands are aligned close to each other such that hydrogen bonds can form between C = O of one β strand and NH on an adjacent β strand. The β sheeps that

Table 5.1. Regualr Secondary structures of proteins

Secondary Structure	*Optimal Dihedral Angles* ϕ	ψ
Right-handed α helix (3.6_{13})	–57	-47
Rihgt-handed 3_{10} helix	–49	-26
Right-handed π helix (4.4_{16})	–57	-70
2.2_7 Ribbon	–78	+59
Left-handed α helix	+57	+47
Collagen triple helix	–51	+153
Parallet β sheet	–119	+113
Antiparallel β sheet	–139	+135

are formed from several β strands are pleated with side chains pointing alternatively above and below the β sheet. The β strands can interact in a parallel or an antiparallel manner. The parallel β sheet has evenly spaced hydrogen bonds that angle across between the β strands while the antiparallel β sheet has narrowly spaced hydrogen bond pairs that alternate with widely spaced pairs. Almost all β sheets in proteins have their strands twisted in a right-hand direction. These repeated local structures are known as *secondary structures,* some of which (with the optimal ϕ and ψ values) are listed in Table 5.1.

The regular secondary structures, α helices and β sheets, are connected by coil or loop regions of various lengths and irregular shapes. A variant of the loop is the β turn or reverse turn, where the polypeptide chain makes a sharp, hairpin bend, producing an antiparallel βturn in the process. The secondary structures are combined with specific geometric arrangement to form compact globular structure known as *tertiary structure.* The fundamental unit of tertiary structure is the *domain,* which is defined as a polypeptide chain or a part of a polypetide chain that can independently fold into a stable tertiary structure. Domains are also units of function, and often the different domains of a protein are associated with different functions. polypeptide chains, especially of regulatory proteins, often aggregate by specific interactions to form oligomeric structures. These oligomeric proteins are said to exhibit *quarternary structure.*

The association of proteins with other biomacromolecules to form complexes of celluar components is referred to as *quinternary structure.* Both internal structure and overall size and shape of proteins vary enormously. Globular proteins vary considerably in the tightness of packing and the amount of internal water of hydration. However, a density of ~ 1.4 g cm^{-3} is typical.

NUCLEIC ACIDS

The nucleic acids, deoxyribonucleic acids (DNA), and ribonucleic acids (RNA) are polymers of nucleotides which are made up of three parts: a purine or pyrimidine base, D-2-deoxyribose for DNA or D-ribose for RNA, and phosphoric acid. The nucleotides are joined through phosphodiesteric linkages between the 5′-hydroxyl of the sugar in one nucleotide and the 3′-hydroxyl of another. Two purine bases, adenine (A) and guanine (G), occur in both DNA and RNA. However, the presence of pyrimidine bases differ such that cytosine (C) and thymine (T) are found in DNA while cytosine and uracil (U) are found in RNA. In addition, some minor bases are present in transfer RNA (tRNA). Deoxyribonucleic acid is the genetic material such that the information to make all the functional macromolecules of the cell is preserved in DNA.

Ribonucleic acids occur in three functionally diffrent classes: messenger RNA (m RNA), ribosomal RNA (rRNA), and transfer RNA (rRNA),Messenger RNA serves to carry the information encoded from DNA to the sites of protien synthesis in the cell where this information is translated into a polypeptide sequnce. Ribosomal RNA is the component of ribosome which seves as the site of protein synthesis. Transfer RNA (tRNA) serves as a carrier of amino acid resideus for protein synthesis. Amino acids are attached as aminoacyl esters to the 3′-termini of the tRNA to form aminoacyl-tRNA, which is the substrate for protein biosynthesis. In the nucleic acids, the furanose ring of ribose or deoxyribose can exist in several envelope and skew conformations.

It is ordinarily C2′- or C3′ that is out of the plane of the other four ring atoms. If this carbon atoms lies above the ring toward the base, the ring conformation is known as *endo*l if the atom lies below the ring away from the base, the conformation is known as *exo,* The C2′-*endo* and C3′-*endo* conformations are most common in individual nucleotides. The orientation of the baee with respect to the sugar is specified by the torsion angle λ of the bond connecting N1 (pyrimidines) or N9 (purines) or base to C1′ of sugar (Figure 5.1 c). The zero ′ λ angle is often assigned to the conformation in which the C2′–O1′ bond of the sugar pyrimidine or the C4-N9 bond of purine is cis to the C1′ – O1′ bond of the sugar. Measured values of χ vary among different nucleotides, a typical value being ~ 127°. In this *anti* conformation the CO and NH groups in the 2 and 3 postions of the pyrimiding ring or the 1, 2, and 6 positions of the purine ring are away form the sugar ring, while in the *syn* conformation they lie over the ring. The *anti* conformation is the one commonly persent in most free nucleotides and nucleic acid. The two torsion angle ϕ (α) and ψ (ζ) are specified around the phosphodiestric bonds, O (5′) – P and P– O (3′), respectively, of the elongated polynucleotide chains.

The B form of DNA has $\phi = -96^\circ$ and $\psi = -46^\circ$. One of the most exciting biological discoveries is the recognition of DNA as a double helix of two antiparallel polynucleotide chains with the base pairing between A and T, and between G and C (Watson and Crick's DNA structure). Thus, the nucleotide sequnce in one chain is complementary to, but not identical to, that in the other chain. The diameter of the double helix measured between phosphorus atoms is 2.0 nm. The pitch is 3.4 nm. There are 10 base pairs per turn. Thus the rise per base pair is 0.34 nm, and bases are stacked in the center of the helix. This form (B form), whose base pairs lie almost normal to the helix axis, is stable under high humidity and is thought to approximate the conformation of most DNA in cells. However, the base pairs in another form (A form) of DNA, which likely occurs in complex with histone, are inclined to the helix axis by about 20° with 11 base pairs per turn.

While DNA molecules may exist as straight rods, the two ends bacterial DNA are often covalently joined to form circular DNA molecules, which are frequently supercoiled. While RNA molecules usually exist as single chains, they often form hairpin loops consisting of double helics in the A conformation (Moore, 1999). The best-known forms of RNA are the low-molecualr weight tRNA molecules. In all of them the bases can be paired to form a cloverleaf structure with three hairpin loops and sometimes a fourth. The colverleaf strucrure of tRNA is further folded into an L-shape conformation with the anticodon triplet and the aminoacyl attachment CCA forming the two ends.

Micro Biomolecules

Vitamins and hormones are minor organic biomolecules, but both of them are required by animals for the maintenance of normal growth and helath. They differ in that vitamins are not synthesized by animals and must be supplied in diets while hormones are secreted by specialized tissues and carried by the circulatiory system to the target cells somewhere in the body to intiate/stimulate specific biochemical or physiological activities. Vitamins can be classified as water-soluble (B vitamins and vitamins C) or fat-soluble (vitamins A, D, E, and K) and act as cofactors for numerous enzyme catalyzed reactions or cellular processes. Hormones can be classified structurally as follows:

(*i*) *amino acid dervatives* – for example, epinephrine, norepinephrine, and thyroxine;

(*ii*) *polypeptides*–for example, adrenocorticotropic hormone and its releasing factor, atrial natriuretic factor, bradykinin, clcitinin, cholecystokinin, chorionic gondotrpin and its releasing factor, enkephalins, folicle-stimulating hormone, gastric inhibitory peptide, gastrin, glucagon, growth hormone and its releasing factor, insulin, luteinzing hormone, oxytocin, parathyroid hormone, prolactin, secretin, somatomedins thyrotropin and its releasing factor, and vasopresssion; or

(*iii*) *steriods* – for example, androgens, estrogens, glucocorticoids, mineralocorticoids,and progesterones.

They function as regulators or meassengers of specific cellular processes.

SPECIFICITY

One of the approaches to understanding biological phenomena has been to purify an individual chemical component form a living organism and to characterize its chemical structure or biological activity. The most commonly used techniques for purifying biomolecules are chromatographic and electrophoretic methods, and the most facile approaches to characterzation of molecular structures are spectroscopic methods.

Chromatographic Purification

Chromatography—in particualr, high-performance liquid chromatography (HPLC)– is an ideal technique for purification and analysis of biomolecules. The Position or retention time of a charomatographic peak is governed mainly by the fundamental thermodynamics of solute partitioning between mobile and stationary phases. The parameters of interest are the capacity factor (k) and the resolution (R). The capacity factor is estimated by

$$k = t_r / t_0 - 1 \text{ or } k = v_r / v_0 - 1$$

where t_r is retention time and v_r is elution volume of the peak measured at the peak maximum,and t_0 is dead time and v_0 is void volumc of the column. The resoultion is calculated according to

$$R = (t_{r1} - t_{r2}) / \{2(W_1 - W_2) \text{ or } R = (v_{r1} - v_{r2}) / \{2(W_1 - W_2)$$

where t_{r1} and t_{r2} are retention times, and v_{r1} and v_{r1} are elution volumes. W_1 and W_2 are full widths at the peak base of the first and second peaks, respectivtly.

The modes of liquid chromatographic separation can be classified as follows:

1. Gel-filtration/Permeation Chromatography

The technique is normally used for the separation of biological marcomolecules and polymers. In separates compounds on the basis of size. Solutes are eluted in the order of decreasing molecular size.

2. Adsorption Chromatography

The process can be considered as a competition between the solute and solvent molecules for adsorpition sites on the solid surface of adsorbent to effect separation. In normal phase or liquid_solid cromatography, relatively nonpolar organic eluents are used with the polar adsorbent to separate solutes in order of incrasing polarity. In reverse-phase chromatography, solute retention is mainly due to hydrophobic interactions between the solutes and the hydrophobic surface of adsorbent. Polar mobile phase is used to elute solutes in order of decreasing polarity.

3. Partition chromatography

A partition packing consists of a liquied phase coated on an inert solid. The separation is effected by the interaction of the solute between the mobile phase and the liquied stationary phase.

4. Ion-exchange chromatography

Ion-exchange chromatography separates compounds on the basic of their molecular chargs. Compounds capable of ionization, particularly zwitterionic compounds, separate well on ion-exchange column. The separation proceeds because ions of opposite charge are reatined to different extents. The resolution is influnced by (a) the pH of the eluent which affects the selectivity and (b) the ionic strength of the buffer which mainly affects the retention.

5. Affinity Charomatography

Affinity chromatography is a type of absorption chromatography in which in the molecule to be purified is specifically and reversibly adsorbed by a complementary binding ligand immolbilized on the matrix. It is used to purify biomolecules on the basis of their biological function or specific chemical structure. Purification is often of the order of several thousandfold, and recoveries of the purified biomolecualr are generally very high.

Electrophoretic Separation

Charged biomolecules– especially amino acids, peptides, proteins (at pH other than their isoelctric points), nucelotides, and nucleic acids (at pH above 2.4) – migrate in an electric field with the rates of migration dependent upon their charge densities. Zone electrophoresis is the separation of charged molecules in a supporting medium resulting in the migration of charged species in discreate zones. Various gels (such as agar, agarose,and polyacrylamide) used as the supporting media may also exert a molecular sieving effect. This allows gel electrophoresis to separate charged biomolecules according to their mobilities (size, shap,e and charge), and the applied current,and the resistance of the medium.

Polyacrylamide gels can be prepared with pores of the same order of size as protein/ oligonucleotide molecules and are hence very effective in the fractionation of proteins and oligonucelotides. On the other hand, agarose gels are used to spearate large molecules or completes such as certain nucleic acids and nucleoproteins. The gel concentration required for polyacrylamide gel electrophoresis (PAGE) to achieve optimal resoultion of two proteins (or nucleic acids) can be determind by measuring the relative mobility of each protein in a searies of gels of different acrylamide concentrations to construct a Ferguson plot ($\log_{10} R_f$ versus % T) according to

$$\log_{10} R_f = Y_o + K_r (\%T)$$

where % T refers to percent total monomer (acrylamide + bisacrylamide used) and R_f is the the distance migrated by protein/distance migrated by marker protein or tracking dye. Each Ferguson plot is characterized by its slop K_r, a retardation coefficient which is related to the molecular size of the protein,and its ordinate intercept Y_0, which is a measure of the mobility of the protein in solution and is related to its charge.

Under appropriate conditions, all reduced polypedides bind the same amount of sodium dodecylslfate (SDS), that is, 1.4 g SDS/g polypeptide. Furthermore, the reduced polypeptide–SDS complexes from rod-like particles with lengths proportional to the molecualr weight of the polypeptides. This forms the basis for the empirical estimation of the molecular weight (M_r) of proteins using SDS-PAGE according to

$$R_f = \alpha + \beta \log M_r$$

where R_f are relative mobilities of the sample and reference proteins in the same run.

Electrofocusing (EF) is a charge fractionation technique that separates molecules predominantly by the difference in their net charge, not by size. Thus EF can be considered as an electrophoretic technique by which amphoteric compounds are fractionated according to their isoelectric points (pIs) along a continuous pH gradient maintained by ampholyte buffers. This is contary to zone elctrophoresis, where the constant pH of the separation medium establishes a constant charge density at the molecule and cause it to migrate with constant mobility. The charge of an amphoteric compound in EF decreases according to tis titration curve, as it moves along the pH gradient approaching its equlibnrium postion at pl where the molecule comes to a stop. Proteins with difference in pls of 0.02 pH units can be resolved by EF.

Spectroscopic Detection

Spectroscopic methods are used at some point in the structural characterization of biomolecules. These methods are usually rapid and noninvasive, generally require small amount of samples, and can be adapted for analytical purposes. Spectroscopy is defined as the study of the interaction of electromagnetic radiation with matter, excluding chemial effects. The electromagnetic spectrum covers a very wide ragne of wavelengths . Interaction of radiation with matter may occur in a number of ways. X-ray diffraction depends on elastic scattering of the radiation.

The molecule may either emit radiant energy at the expense of its internal energy or it may absorb radiant energy, being promoted to an excited state. According to the quantum theory, the energy content of a molecule is confined to certain discrete values (*energy level*). Furthermore, a molecule will absorb radiation only when the frequency (υ) of the radiations related to the energy difference (ΔE) between two energy levels by the equation $\Delta E = h\upsilon$, where h is Plank's constant (6.67×10^{-27} erg sec). Because the relative positions of the energy levels depend characteristically on the molecular structure, absorption spectra provide subtle tools for structural investigation. A molecule can become excited in a variety of ways, corresponding to absorption in different regions of the spectrun.

Ultraviolet (UV) and visible spectra, also known as electronic spectra, involve transitions between different electronic states. The accessible regions are 200–400 nm for UV and 400–750 nm for visible spectra. The groups giving rise to the electronic transtitons in the accessible regions is termed *chromphores,* which include aromatic amino acid residues in proteins nucleic acid bases, NAD (P)H, flvins, hemes, and some transition metal ions. Two parameters characterize an absorption band, namely the position of peack absorption (λ_{max}) and the extinction coefficient (ε), which is related to concentrations of the sample by the Beer_Lambert law:

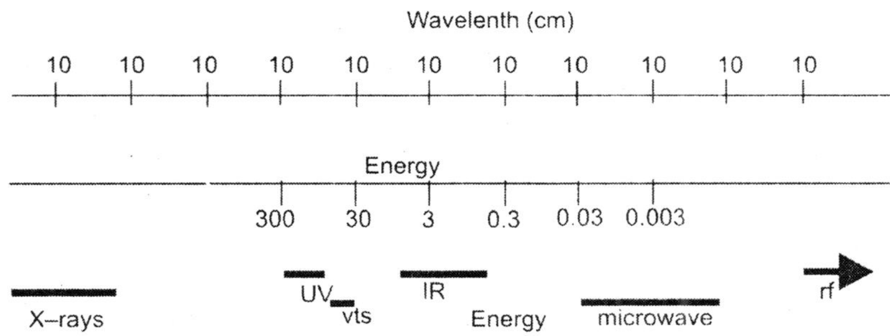

Fig. 5.2. Electromagnetic radiation and its corresponding spectra.

weher l_o and l_s and are the incident and transmitted radiation, respectively, and i is the length of the cell through which radiation travels. The biochemical applications of UV and visible spectroscopy are determination of concentration and interaction of ligand with biomacromolecules. The sensitivity of UV and visible spectra to the solvent environment of the chromophore leads to shifts in λ_{max} and ε, and it is the basic of solvent perturabation spectra in the structural studies of biomacromolecules.

Infrared (IR) spectra provide information on molecular vibration. the common region of IR spectrum is 1400–4000 cm^{-1}. The main experimental parameter is the frequencies (*e.g.*υ for stretching, δ for in plane bending, and γ for out-plane bending) of the absorption bands characterzing functional groups– in particular, υ_c = o and υ_n – H for biomolecules. Water interferes with IR spectrum. The main biochemical application involve studying ligand binding to macromolecules, probing hydrogen bonds, and molecular conformation in nonaquous or oriented samples.

Fluoresence is the emission of radiation that occurs when an excited molecule returns to the ground stat. The molecular group giving rise to fluorescence is termed *fluorophore,* which includes tryptophan residue in proteins, NAD (P) H, and cholorophyll. The measurable parameters are the quantum yield, the intensity, and the position of peak emission (λ max),

Table 5.2. Spectroscopic Methods .

Spectroscopic Method	*Principle*	*Approximate Sample Size*	*Information Obtained and Applications*
Ultraviolet and visible spectroscopy	Absorption of UV and visible radiation leading to electronic excitation	0.1– 10 mg	Presence and nature of unsaturation, especially conjugated double bonds and aromatic systems. Aqueous solutons can be used . Quantitative analysis of proteins and nucleic acids DNA conformations.
Infrared spectroscopy	Absorption of IR radiation leading vibrational excitation	1–10 mg	Presence and enviornment of functional groups, especially X_H or multiple bonds for example, C= O. Diagnosis of finer structural detail such as conformation, intramolecular hydrogen bonding. Identification of unknown by fingerprinting. Studies of macromolecular dynamics by H-D exchanges.
Raman spectroscopy	Scattering of visible or UV light with abstraction of some of the energy leading to vibrational excitation.	0.05 –10 g	AS for IR but with sampling restriction However, aqueous solution can be used. Functional groups that have weak IR Absorption oftern give strong Raman spectra.
Fluorescene spectroscopy	Emission of radiation when a molecule in an excited state returns to the ground state	1 μ M solution	Enviornment, relative abundance, and interactions of fluorphore. Quantitation of proteins and flurophoric compounds. Ligand binding studies.
Nuclear magnetic resonance	Absorption of radiation giving rise to transitions orientations between different spin of nuclei in a magnetic field.	0.1– 1 M solutin	Environment of nuclei, hence unambiguous detection of certain functional groups and information about the environment. Determination of proton sequences and hence of relative points of attachement of functional groups. inference to moleculr configuration and conformation. Studies of macrtomolecular structures and ligand interactions.
Electron spin resonance	Absorption of radiation giving rise to transitions	10^{10} mole of Radial	Detection and estimation of free radicals and diagnosis of their

Spectroscopic Method	*Principle*	*Approximate Sample Size*	*Information Obtained and Applications*
	between opposite spin orientations of of unpaired electrons in a magnetic field.		structure and electron distribution.
Optical rotatory dispersion	Rotation of the plane polarized light by asymmetric molecules in solution without and with variation in wavelngth	1–200 mg/0.5 –20 mg	Determination of relative and absolute configurations of asymmetric centers. Location of functional groups in certain types of compounds. Information about conformation.
Circular dichroism	Difference in intensity of absorption of right and left-circularly polarized light by functional groups in asymmetric environment	1–15 mg	Applications similar to but more powerful than ORD especially for functional groups such as C = O. Conformational analysis of biomacromolecules in solutions.
X-ray diffraction	Interference between scattered X rays caused by atomic electrons	1–10 mg	Determination of complete molecular structure and stereochemistry. Such structural analysis gives bond lengths and angles as well as distances between nonbonded atoms in crystals.
Electron diffraction	Interference between scattered electron beam caused by electrostatic field	1–10 mg	Determination of complete molecular structure of fairly simple molecules.
Neutron diffraction	Interference between neutron beam caused by atomic nuclei	0.1– 1 g	Particularly useful for the location of hydrogen atoms in a molecule.
Mas spectrometry	Determination of mass/ charge ratio and relative abundance of the ions formed upon electron bombardment	0.01–1 mg	Accurate molecular weight determination. Elucidation of structure especially the nature of the skeleton and the length of side chains. Sequence determination.

which are sensitive to environment. The quantum yiled (ϕ_F) is the fraction of molecules that becomes de-excited by fluorescence and is defined as:

$$\phi_F = t/t_F$$

where τ is the observed lifetime of the excited state and τ_F is the theoretical lifetime. Applications include ligand binding, probing of environment and measurement of distance between fluorphores.

Nuclear magnetic resonance (NMR) is a technique that detects nuclear spin reorientation in an applied magnet field. The system requires that the nuclei display unpaired spin state, S such as ^1H,, ^{13}C, and ^{31}P of biochemical interest. The spinning nucleus generates a magnetic field and thus has an associated magnetic moment which interacts with the applied field. The parameters of NMR are chemical shift, spin_spin coupling, area or intensity of the signal, and

$$A = \log (l_0/I_s) = \varepsilon cl$$

two relaxation times (T_1 and T_2) with T_2 related to the line width for freely timbling molecules in solution. Chemical shifts (δ) that measure the resonance frequencies of nuclei are affected by the magnetic shielding for their circulating electrons and are dependent on the molecular environment. The chemical shifts are normally recorded in the NMR spectra in field-independent units with a reference to the frequency (υ_{ref}) of the common reference compound:

$$\delta\,(\text{ppm}) = (\upsilon - \upsilon_{ref})\ \text{Hz/operating frequency, MHz} \times 10^6$$

They are useful for structural inference. The magnetic interaction between neighbouring nuclei is called spin-spin coupling. Its magnitude, known as the spin-spin coupling constant (j), is dependent on the degree of delocalization of electrons between the interacting nuclei, and thus it is informative of the neighboring groups, The intensity gives the relative quantities of the resonance nuclei. The dynamic state of the resonance nucleus can be inferred from its T_2 2D-NMR has yeilded valuable structural information on biomacromolecules in solutions.

Various applications of NMR in biochemistry include structural identification of biomolecules, chemistry of individual groups in macromolecules, structural and dynamic information of biomacromolecules, metabolic studies, and kinetic and association constants of ligand bindings to macromolecules.

The optical activity arises from the chiral centers of chemical compounds or interactions of asymmetrically placed neighboring groups in macromolecules. An optically active molecule interacts differentially with left-and right-cricularly polarized light. This interaction can be detected either by optical rotatory dispersion (ORD) or by circular dichroism (CD). ORD spectrum records a differentially change in velocity of the two beams of the polarized light and is characterized by $[\alpha]_\lambda$ which is the specific rotation at a given wavelength or the molar rotation $[\phi]_\lambda$. CD spectrum recoreds a differential absorption of each beam of the polarized light and is characterized by ΔA (the differential absorption of the two beams) or the molar ellipticity θ_m. Main applications of ORD or CD spectroscopy are the determination of the secondary structure of biomacromolecules and detection of their-conformational changes.

X-ray and neutron diffraction patterns can be detected when a wave is scattered by a periodic structure of atoms in an ordered array such as a crystal or a fiber. The diffraction patterns can be interpreted directly to give information about the size of the unit cell, information about the symmetry of the molecule,and in the case of fibers, information about periodictiy. The determination of the complete structure of a molecule requires the phase informatoin as well as the intensity and frequency information. The phase can be determined using the method of multiple isomorphous replacement were heavy metals or groups containing heavy element are incorporated into the diffracting crystals. The final cooordinates or biomacro-molecules are then deduced using knowledge about the primary structure and are refined by processes that included comparisons of calculated and observed diffraction pattersn. Three-dimensional structures of proteins and their complexes, nucleic acids, and viruses have been determined by X-ray and neutron diffractions.

The mass spectrometry does not involve an interaction between electromagnetic radiation and sample molecule. The functions of a mass spectrometer are to produce positive ions from the sample under investigation, to resolve these ions into a series of ion beams that are homogeneous with respect to their mass/charge ratio (m/e), and to measure the relative abundance of the ions in these beams. The main applications include the molecular weight and structural determinations. Mass spectrometry has emerged as the method for rapid analyses of protein sequences and annotation of their databases.

SEACRHING BIOMOLECULAR DATA

Chromatographic and Spectroscopic Peak Fitting

The success of chromatographic and spectroscopic techniques depends largely on the resolution and analysis of chromatographic/spectroscopic peaks. Automatic peak fitting software, Peak Fit,uses the following routines for finding hidden peaks, thus enhancing the resoultion and facilitating the analysis of

I. Auto Fit Peak I, *Residual method:* A residual is the diffrence in y value between a data point and the sum of component peaks evaluated at the data point's x value. Hidden peaks are revealated by positive residuals.

II. Auto Fit Peaks II, *Second derivative method:* A smooth second dervative of the data will contain local minima at peak positions. The second derivative method requires a constant x-spacing operated in the time domain.

III. Auto Fit Peaks III, *Deconvolution method:* Deconvolution is a mathematical procedure that is used to remove the smearing or braadening of peaks arising because of the imperfection in an instrument's measuring system. Hidden peaks that display no maxima may do so once the data have been deconvoluted and filtered. This method requires a uniform x–specing operated in the frequency domain.

Some of the common Peak Fit functions that can be selected from the program are as follows:

Charomatography:	Haarhoff–Van der Linde (HVL)
	Gidding
	Nonlinear chromatography (NLC)
	Exponentally modified Gaussian (EMG)
Spectroscopy:	Gaussian, amplitude and area
	Lorentizian, amplitude, and area
	Voigt, amplitude and area

The hierarchy of processing in Peak Fit generally follows the order listed below:

1. *Baseling* fitting
2. *Width and shape*
3. *Smooth option*–Savitzky–Golay as the default is adequate. For start, choose Al Expert to let the program seeks an optimum smoothing level.
4. *Peak function family*–Specify either chromatography or spectroscopy.
5. *Peak function type*–Select one of the peak functions appropriate for chromatography or spectroscopy.
6. *Amp %*
7. *Auto Fit*–Start Peak Fit with full graphical update or fast numerical update.

Prepare chromatographic/spectroscopic data files in fileneme.dat or filename. txt (ascii), preferbly filename.xls (Excel) format and save to your disk Launch Peak Fit. from the file menu, invoke Import than enter (or browse) A:/filename.xls. Select columsn for *x–y* data by highlighting column data for *x* and *y*, respectively. Enter titles and then click OK. Click toolbar "AutoFit Peaks I: Residual" initially then follow the hierarchy of fitting process. Click toolbar Auto Fit Peaks I, II, or III of your choice. Select default "Savitzky_Golacey smooting alogrithm" and click Al expert. Invoke Review Fit to view and save the fitted data.For the peak function family, choose either chromatography or spectroscopy and an appropriate fitting function.

Biochemical Databases

Most biochemical databases request the user to enter keywords to search/ retrieve informaion concerning biomolecules (unless the indentifers of the compounds are known). The keyword is normally the IUBMB (international Union of Biochemistry and Molecular Biology) name or common name of the compound. In particular, the linkage and conformational desigantions of oligomeric compounds needs to be specified. The IUBMB nomenclature for biochemical compounds can be accessed from the IUBMB site.

The collection and categorization of biomolecules can be found at Klotho server. On the Klotho home page, click Compound Listing to open the alphabetical compound list. Enter the compound name and click Search Klotho or scan the list for the desired compound that is displayed (in the gif format). Activate (requiring PDB plug-in on the user's computer) the molecular view by clicking on Interactive Viewer. Left press the mouse to move/ rotate the molecular view. Right click the view window to bring up to pop-up option menu. The molecule can be viewed in wireframe (default) sticks, ball and sitck, or space -fill model by choosing the Display option. Select File → Save Molecule As to save the molecule in PDB (.pdb) format (which is the format for most molecular modeling programs) or MDL (.mole) format. This is a handy way to obtain 3D structurs of biochemical molecules for modeling. To save 2D structure, selecte Edit → Transfer to ISIS draw to convert 3D view into 2D view as molecule.skc. The Options toggle-check for displays of hydrogen bond, disulfide bond, and dot surface according to van der Waals radii or Connolly/Richards solvent. Select Rotation and then Start to initiate an automatic rotation until Stop command is issued.

Table 5.3. Some Databases for Biochemical Compounds.

Resorce	*URL*
Chem Finder: Chemical structures, properties	http://www.chemfinder.com
Klotho:General,metabolites	http://ibc.wustl.edu/klotho
Monosaccharide database	http://www.cermav.cnrs.fr/databank/mono/
Glyco Suite DB: Glycan structures	http://www.glycosuite.com/
Lipid Bank: comperhensive lipid information	http://lipid.bio.m.u-toky.ac.ip/
LIPID: Membrane lipid structures	http://www.biochem.missouri.edu/:LPIDS/membrane lipid-html
Mptopo:Membrane protein topolgy	http://blanco.biomol.uci.edu/mptopo/
AAindex: Parameters for amino acids	http://www.genomead.jp/dbget-bin/
Entrez: Sequences of Proteins/ nucleic acids	http://www.ncbi.hlm.nih.gov.Entrez
EBI: Sequences of proteins/nucleic acids	http://www.ebi.ac.uk/
PDB: 3D structures of biomacromolecules	http://www.rcsb.org/pdb/
Histone database	http://genome.nhgri.nih.gov/histones/
RNA structure databse	http://rnabase.org
RNA modification database	http://medlib.med.utah.edu/RNAmods
European large subunit rRNA database	http://rrna.uia.ac.be/isu/indx.html
European samll subunit rRNA database	http://rrna.uia.ac.be/isu/inde.html
tRNA sequence database	http://www.uni-bayreuth.de/departments/biochemie/trna/
Merck manual: Vitamin/hormone function	http://www..merck.com/pubs/manual/

To obtain useful informatoin on chemical (including biochemical) compounds from chem Finder enter the compound name and click Search. The serve returns with synonymous names, chemical formula and structure, and onsite information, as well as links for further biochemical information. The structure can be saved in.cdx format (Chem Draw) and viewed with Chem 3D.

The monosaccharide database the resource site for monosaccharides. Click Databank to open the query/result windows. Select the desired compound from Choose sugar type (*e.g,.* Ribo-,Gluco-) and the subsequent pop-up list (of the sugar type). The search result returned with the 2D molecular structure and the choices for 3D structural view (double click to enalrge the view window), along with atomic coordinate files in PDB or MDL format. Glysco Suite DB provides annotated glycan (polysccharide) structures that can be queried by:

I. mass or mass range,

II. attached protein by protein name, keyword or Swiss-Prot accession number,

III. taxonomy via selection from the scrolling list box,

IV. composition by entering the number of units for each monosaccharide field (*i.e.*, Hex, Hex NAc, dhex, Pent, Neu Ac, and Neu Gc for hexose, N acetyl- hexosamine, deoxyhexose, N-acetylneuraminic acid , and N-glycolyleuramic-nic acid, respectively),

V. tissue/cell type via selection from the scrolling list box,

VI. linkage (*i.e.*, N-linked, O-linked or C-linked),

VII. Glysco Suit DB accession number, or

VIII. advance query (all categories selected by the user).

To query by "composition" enter the desired number for monosaccharide units and click the "*perform search*" button. From the returned list (arranged by taxanonmy and tissue/cell types), choose the entry (or entries) with the matched compositions (s) from the desired biological sources (s) and then click either the "refine selection"button to trim the hit list or the '*show glycan entries*" button to display the search results : clicking the '*show glycan entries*' button returns the list of glycan structures with pertaining information

Table. 5.4. Lipid Menu from lipid Bank.

Click at the square box to open the query from the lipid class of compounds	
ACYLGLYCEROLS	HOPANOIDS
BILE ACIDS (CHLANOIDS)	ISOPERENOIDS
DERIVED LIPIDS	LIPOAMINO ACIDS
LONG-CHAIN ALCOHOL	LIPOAMINO ACIDS
LONG-CHAIN ALDEHYDE	LIPOROTEINS
LONG-CHAIN BASE and CERAMIDE	
ETHER TYPE LIPIDS	MYCOLIC ACIDS
FAT SOLUBLE VITAMINS	PHOSPHOLIPIDS
CAROTENOID	GLYCEROPHOSPHOLIPID
COENZYME Q	PAF
VITAMIN A	SPHINGOPHOSPHOLIPID
VITAMIN D	
VITAMIN E	PROSTANOIDS
VITAMIN K	STEROIDS
GLYCOLIPIDS	WAXES
glycoSPHINGOLIPID	
glycoGLYCEROLIPID and OTHERS	

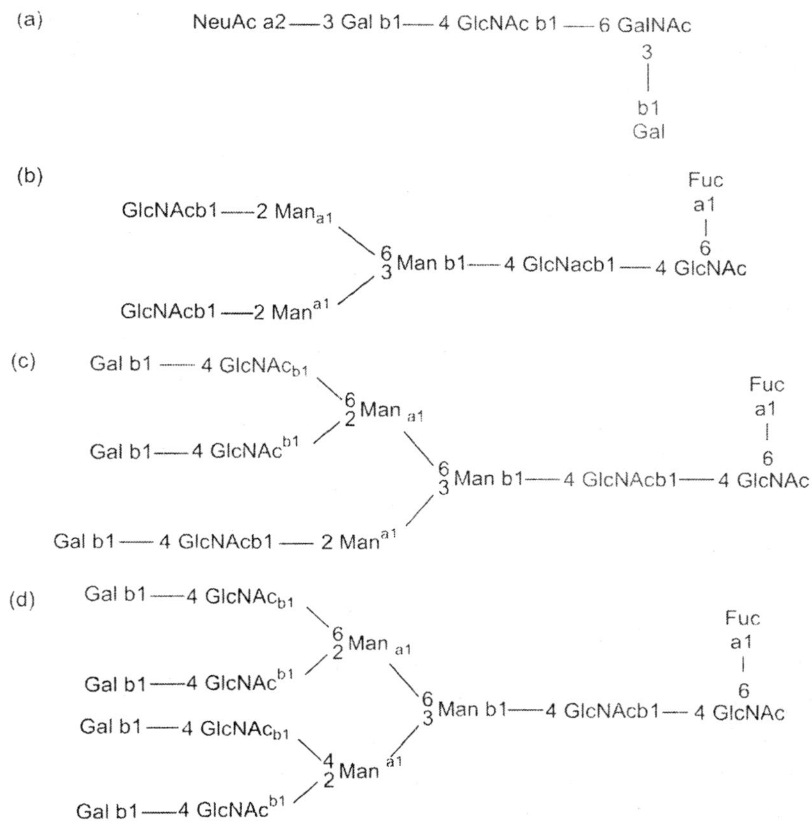

Fig. 5.3. Examples of glycans retrieved from GlycoSuite DB. Examples of various anternnary sturctures are illustrated for human glycans retrieved from GlycoSuite DB. (a) Monoantenary Hex_2 Hex NAC_2 $NeuAc_1$, O-linked to mucin of intestine, colon, and mucosa. (b) Bianternary Hex_3 Hex NAc_4 $dHex_1$, N-linked to saposin A of spleen lymphatic system. (c) Trianternary Hex_6 Hex NAc_5 $dHex_1$,N-linked to plasma coagulation factor X. (d) Tetrantennary Hex_7 hex_6 dHex1, N-linked to saposin B of liver. Abbreviations used are: Ac, acetyl; Fuc, L-fucose (6-dexy-Lgalactopranose); Gal, D-galactopyranose; GaLNAc, 2-acetamido-2 dexoy-d galactopyranose; GlcNAC, 2-acetamido-2deoxy-D-dexy-D glucopyranose; Manm D-Mannopyranose;and NeuAc, N-acetylneuraminic acid.

Select lipid category from the Lipid Menu to open the query form. The databse serch can be conducted via Keyword (lipid name, source, biological activity, metabolism, or genetic information), Classification (lipid class from classification list), Numeric Attributes (number of carbons, double bonds, etc.), and Linkage Position (carbon position attached to specific residue). Choosing the desired lipid class from the Classification list (the easiest approach) returns a hit list of compounds. Select the desired compound by clicking the View button on the left-hand side of the compound. Tabulted information including names, formula, molecular structure (which can be downloaded in Chem Draw fromat), biological activity, physical and chemical properties, spectra data, source chemical synthesis, metabolism, genetics and cited references is returned.

The AAindex database of DBget, which can be accessed from provides physcochemical properties, conformational propensities, and mutation matrices of amuno acids. To search AAindex database enter keywords (*e.g.*,chemical shifts, volume), choose the *bfind* mode and click Submit. The search returns a list of hits from which you select the desired entry. The amino acid index data for each entry starts with an index I in the following alphabetical order: Ala,Arg, Asan, asp, Cys, Gln, Glu, Gly, His, Ile (in the first line), and Leu, Lys, Met, Phe, pro, Ser, Thr, Trp, Tyr, Val (in the second line). Alternatively, entering accession number in the *bget*mode will reutrn the index data.

RNA modification database at furnishes information concerning naturally modified nucleosides in RNA. After clicking Search, select individual files (pick corresponding radios under categories: base type, RNA source, and phylogenetic occurrence) to be accessed,and output options (common name structures and/or mass values). `Enter partial name (*i.e.,*substituent name such as methyl, thio, etc). and choose the parent base (s) from which modified unceloisdes are derived. Click Search button. The tabulated search results based on selected files are returned according to the requested output options.

The amino acd sequences of proteins and nucleotide sequences of DNA can be retrieved from the intergrated database retrieval systems: Entrez of NCBI, EBI and DDBj. The three-dimenstional structures of proteins, nucleic acids,and polysaccharides can be retrieved from PDB. The structural information on protein toplogy of biomembrane can be obtained from the Mptopo site. The biomedical information for vitamins and hormones can be obtained from the Merck manual online at. For example, entring vitamin returns a list of vitamins from which an entry describing deficincy, dependence and/or toxicity of the vitamin can be selected, viewed, and saved.

Spectral Information

National Institute of Materials and Chemical Research, Japan maintains Spectral Database Systems (SDBS) of organic compounds including a number of biochemical compounds. Enter compound name and/or molecular formula. The wild card% can be used to search spectra data by molecular formula. Clicking the SDBS number that corresponds to the desired compound on the response page returns the information page from which the desired spectrum is selected.

The infared (IR) spectra (in liquid film, KBr dist, nujor mull, or CCI_{14}/CS_2 solution) are recorded with an FT spectrometer in the region of 4000–400 cm^{-1} with an optical resoultion of 0.5 cm^{-1}. The proton magnetic resonace (^{1}H-NMR) spectra in ($CD.I_3$ or D_2O/DMSO at the sample concentration of less than 100 *mg/ml*) are obtained at the resonance frequency of 90 MHz with the digital resolution of 0.0625 Hz or 400 MHz at the digital resoultion of 0.125 Hz. The carbon 13 unclear magnetic resonance (^{13}C-NMR) spectra (in $CDCI_3$ or D_2o/DMSO at the sample concentration of about 25 mol%) are obatined under ^{1}H decoupled condition. The flipangle is 22.5 –45.0 degree, and the pulse repetion time 4–7 sec with the digital resolution of 0.025 – 0.045 ppm. The mass (MS) spectra are obtained by the electron impact method with the electron accelerating voltage of 75 V and the ion accelerating voltage of 8–10 V. The NMR and IR spectra are appended with information on peak and frequency analyses. Clicking *peck data* on the mass spectral page returns the table of mass numbers and relative intersities.

Alternatively, a search for the structure of an unknown compond with spectral information can be conducted by entering the peak positions in ppm of C^{13} and/or ^{1}H NMR spectra or sets of mass number and relative intensity of MS spectrum. NMR data for some proteins can be obtained from Bio MagResBank (BMRB)

Fitting and search of Biomolecular Data and information

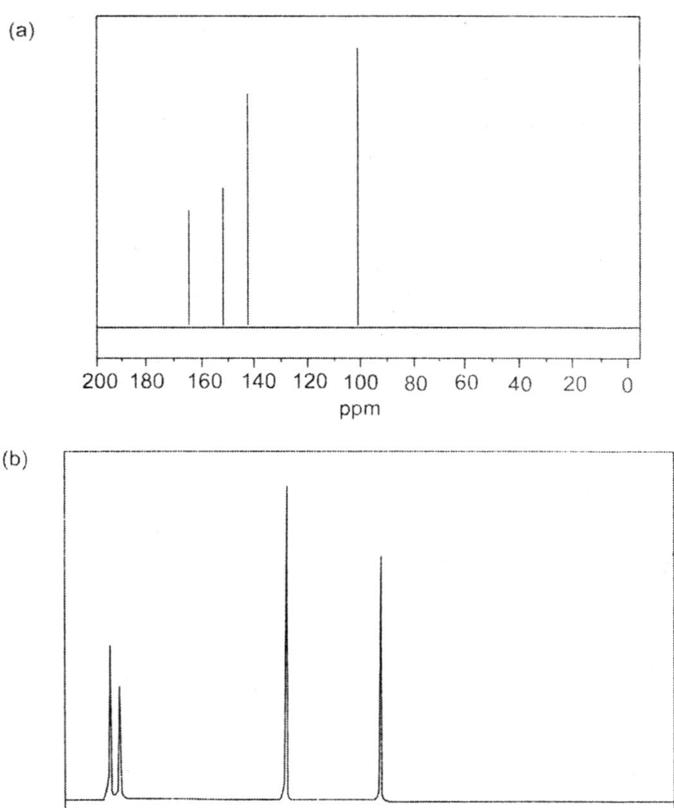

Fig. 5.4. Sample spectra retrieval from SDBS (a) ^{13}C-NMR spectrum in DMSo-d_6. (b) ^{1}H-NMR (400 MHz) spectrum in DMSO-d_6. (c) Mass spectrum. (b) infrared spectrum in KBr. Sample spectra (including spectral analysis) or uracil are retrieved from Spectral Database Systems. The structure of uracil (molecular weight = 112) is represented with the number corresponding to the postion carbons and the alphabet denoting the postion of protons to facilite NMR assignments;

Carbon -13		Proton NMR	
C-Position	chemical shiff (ppm)	H Position	Chemical shift (ppm)
1	165.09	a	11.02
2	152.27	b	10.62
3	142.89	c	7.41
4	101.01	d	5.47

The major mass spectral peaks and their relative intensities based on 100 for the molecular ion (m+) are m/e = 18.0 (12), 28.0 (34), 42.0 (47), 69 (64), and 112.0 (100). The retrieved uracil spectra c-13 NMR, proton NMR, mass, and infrared) are depicted.

WORKSHOPS

1. Perform peak fits for the liquid charomatographic sepeation of proteins shown below:

Fraction	*A 280*	*Fractions*	*A280*	*Fraction*	*A 280*	*Fractions*	*A 280*
20	0	55	1.851	87	0.117	119	0.768
22	0	56	1.068	88	0.235	120	0.832
24	0	57	0.695	89	0.478	121	1.016
26	0	58	0.268	90	0.760	122	1.272
27	0.04	59	0.102	91	0.996	123	1.668
28	0.08	60	0.066	92	1.122	124	1.962
29	0.22	61	0.034	93	1.016	125	1.855
30	0.54	62	0.026	94	0.836	126	1.792
31	0.28	63	0.014	95	0.545	127	1.662
32	0.11	64	0.032	96	0.482	128	1.536
33	0.06	65	0.064	97	0.482	129	1.472
34	0.04	66	0.141	98	0.482	130	1.407
35	0	67	0.256	99	0.482	131	1.343
36	0	68	0.442	100	0.456	132	1.279
37	0.006	69	0.665	101	0.249	133	1.151
38	0.088	70	0.908	102	0.128	134	0.962
39	0.326	71	1.139	103	0.064	135	0.896
40	0.698	72	1.304	104	0.026	136	0.768
41	1.193	73	1.096	105	0.23	137	0.602
42	1.624	74	0.832	106	0.71	138	0.489
43	1.824	75	0.603	107	0.128	139	0.382
44	1.647	76	0.424	108	0.332	140	0.297
45	1.195	77	0.298	109	0.639	141	0.212
46	0.726	78	0.192	110	0.984	142	0.135
47	0.382	79	0.137	111	1.322	143	0.086
48	0.218	80	0.098	112	1.842	144	0.047
49	0.197	81	0.064	113	1.458	145	0.023
50	0.355	82	0.045	114	1.088	146	0.011
51	0.729	83	0.030	115	0.924	147	0.006
52	1.426	84	0.024	116	0.832	148	0.002
53	1.943	85	0.319	117	0.768	149	0
54	2.034	86	0.053	118	0.703	150	0

2.Perform peak fit for the spectrum of FAD chromophore of a flavoprotein:

Wave Number	*A*	*Wave Number*	*A*	*Wave Number*	*A*
14025.0	0.000	208.25.0	0.448	27625.0	0.463
14425.0	0.000	21225.0	0.513	28025.0	0.454
14825.0	0.000	21625.0	0.565	28425.0	0.422
15225.0	0.000	22025.0	0.584	28825.0	0.394
15625.0	0.000	22425.0	0.546	29225.0	0.362
16025.0	0.000	22825.0	0.498	29625.0	0.325
16425.0	0.000	23225.0	0.430	30025.0	0.298
16825.0	0.004	23625.0	0.357	30425.0	0.271
17225.0	0.008	24025.0	0.281	30825.0	0.214
17625.0	0.026	24425.0	0.253	31225.0	0.178
18025.0	0.054	24825.0	0.224	31625.0	0.153
18425.0	0.112	25225.0	0.201	32025.0	0.130
18825.0	0.187	25625.0	0.198	32425.0	0.117
19225.0	0.195	26025.0	0.235	32825.0	0.094
19625.0	0.198	26425.0	0.292	33225.0	0.072
20025.0	0.252	26825.0	0.377	33625.0	0.052
20425.0	0.375	27225.0	0.458	34025.0	0.034

3. A dehydrogenase is purifed to homogeneity from fish liver. The atomic absorption spectrometry shows that the enzyme contains 0.316% (by weight) of zinc. The enables calculation of the minimal molecualr weight (M_{min}) of the metalloenzyme according to

$$M_{min} = \{(\text{atomic weight of Zn } (\% \text{ Zn})\} \times 100$$

Two empirical approaches, SDS polyacrlamide electrophoresis (SDS-PAGE) and gel filtration chromatography, are used to determine the subunit molecualr weight (M_{su}) and Stokes' radius (r) of the dehydrogenase. In the presence of sodium dodecylsulfate (SDS), protein undergo disscoiation and unfolding, therefore their electrophoretic mobities (x) are related to the dissociated subunit weight (M_{su})

$$\log M = \alpha - \beta x$$

where α and β are constants for a given gel at a given electric field. In gel filtration chromatography the elution constant, K_e, of protein is related to its radius of an equivalent hydrodynamic sphare known as the Stokes' radius (r) by

$$(-\log K_e)^{1/2} = A + Br$$

where A and B are constants and K_e is determined experimentally from the elution volume (V_e) in relation to the total bed volume (V_t) and the void volume (V_0) of the column according to $K_e = (V_e - V_0)/(V_e - V_t)$. The electrophoretic mobilities and elution constants of the dehydrogenase and marker proteins of known subunit weights and Stokes' radii are given, respectively, on next page:

SDS-PAGE in 10% Separating Polyacrylamide Gel

Protein	*Electrophoretic Mobility*	*Subunit Weight (kdal)*
Lysozyme	0.81	14.4
Hemoglobin	0.77	16.0
Chymotrypsinogen	0.56	25.7
Trisephosphate isomerase	0.54	26.5
Yeast alcohol dehydrogenase	0.43	35.0
Aldolase	0.37	40.0
Horse liver alcohol dehydrogenase	0.34	42.0
Hexokinase	0.26	51.0
Catalase	0.22	57.5
Purified dehydrogenase	0.36	?

SephadeX G200 Filtration Chromatography

Protein	*Elution Constant K_e*	*Stokes' Radius (nm)*
Cytochrome c	0.59	1.64
Ribonuclease	0.56	1.92
Tripsin	0.51	0.94
Chymotrypsin	0.45	2.09
Monomeric serum albumin	0.19	3.50
Yeast alcohol dehydrogenase	0.062	4.53
Catalase	0.038	5.20
Pufified dehydrogenase	0.22	?

Perform regression analyses of the SDS-PAGE and the gel filtration chromatographic results to evaluate the subunit weight (M_{su}) and the Stockes radius (r) of the purified dehydrogenase.

The native molecular weight (M) of 83.0 kdal is determined independently with electron spary ionization mass spectrometry. If the protein is spherical and unhydrated its radius (r_o) can be estimated by

$$r_0 = [3\bar{v}M)/(4\pi N)]^{1/2}$$

where π = 3.1416 and N is the Avogadro's number (6.022×10^{23} mol^{-1}). From the molecular weight and the Stokes' radius a useful shape parameter, the fractional ratio (f/f_o), which expresses the deviation of the protein molecule from the unhydrated spherical shape, can be claulated according to

$$f/f_0 = [3\bar{v}\,M)/(4\pi N)]^{1/3}$$

Evaluate the frictional ratio by taking the partial specific volume to be 0.735 $g\ cm^{-3}$.Deduce the structural features of the dehydrogenase.

4. Search for oligo-/polysaccharide structures of glycoproteins from human with following saccharide compositions:

(*a*) Hex_1 $HexNAc_2dHex_1NeuAc_1$
(*b*) Hex_2 $HexNAc_2dHex_2NeuAc_1$
(*c*) Hex_3 $HexNAc_2dHex_1NeuAc_1$
(*d*) Hex_4 $HexNAc_2dHex_2NeuAc_1$
(*e*) Hex_5 $HexNAc_3dHex_2NeuAc_1$
(*f*) Hex_6 $HexNAc_5dHex_1NeuAc_1$

Depict their linkage and antennary structures using abbreviations for monosaccharide units.

5. Physicochemical properties of amino acids are very useful descriptors for understanding the structurs and properties of proteins. These properties are expressed numerically in indexes that can be retrieved from the A Aindex database. Desing an index database of physicochemical properties of amino acids with Microsoft Access that may facilitate the data retrieveal according to their chemical similarities:

Chemical Characteristics	*Residues Groups*
Small	Ala, Gly, Pro
Small/relatively polar	Cys, SEr, Thr
Acidic/amide	Asp, Asn, Gln, Glu
Basic	Arg, His, Lys
Apiphatic/hydrophobic	Ile, Leu, Met, Val
Aromatic	Phe, Trp, Tyr

6. The carbonyl functionality at C1 and the primary hydroxyl group at C6 of D-glucose are readily oxidized/reduced to yield D-fucose, D-glucaric acid, D-glucitol,

D-gluconic acid, and D-glucuronic acid. Search Klotho and/or Chem Finder sites fro their structurs. How would you differentiate these redox derivatives by spectro-scopic methods.

7. Mass spectromety is one of the best techniques for identifying fatty acids, which are the major component of lipids. its application revelas the following:

(*a*) The molecular ion (M^+) of fatty acids and their esters are relatively abundant, thus it is easy to identify them and determine their molecular weight.

(*b*) The fragmentation process leading to intense peaks follows the simple C-C cleavage.

(*c*) The rearrangement event leads to the intense peak.

H C C C O C OH → H C C C C C OH +

(*d*) Most intense peaks are oxygen-containing fragments.

(*e*) The periodicity C-C cleavage of $-(CH_2)_4^-$ is favored.

Search Lipid Bank for structures and SDBS for mass spectra of common staurated fatty acids and identify the intense peaks with the above-mentioned characteristics.

8. Nuclear magnetic resonace spectgroscopy is a powerful technique for investigating structure of biomolecules. The ^{1}H and ^{13}C-NMR spectra of L–α amino acids have been compiled (wuthrich, 1986) and can be retrieved from SDBS. Design a database for 1H or ^{13}C-NMR data that can be used in the identification of amino acids.

9. The characteristic ^{13}C-NMR data for oligosaccharides of glucopyranoses have been complied and the chemicl shifts (δ in ppm) for glucobioses are listed below:

Glcobiose	*Nonred C1*[a]	*Red C1*[a]	*αGlucosid C2 to C4, C6*[b]	*βGlucosid C2 to C4, C6*[b]	*Other C2 to C5*[c]	*Other C6*[c]
αGlc (1 → 1) α-Glc	94.0				72.0 ± 1.5	61.5
αGlc (1 → 1) β-Glc	103 ± 1				73.9 ± 3.5	62.0 ± 0.4
βGlc (1 → 1) β-Glc	100.7				74.2 ± 3.1	62.5
αGlc (1 → 2) α-Glc	97.1	90.4	76.7		72.4 ± 1.7	61.6
αGlc (1 → 2) α-Glc	98.6	97.1	79.5		73.7 ± 3.0	61.6
βGlc (1 → 2) α-Glc	104.4	92.4		81.4	73.5 ± 3.1	61.7
βGlc (1 → 2) β-Glc	103.2	95.1		82.1	73.5 ± 3.1	61.7
αGlc (1 → 3) α-Glc	99.8	93.1	80.8		72.4 ± 1.8	61.8
αGlc (1 → 3) α-Glc	99.8	97.0	83.2		73.6 ± 3.0	61.8
βGlc (1 → 3) β-Glc	103.2	92.7		83.5	72.7 ± 3.7	61.7
βGlc (1 → 3) β-Glc	103.2	96.5		86.0	72.7 ± 3.7	61.7
αGlc (1 → 4) α-Glc	100.7	92.8	78.5		72.3 ± 1.9	61.6
αGlc (1 → 4) β-Glc	100.7	96.8	75.2		73.8 ± 3.4	61.7 ± 0.1
βGlc (1 → 4) α-Glc	103.6	92.9		79.9	73.8 ± 3.2	61.4 ± 0.4
βGlc (1 → 4) β-Glc	103.6	92.8		79.9	73.8 ± 3.2	61.5 ±0.3
αGlc (1 → 6) α-Glc	98.5	92.9	66.5		72.3 ± 1.9	61.6
αGlc (1 → 6) β-Glc	98.5	96.8	66.5		73.3 ± 2.9	61.6
βGlc (1 → 6) α-Glc	103.0	92.5		69.4	73.3 ± 3.0	61.7
βGlc (1 → 6) β-Glc	103.0	96.4		69.4	73.3 ± 3.0	61.7

[a] Nonred C1 and Red C1 refer to C1 of nonreducing and reducing glucopyranose units.

[b] αGlucosid C2 to C4, C6 and Glucosid C2 to C4, C6 refer to C2, C3, C4 of glucopyranose unit involved in the
or β-(1 → 2), (1 → 3) α (1→ 4) or (1 → 6) glucosidic linkages, respectively.

[c] Other refers to glucopyranose Cs with free OH groups.

Desing the ^{13}C–NMR database (Access)for glucobioses and create a query form to identify the structur of glucobiose with 13 C-NMR signals at δ103.65, 92.94, 79.88, 77.08, 76.66, 74.29, 72.47, 72.38, 71.20, 70.61, 61.74, and 61.08 ppm.

10. The anticodon of tRNA molecule interacts with codon of mRNA. The degeneracy of the genetic code means that several codons can specify a single amino acid. To reduce the number of tRNAs necessary to read the codons, the first nucleotide (5′ end) of the anticodon which pairs with the third base (3′ end) of the codon is wobbly such that it can pair with two or three different bases at the 3′ end of the codon. This reduces the specificity of interaction between the rRNA molecule and the transcript. It appears that one approach taken by anture to enhance the specific base pairing interaction between tRNA and mRNA is to use modified bases at postions flanking the antocodon in the tRNA molecule. Thus tRNA is noted for the number and variety of modified bases in its structure. The modification of purine and pyrimiding bases is rquired to generate mature tRNA Examples of the modified bases are as follows:

Adenosin (A)	*Cytidine (C)*	*Guanidine (G)*	*Uridine (U)*
1-Methyl (m^1A)	3-Methyl (m^3C)	1-Methyl (m^1G)	Dihydro (D/hU)
N^6-Methyl (m^6A)	5-Methyl (m^5C)	7-Methyl (m^7 G)	5-Methoxycarbonylmethlyl (mcm^5U)
N^6-Isopentlyl (t^6A)	N^4–Acetyl (ac^4C)	N^2-Methyl (m^2 G)	4- Thio (s^4U)
2-Methylthio-t^6A	2-Thio (s^2C)	N^2, N^2-Dimetly^l ($m^2{}_2$G)	5-Methylaminomethlyl-s^2U

Search for the structures of methyl nucleosides in tRNA and suggest spectral characteristics that distinguish these minor bases from their corresponding parent (major) bases.

CHAPTER 6 BIOMOLECULAR INTERACTIONS

Understanding the structure-function relationship of boomacromolecules furnishes one of the strong motives for investigating biomacromolecule-ligand interactions that constitute an initial step in their biological functions. Various situations arising from multiple equilibria including cooperativity and allosterism are introduce. Binding equilibrium is analyzed with Dyna Fit. Concepts and databases for receptor biochemistry and signal transduction are presented.

Most physiological processes are the consequences of an effector interaction with biomacromolecules, such as interactions between enzymes and their substractes, between hormotes and hormone receptors, between antigens and antibodies, between inducer and DNA, and so on. In addition, there are macromolecule-macromolecule interactions such as between proteins, between protein and nucleic acid and between protein and cell-surface raccharide. The effector of small molecular weight is normally referred to as the ligand, and the macromolecular combinant is known as the receptor.

The biochemical interaction systems are characterized by general ligand interactions at equilibrium, site-site interactions, and cooperativity as well as linkage relationships regarding either (a) two different ligands binding to the same macromolecule of (b) the same ligand binding to the different sites of the macromolecule.

Consider a macromolecular receptor, R, which contains n sites for the ligand L. Each site has the microscopic ligand association constant K_i for the i-site.

$$R + L \rightleftharpoons RL \qquad K_1 = [RL]/[R][L]$$

$$RL + L \rightleftharpoons RL_2 \qquad K_2 = [RL_2]/[RL][L]$$

$$\cdots \qquad \cdots$$

$$RL_{n-1} + L \rightleftharpoons RL_n \qquad K_n = [RL_n]/[RL_{n-1}][L]$$

Equilibrium measurement of ligand bindig typically yields the moles of ligand bound per mole of macromolecule, v, which is given by

$$v = \frac{K(L)}{1 + K(L)} \qquad \text{for the single equilibrium}$$

and

$$v = \frac{\sum\{i \prod K_i\}(L)^i}{1 + \sum\{\prod K_i\}(L)^i} \qquad \text{for the multiple equilibrium}$$

The solution of v for n-sites give different expressions uder various situations, such as:

1. ***n*- Equivalent Sites:** The macromolecular receptor contains n sites that are thermodynamically equivalen. The equilibrium treatement of the ligrand binding to a receptor with n- equivalent sites which may be either independent (noninteracting) or interacting gives rise to

   a. Noninteracting n-equivalent sites

$$v = v\frac{nK(L)}{1+K(L)}$$

   The linear transformation of this expression gives

   Klotz's equation: $1/v = 1/n + 1\{n\,K(L)\}$

   or

   Scatchard's equation: $v/(L) = nK - vK$

   Bothe equations are used in the linear and graphical analysis of binding data.

   b. interacting n-equivalent sites

$$v = \frac{nK(L)^h}{1+K(L)^h}$$

   This expressin is known as Hill's equation and undergoes linear transformation to yield

$$\log\{v/(n-v)\} = \log K + h\log(L)$$

   The Hill's interaction coeffcient, h, is a measure of the strength of interaction among *n-sites. If the* n-equivalent sites are noninteracting, then $h = 1$, wheres h approaches n (number of equivalent sites0 for the stronglyb interacting n-equivalent sites.

2. ***n*-Nonequivalent Sites:** The macromolecular receptor contains n sites that bond lagrand with different equilibrium constaints. For a macromolecule with m classes of independent sites, each clas i has n_i sites with an intrinsic association constant, K_i.

$$v = \sum^{m}\frac{n_i\,K_i(L)}{1+K_i(L)}$$

   The receptors with n-nonequivalent sites are generally oligomeric, consisting of heterooligomers or hommoligomers with different conformations. These receptors may display cooperativity in the ligand-recptor interactions.

ALLOSETERISM

The ligand-induced Conformational changes appear to be an important feature of recptors with regulatory functions. The concepts of cooperativity and allosterism in proteins have been applied to explain various ligand-r4eceptor interaction phenomena. The observed changes in successive association constants for a multiple ligand binding system suggest the cooperativity in the interaction betwen binding processes. The interaction that causes an increase insucessive association constants is called positive cooperativity, while a decrease in successive association constants is the consequence of negative cooperativity. The *cooperativity* refers to interaction between binding sites in which the binding of one ligand modifies the ability of a subsequent ligand molecules to bind to its binding site, whereas the *allosterism* refers to the binding of ligand molecules to different sites.

Because a single binding site cannot generate cooperativity, a cooperative binding system must consist of two or more binding sites on each molecule of protein. Although there is no need for such proteins to be oligomeric, nearly all known cases of cooperativity at equilibrium are found in proteins with separate binding sites on different subunits. Furthermore, the cooperativity can be observed with interactions of a single kind of ligand in homotropic (same ligand) interactions or those involving two (or more) different kinds of ligands in heterotropic interactions. Two models have been proposed for the cooperativity of homotropic interactions between ligand and proteins:

The *symmerty model* of Monond, Wyman, and Changeux. This model was originally termed the allosteric model.The model is based on three postulates about the structure of an oligomeric protein (allosteric protein) capable of binding ligands (allosteric effectors):

1. Each subunit (protomer) in the protein is capable of existing in either of two conformational states, namely, T (tight) and R (relaxed) states.
2. The conformational symmetry is maintained such that the all protomers of the protein must be in the same conformation, either all T or all R at any instance. Therefore, a protein with n subunits is limited to only two conformational states, T_n and R_n. Thus only a single equilibrium constant $L = [T_n]/[R_n]$ is sufficient to express the equilibrium between them.
3. The dissociation constants, K_T and K_R for binding of ligand S to protomers in theT and R conformations, respectively, are different, with the ratio $K_R/K_T = C$ that is assumed to favor binding to the R conformation.

The fractional saturation has the following expression:

$$v = \frac{\alpha(1+\alpha)^{n-1} + LC\alpha\,(1+\alpha)^{n-1}}{(1+\alpha)^n + L(1+\alpha)^n}$$

where = αSK_R and $C\alpha = SK_T$.

This *sequential model* of Koshland, nemethy and Filmer. The sequential model Proposes that the conformational stability of each subunit is determined by the conformations of the subunits with which it is in contact. The model is based on three postulates:

1. Each subunit in the protein is capable of existing in either or two conformational states A and B.
2. There is no requirement for conformational symmetry; therefore mixed conformaions are permitted. The stability of any particular state is determined by a product of equilibrium constants including a term K_t for each subunit in the B conformation expressing the free energy of the conformational transition from A to B of an isolated subunit, a term K_{AB} for each contact between a subunit in the A conformation with one on the B conformation, and a term K_{BB} for each contact between a pair of subunits in the B conformation.
3. Ligand bonds only to conformation B, with association constant Ks. The binding to conformation A does not exist.

Because the sequential model deals with subunit interactions explicitly and referes to contacts between subunits, it is ncecssary to consider the geometry of the molecule; for example, a terameric protein may exist as square or terrahedral of which tetrachedral is the preferred arrangement. Treatment of the tetrahedral tetrameric protein gives rise to the folliwng expression;

$$v = \frac{K_S K_t K_{AB}^3 [S] + 3K_S^2 K_t^2 K_{AB}^4 K_{BB} [S]^2 + 3K_S^3 K_t^3 K_{AB}^3 K_{BB}^3 [S]^3 + K_S^4 K_t^4 K_{BB}^6 [S]^4}{1 + 4K_S K_t K_{AB}^3 [S] + 6K_S^2 K_t^2 K_{AB}^4 K_{BB} [S]^2 + 4K_S^3 K_t^3 K_{AB}^3 K_{BB}^3 [S]^3 + K_S^4 K_4' K_{BB}^6 [S]^4}$$

Treatment of other geometries leads to the same expressions for the conecntrations except for the subunit interaction terms, which are different for each geometry.

The two models further differ in that the symmerty model predicts only positive cooperativity wheres both positive and negative cooperativities are possible according to the sequential model. it is noted that the application of the multiple equilibrium to cooperativity, thouth fundamentally important, is not practical because it requires knowledge of the number of binding sites and the values of the ndividual association constants. it is often convenient to define cooperativity with reference to the shape of saturation curve. Operationally, the Hill equation (or the Hill plot) provides a measure of cooperativity. The Hill coefficeint, h (or the slope of the Hill plot), is large than 1 for positive cooperativity, less than 1 for negative cooperativity, and equal to 1 for noncooperativity. Thus the ligand-biomacromolecule interaction is positively cooperative, negatively cooperative, or noncooperative according to the sign of $(h - 1)$

SIGNAL TRANSDUCTION

A receptor is a molecule (commonly biomac-romolecule) in/on a cell that specifically recognizes and binds ia ligand acting as a siganl molecule. Ligand-receptor interactions constitute important initial steps in various celluar processes. The ligands such a hormoune and neurotransmitters bind to plasma membrane receptors, which are transmembrane glycoproteins. These ligands include bioactive amines (acetylcholine, adrenaline, dopamine, histamine, serotonin,) peptides (clactonin, gluacgon, secretin, angiotensin, bradykinin, interleukin, chemokine, endothelin, melanocortin, neuropetide Y, neurotensin, somatostatin, thrombin, galanin, orexin), hormone proteins, prostaglandin, adenosine, and platelet activating fractor. Other ligands such as steroids and thyroid hormones bind soluble DNA-binding proteins. The ligand-receptor interactions initiate various *singal transduction* (pathways that mobilize second messengers which activate/inhibit cascade of enzymes and proteins involved in specific cellular processes. Three membnrance receptor groups that mediate eukaryotic transmembrane signalig processes are as follows:

1. Group 1, *Single-transmembrane segment catalytic receptors:* Proteins consisting of a single transmembrane segment with (a) a globular extracellular domain, which is the ligand recognition site, and (b) an intracellular catalytic domain, which is either tyrosine kinase or guanylyl cyclase.
2. Group 2, *Seven-transmembrane segment receptors:* Intergral membrance proteins connsisting of seven transmebrance helical segments with extracellular reognition site for ligands and an intracellular recognition site for a GTP-binding protein.
3. Group 3, *Oligomeric ion channels:* Ligand-gated ion channels consisting of associated protein subunits that contain several transmembrane segments. Typically, the ligands are necurotransmitters that open the ion channls upon binding.

The binding of these receptors with may ligands stikulates A G-protein (GTP-binding protein), which in turn activates an effector enzyme (*e. g.,* adenylyl; cyclae/phospholipase C). Typically, G-proteins are heterotrimers ($G_{\alpha\beta\gamma}$) consisting of α (G_α), β (G_β), and γ (G_γ), subunits. Binding of the ligand to receptor stimulates an exchange of GTP for abound GDP on G_α causing G_α to dissociate from $G_{\alpha\beta\gamma}$ and to associate with an effector enzyme which synthesizes the second messenger (Ross and Wilkie, 2000). G-Proteins are a universal menas of signal transduction in higher crganisms, activting many hormone-receptro-initiated cellular processes via activation of adenlyl cyclases, phospholipases A/C phosphodiesterases, and ion (Ca^{2+}, Na^+, K^+) channels. Each hormone receptor protein interacts specifically with either a stimulatory G protein or an inhibitory G protein. Some of G proteins and their physioligical effects are given in Table 6.1.

Table 6.1. Some Proteins and Their Physiological Effects.

G Protein	*Ligand*	*Effector*	*Effect*
G_s	Epinephrine, glucagon	Adenylyl cyclase	Glycogen/fat breakdown
G_s	Antiduretic hormone	Adenylyl cyclase	Conservations of water
G_s	Luteinizing hormone	Adenylyl cyclase progesterone synthesis	Increase estrogen, Progerter one synthesis
G_l	Acetyl choline	Potassium channel	Decrease heart rate
G_i/G_o	Enkephalins, endorphins, opioids	Adenylyl cyclase, ion channel	Changes in neuron electrical activity
G_q	Angiotenisn	Phospholipase C	Muscle contraction, blood presurre elevation
G_{olf}	Odorant molecules	Adenlyl cyclase	Odoratn detection

Two stages of amplification occur in the G-protein-mediated signal response. First, a signle ligand-receptor complex can activate many G proteins. Second, the G-protein-activated adenylyl cyclase or phospholipase synthesizes many second messenger molecules. Some of these intracellular second messengrers and their effects are given in Table 6.2

Table 6.2. Some Second Messengers and Their Effects.

Second Messenger	*Effect*	*Soruce*
cAMP	Protein kinase activation	Adenylyl cyclase
c GMP	Protein kinase activation, ion channel regulation	Guanylyl cyclase
Ca^{2+}	Protein kinase and Ca^{2+} – regulatory protein activation	Membrance ion channel
Inositol-1, 4, 5-trip	Ca^{2+} -channel activation	Phospholipase C on P-inositol
Diacylglyceorl	Protein kinase C activation	Phospholipase C on P-inositol
Phosphotidic acid	Ca^{2+} -channel activation, adenylyl cyclase inhibition	Phospholipase D products
Ceramide	Protein kinase activation	Phospholipase C on sphingomyelin
Nitric oxide	Cyclase activation, smooth muscle relaxation	nitric oxide synthase
cADP-ribose	Ca2+ -channel activation	cADP-ribose synthase

Cyclic AMP (cAMP) is produced from ATP by the action of an integral membrance enzyme, adenylyl cyclase. Various second messengers, such as inositiol 1, 4, 5-triphosphate, diacylglycerol, and arachiodonic acid, are generated via breakdown of membrance phospholipids by the action of phospholipases. Calcium ion is an important intracellular siganl that is affected by either cAMP or inositol-1, 4, 5- triphosphate. The Ca^{2+} signals are translated into the desrirde intracellular response by calcium binding proteins (*e.g.*, calmodulin, parvalbumun), which in turn regulate cellular processes via protein kinase C (PKC). PKC is a cellular transducer, translationg the ligand message and the signals of second messengers to phosphorylate serine and threonine residue of wide range of proteins that control growth and develpment. Nitric oxide (NO) which is synthesized from arginine by nitric oxide synthetase, acts as a neurotransmitter and as a second messenger. Cyclic GMP generated by the No-stimulated solube guanylyl cyclase acts to regulate ion channel conductivity and to block gap junction conductivity. Srteroid hormones and carcinogens exert their effect as transcription regulators to mediate gene expression.

RECEPTOR DATABASES

Binding Data

Dyna Fit is a versatile program for statistical analysis and fitting of the equilibrium binding data, the intitial velocties, and the time course of enzymatic reactions according to the user defined mechanisms. The program performs nonlinear least-squares regression of ligand-receptor binding and enzyme kinetic data. To run DynaFit, you must first create a user subdirectory into which you should place a script file and a data file or a data sub-subdirectory containing several data files. The script file is an ascii text file that contains all the information required for the program to perform analysis and simulation.

The general format of the script file for ligand (L) binding to two-site receptor (R) is

```
[task]
  data = equilibrium
  task = fit
  model = fixed?
[mechanism]
  L + R <=> LR        :    K1 dissoc
  L + LR <=> L2R :   K2 dissoc
[constants]
  K1 = 5.0 ?,          ; mM
  K2 = 20.0 ?          ; mM
[response]
  LR = 1.0
  L2R = 2.0
[concentrations]
  R = 20.00   ; enzyme concentration
[data]
  variable  L
  file                 user/1 igand / data. txt
[otuput]
  directory    user/ 1ligand/output
[end]
```

The student should consult the script files in the example directory of the distribution program for the requisite format in order to run Dyna Fit successfully.

The data files are ascii text files arranged in columns separated by space, tabs, or commas with header lines. *Equilibrium files* contain total ligand concentrations in the first column and free ligand concentrations in the second column *Initial velocity files* contain subtrate concentrations in the first column and initial velocities of duplicate assays inthe subsequent columns. *Progress curve files* contain time in the first column and the measured experimental siganl in the second column.

Three-component windows open at the start of Dyna Fit (double click Dyna Fit. exe orDyna Fit icon). Status window provides messages for the action to be taken or in progress. Text window gives mechanism, analytical, and fitted text results, while the fitted and simulated plots are displayed in Graphic window. From the File menu, load the script file which processes and analysis and fitting of input data in the data files. An intial plot based on the input data and parameters in displayed on the Graphic window, and the mechanism used is listed on the Text window. After the Stutus window requsting "Ready to run.....,' choose Run from the File menu. The analytical and fitting results are displayed continuously on the Graphic and the Text windows with the Status window providing messages of progress. AT the end of an analytical/fitting cycle, answer *Yes* to "Is the initial estimate good enough/" of the pup-up dialogue box to view result. Answer *Yes* to "Terminate the least square minimization?" of the pop-up dialogue box in order to terminate analysis. otherwise, answer No to continue with an addition cycle of analysis and fitting. AT the end of fitting (the Status window shows a message, "All tasks completed. Examine output files)", the final fitting is displayed on the Graphic window and the summary results are given in the Text window. The details of analysis and results can be viewed via Notepad by invoking Results from the Edit menu and saved as result. txt.

Signal Transduction Pathways

The receptors are classified according to membrane receptors and nuclear recptors, which are also listed alphabetically for retreieval at Receptor database. A keyword query can be initiated to search for a specific receptor, its link to sequence database (Gen-Bank, PIR, and SWISS- PROT), and multiple alignment of a receptor family via MView option. GPCRDB and Nuclea RDB are Information Systems for G protein-coupled receptors and nuclear receptors, respectively (Horn et.al., 2001). both sites provide links to sequence databases, bibliographic references, alignments, and phylogenetic trees based on sequence data. In addition, GPCRDB also provides mutation data lists of available pdb files, and tables of ligand binding constants to receptors of aceptylcholine, adrenaline, dopamin,e histamine, serotonin, opioid, adenosine, cannabis, melatonin, and γ aminobutyric acid.

The Bimolecular Interaction Network Database (BIND) provides a text query for information, on biomolecular interactions relating to cellular communication, differentiation, and growth. Enter the name of receptor or second messenger but not ligand (*e.g.*, the text query accepts dopamine receptor but not dopamine) and click Find. This returns a list of hits. Select the InteractionID (full record) or desired information (Full record, Pub Med abstract, BIND publication, View ASN report, or View XML report). Click Visualize Interaction to open the Interaction Viewer window showing the interaction between two components in the signaling pathway. Further interactions involving the pathway members can be viewed by double-clicking the highlighted boxes.

Reli Base is the resource site for receptor-ligand (Reli) interactions and provides facilites for similarty search/analysis of the Reli database for similar ligands, similar binding sites, and Reli complex. The query at Reli Base can be conducted via Text (*e,g.* chemical name of ligand),

Sequence (of receptor), Smiles (strings), or 2D/3D (fragments of structure) of ligand by clifcking respective menu button. The instructions on how to use Relibase can be obtained by clicking the Help button. A ligand and its receptor complex can be searched/retrieved by entering the ligand name after clicking text button or entering SMILES string (by clicking the Smiles button). to save the atomic coordinate file (pdb format) of a ligand-binding complex, click Complex PDB file. To search for similar ligands with a receptor sequence, click Sequence on the Relibase homepage.

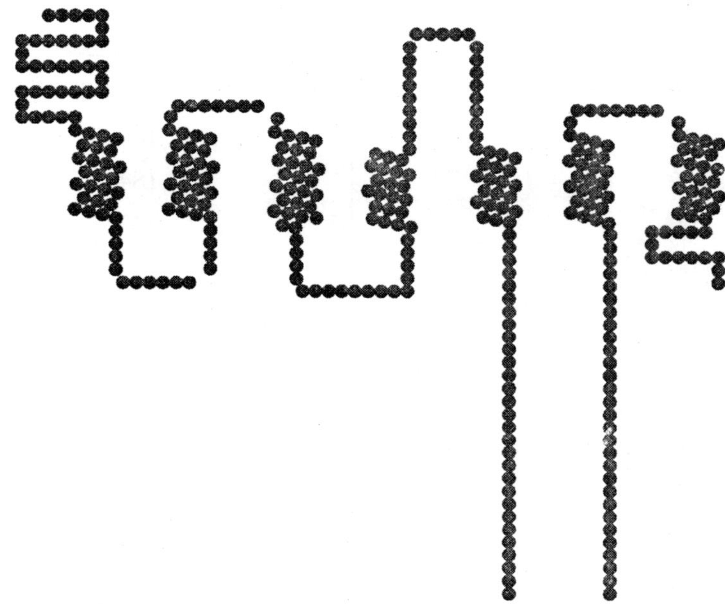

Fig. 6.1. G Protein-coupled receptor retrieved from GPCRDB. Seven transmembrane segment receptor for serotonin retrieved from GPCRDB is visualized in a snake-like plot. The colored groups of residues show a correlated behavior as determined by correlated mutation analysis (CMA). The colored positions are hyperlinked to their corresponding residue locations in the multiple sequence alignments.

Copy and paste the amino acid sequence into the query box and click Show Ligands. Clicking the 2D/3D button opens the structure window for drawing a structure component used to search the database by substructure.

WORKSHOPS

1. The data below describe the binding of ligand L to its receptor R(100 μM). Build a script file and data file to solve for the association constant and the number of non interacting equivalent sites.

Total Ligand (mM)	*Bound Ligand (μM)*
1.00	24.58
1.25	29.59
2.00	43.62
2.50	51.51
5.00	81.94
10.00	166.23

2. The data below describe the binding of ligand L to the oligomeric receptor R(100 μM). Create a script file and data life to evaluate the association constant and the number of equivalent sites. Calculate Hill's interaction coeffcient by perform regression analysis.

Total Ligand (mM)	Bound Ligand (μM)
1.00	98.22
1.25	108.12
1.60	130.90
2.00	150.32
2.50	184.07
4.00	245.71
5.00	279.36
8.00	347.78
10.00	378.92

3. The binding data of ligand L with its receptor protein (100 μM) are given below. Estimate the association constants by assumung two classes of equivalent sites and deduce their cooperativity.

Total Ligand (mM)	Bound Ligand (μM)
1.00	10.30
1.250	15.90
1.425	17.78
1.600	19.55
2.000	22.94
2.500	26.42
3.125	28.73
4.000	30.86
5.000	32.48
8.000	35.05
10.000	36.45
20.000	38.72
40.000	39.96

4. The binding data of ligand L iwth its receptor protein (100 μM) are given below. Estimate the association constants by assuming two classes of equivalent sites and deduce their cooperativity.

Total Ligand (mM)	Bound Ligand (μM)
1.000	11.42
1.250	13.07
1.425	14.05
1.600	15.06
2.000	16.47
2.500	18.07
3.125	19.58
4.000	21.58
5.000	23.69
8.000	19.80
10.000	33.19
14.000	38.40
20.000	41.61
40.000	48.74

5. Search receptor database for the classification of receptors according to the chemical nature of their ligands.
6. List hormones according to their receptors being membrane (G-protein coupled)proteins or soluble (nuclear bound) proteins.
7. Retrieve the sequence and plot of one of-7 transmembrane segement receptors and discuss its structrural features.
8. Retrieve the sequence and 3D model of a neurotransmitter or hormone receptor that transduces nuclear signaling.
9. Depict one signal transduction pathway each for
   a. G-Protein- coupled, adenylylate activating (via cAMP) pathway
   b. G-Protein-coupled, phospholipase C (via inositol triphosphate) pathway
   c. Nuclear protein mediating pathway
10. List candidate signal molecules for the receptor with the amino acid sequence given below.

MRTLNTSAMD GTGLVVERDF SVRILTACFL SLLILSTLLG NTLVCAAVIR FRHLRSKVTN
FFVISLAVSD LLVAVLVMPW KAVAEIAGFW PFGSFCNIWV AFDIMCSTAS ILNLCVISVD
RYWAISSPFR YERKMTPKAA FILLISVAWTL SVLISFIPVQ LSWHKAKPTS PSDGNATSLA
ETIDNCDSSL SRTYAISSSV ISFYIPVAIM IVTYTRIYRI AQKQIRRIAA LERAAVHAKN
CQTTTGNGKP VECSQPESSF KMSFKRETKV LKTLSVIMGV FVCCWLPFFI LNCILPFCGS
GETQPFCIDS NTFDVFVWFG WANSSLNPII YAFNADFRKA FSTLIGCYRL CPATNNAIET
VSINNNGAAM FSSHHHEPRGS ISKECNLVYL IPHAVGSSED LKKEEAAGIA RPLEKLSPAL
SVILDYDTDV SLEKIQPITQ NGQHPT

CHAPTER 7

MOLECULAR DYNAMICS

An understanding of a wide variety of phenomena concerning conformational stabilities and molecule-molecule association(protein-protein, protein-ligand, and protein-nucleic acid) requires consideration of solvation effects. In particular, a quantitative assement of the relative contribution of hydrophobic and electrostatic interactions in macromolecular recoganition is a problem of central importance in biology.

There is no doubt that molecular dynamics simulations in which a large number of solvent molecules are treated explictily represent one of the most detailed approaches to the study of the influence of solvation on complex biomolecules. The approach, consists in constructing detailed atomic models of the solvated macromolecular system and having described the microscopic forces with a potential function, applying Newton's classical equation F = ma to literally, "simulate" the dynamic motions of all the atoms as a function of time. The calculated calssical trajectory, though an approximation to the real world, provides ultimate detailed information about the etime course of the atomic motions, which is difficult to access experimentally. However, statistical convergence is an important isssue because the net influence of solvation results form an averaging over a larg number of configurations. In addition, a large number of solvent molecules are required to relistically model a dense system. Thus in practical stituations a significant fraction of the computer time is used to calculat the detailed trajectory of the slovent molecules even though it is often the solute that is of interest.

An alternativ approach, consists in incroporating the influcence of the solvent implicitly. Such approximate schemes can provide useful quantitative estimates of solvation free energies while remaining computationally tractable. Implicit solvent approaches avoid the statistical errors assoiciated with averages extracted from simulations with a large number of solvent molecules. Furthermore, implicit solvent models are sometimes better suited for particularly complex situations. For example, an explicit representation of the cellular membrane potential would require prohibtively large atomic simulation systems and is currently inparactical. Finally, implicit solvent representations can be very useful conceptual tools for analyzing the results of simulations generated with explicit solvent molecules and to better undrstand the nature of solvation phenomena in general. The complexity of the environment in which biomolecules must perform their functions is such that information extracted form simple theroretical models may be helpful to further our understanding of these systems.

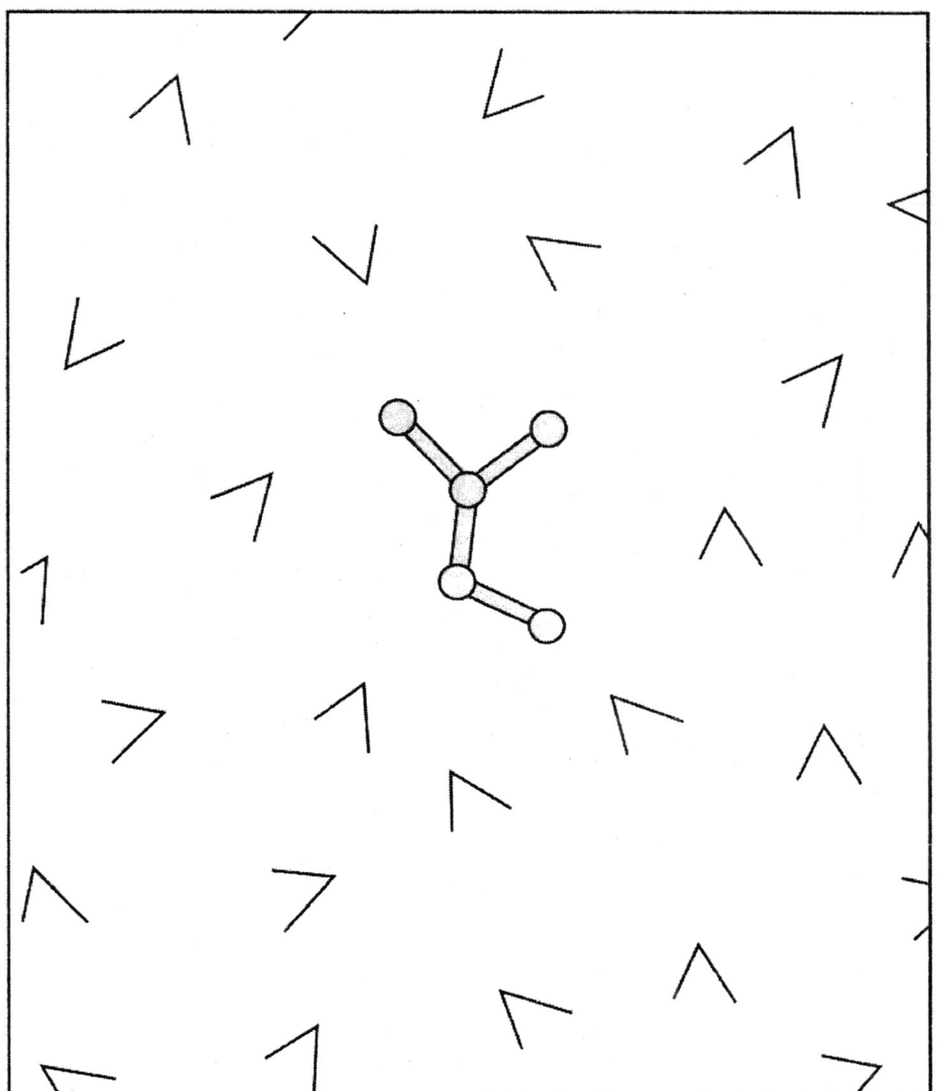

Fig. 7.1. Schematic representation of an atomic model of a biomolecular soulte surrounded by explicit water molecules.

In this chapter we provide an introductory overview of the implicit solvent models commonly used in biomolecular simulations. A number of questions concerning the formulation and development of implicit solvent models are addrssed. In Section II, we begin by providing a rigorous formulation of implicit solvent from statistical mechanics. In addition, the fundamental cocept of the potential of mean force (PMF) is introduced. A decomposition of the PMF in terms of nonpolar and electrostatic contributions is elaborated. For the sake of completeness, other computational schemes such as statistical mechanical integral equations, implicit/explicit solvent boundary potential, solvent-accessible surface area (SASA), and knowledge-based potentials are briefly reviewed.

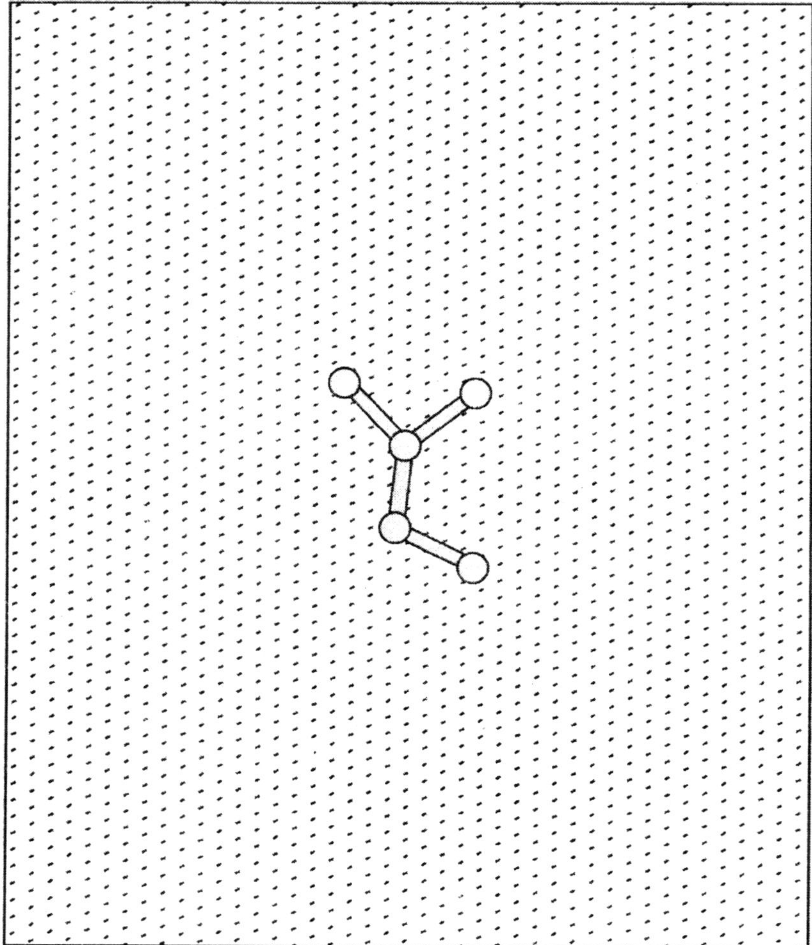

Fig. 7.2. Schematic representation of a biomolecular solute in a solvent enviornment that is taken into account implicitly.

THE POTENTIAL OF MEAN FORCE

As a first step, it is important to establish implicit solvent models on fundamental principles. For the sake of concreteness,, let us consider a solute **u** immersed in a bulk solution **v**. The configuration of the solute is represented by the vector $\mathbf{X} \equiv \{x_1, x_2, \ldots\}$. All other degrees of freedom of the bulk solution surrounding the solute, which may include solvent molecules as well as mobile counnterions. are represented by the Vector **Y.** It is expected that the system is fluctuating over a large number of configurations. It is therefore nceessary to consider the problem for a statistical point of view. For a system in equilibrium with a thermal bath at temperature T, the probability of a given configuration (**X, Y**) is given by the function P (**X, Y**)

$$P(\mathbf{X}, \mathbf{Y}) = \frac{\exp\{-U(\mathbf{X},\mathbf{Y})/K_B T\}}{\int d\mathbf{X}\, d\mathbf{Y} \exp\{-U(\mathbf{X},\mathbf{Y})/k_B T\}} \qquad \ldots(7.1)$$

where U (**X, Y**) is the total potential energy of the system. For the sake of simplicity, we neglect nonadditive interactions and assume that the total potential energy can be written as

$$U(\mathbf{X}, \mathbf{Y}) = U_u(\mathbf{X}) + U_{VV}(\mathbf{Y}) + U_{uv}(\mathbf{X}, \mathbf{Y}) \qquad \text{...(7.2)}$$

where $U_u(\mathbf{X})$ is the intramolecular potential of the solute, $U_{vv}(\mathbf{Y})$ is the solvent-solvent potential, and $U_{uv}(\mathbf{X}, \mathbf{Y})$ is the solute-solvent potential. All observable properties of the system are fundamentally related to averages weighted by the probability function P (**X, Y**). For exmple , the average of any quanitity Q(**X**) depending on the soulte configuration is given by

$$\langle Q \rangle = \int d\mathbf{X}\, d\mathbf{Y}\, Q(\mathbf{X}) P(\mathbf{X}, \mathbf{Y}) \qquad \text{...(7.3)}$$

An important question is whether one can rigorously express such an average with out referring explicitly to the solvent degree of freedom. In other words, Is it possible to avoid explicit reference to the solvent in the mathematical description of the molecular system and still obtain rigorously correct properties? The answer to this question is yes. A reduced probability distribution $\bar{p}$ (X) that depends only on the solute configuration can be defined as

$$\bar{P}(\mathbf{X}) = \int d\mathbf{Y}\, P(\mathbf{X}, \mathbf{Y}) \qquad \text{...(7.4)}$$

The rqduced probability distribution does not depend explicitly on the solvent coordinates **Y**, although it incorporates the average influence of the solvent on the solute. The operation symbolized by Eq. (7.4) is commonly described by saying that the solvent coordinates **Y** have been "integrated out." In a system at temperature T, the reduced probability has the form

$$\bar{P}(\mathbf{X}) = \frac{\int d\mathbf{Y} \exp\{-[U_u(\mathbf{X}) + U_{vv}(\mathbf{Y}) + U_{uv}(\mathbf{X}, \mathbf{Y})]/k_B T\}}{\int d\mathbf{X}\, d\mathbf{Y} \exp\{-[U_u(\mathbf{X}) + U_{vv}(\mathbf{Y}) + U_{uv}(\mathbf{X}, \mathbf{Y})]/k_B T\}} \qquad \text{...(7.5)}$$

$$= \frac{\exp\{-W(\mathbf{X})/k_B T\}}{d\mathbf{X} \exp\{-W(\mathbf{X})/k_B T\}}$$

The function W(**X**) is called the potential of mean force (PMF). The fundamental concept of the PMF was first introduced by Kirkwood to describe the average structure of liquids. It is a simple matter to show that the gardient of W(**X**) in Cartesian coordiantes is related to the average force,

$$\frac{\partial W(\mathbf{X})}{\partial \mathbf{x}_i} = \left\langle \frac{\partial U}{\partial \mathbf{x}_i} \right\rangle = -\langle F_i \rangle_{(\mathbf{X})} \qquad \text{...(7.6)}$$

where the symbol $\langle \ldots \rangle_{(\mathbf{x})}$ represents an average over all coordinates of the solvent, with the solute in the fixed configuration specified by X. All solvent effects are included in W (X) and consequently. In the reduced distribution function $\bar{P}(\mathbf{X})$. The PMF is an effective configuration-dependent free energy potential W(**X**) that makes no explicit reference to the solvent degrees of freedom, such that no information about the influence of solvent on equilibrium properties is lost.

As long as the normalization condition given by Eq. (7.15) is satisfied an aribitrary offset constant may be added to W(**X**) without affecting averages in Eq. (7.3). The absolute value of the PMF is thus unimportant, For convenience, it is possible to choose the value of the free energy W(**X**) relative to a reference system from which the solute-solvent interactions are absent. The free energy W(**X**) may thus be expressed as

$$\exp\{-W(\mathbf{X})/k_B T\} = \frac{\int d\mathbf{Y} \exp\left\{-\left[U_u(\mathbf{X}) + U_{vv}(\mathbf{Y}) + U_{uv}(\mathbf{X}, \mathbf{Y})\right]/k_B T\right\}}{\int d\mathbf{Y} \exp\{-U_{vv}(\mathbf{Y})/k_B T\}} \quad ...(7.7)$$

It is customary to write $W(\mathbf{X}) + U_u(\mathbf{X}) + \Delta W(\mathbf{X})$, where $U_u(\mathbf{X})$ is the intramolecular solute potential and ΔW (X) is the solvent- induced influence. In practice, ΔW depends on **X**, the configruation of the soulte, as well as on thermodynamic variables such as the temperature T and the pressure p.

Relative and Absolute Values

As shown by Eq. (7.6) the PMF is the reversible work done by the average force. It is possible to express relative values of the PMF between different solute configurations $\mathbf{X}_1$ and $\mathbf{X}_2$, using Eq. (7.6) and the reversible work theorem.

$$W(\mathbf{X}_2) = W(\mathbf{X}_1) + \int_{\mathbf{X}_1}^{\mathbf{X}_2} \sum_i d_{\mathbf{X}_i} \cdot \frac{\partial W(\mathbf{X})}{\partial_{\mathbf{X}_i}} \quad ...(7.8)$$

$$= W(\mathbf{X}_1) \int_{\mathbf{X}_1}^{\mathbf{X}_2} \sum_i d_{\mathbf{X}_i} \cdot \langle \mathbf{F}_i \rangle_{(\mathbf{X})}$$

This relationship makes it clear that the PMF is not equal to an average potential energy because one needs to compute a reversible work against an average force to get W (**X**). It is also possible to express the free energy in terms of a thermodynamic integral. Introducing the thermodynamic soulte-solvent coupling parameter λ we write the potential energy as

$$U(\mathbf{X}, \mathbf{Y}; \lambda) = U_u(\mathbf{X}) + U_{vv}(\mathbf{Y}) + U_{uv}(\mathbf{X}, \mathbf{Y}; \lambda) \quad ...(7.9)$$

constructed such that $\lambda = 0$ corresponds to a noninteracting refrence system with $U_{uv}(\mathbf{X}, \mathbf{Y}; 0) = 0$ and $\lambda = 1$ corresponds to the fully interacting system. As long as the end points are respected, any form of thermoynamic coupling is correct. Therefore, we have

$$\Delta W(\mathbf{X}) = \int_0^1 d\lambda \left\langle \frac{\partial U_{uv}}{\partial \lambda} \right\rangle_{(\mathbf{X}, \lambda)} \quad ...(7.10)$$

where the symbol $\langle \cdots \rangle_{(\mathbf{X}, \lambda)}$ represents an average over all coordinates of the solvent for a solute in the fixed configuration **X** with thermodynamic couplig λ. It may be noted that $\partial U_{uv}/\partial\lambda$ in Eq. (7.10) plays the role of a generalized thermodynamic force similar to that of $\partial U/\partial_{\mathbf{X}_i}$ in Eq. (7.8)

DECOMPOSITION

Intermolecular forces are dominated by short-range harsh repulsive interactions arising from Pauli's exclusion principle, van dar Waals attractive forces arising form quantum dispersion, and long-range electrostatic interactions arising from the nonuniform charge distribution. It is convenient to express the potential energy $U_{uv}(\mathbf{X}, \mathbf{Y})$ as a sum of electrostatic contributions and the remaining nonpolar (nonelectrostatic) contributions,

$$U_{uv}(X, Y) = U_{uv}^{(np)}(\mathbf{X}, \mathbf{Y}) + U_{uv}^{(elec)}(\mathbf{X}, \mathbf{Y}) \quad ...(7.11)$$

Although such a representation of the microscopic non-bonded interactions does not follow directly from a quantum mechanical description of the Born-Oppenheimer energy surface, it is commonly used in most force fields for computer simulations of biomolecules (*e.g.*, *AMBER*, *CHARMM*, *OPLS*. The sepration of the solute-solvent interactions in Eq. (7.11) is useful for decomposing the reversible work that defines the function W(**X**). The total free energy of a solute in a fixed configuration **X** may be expressed rigorously as the revesible that modynamic work needed to construct the system in a step by step process. In a first step, the non-polar solute solvent in tractions are switched "on" in the absence on any solute-solvent electrostatic interactions; in a second step, the solute-solvent electrostatic interactions are switched "on" in the presence of the solute-solvent nonpolar interactions. The solute is kept in thefixed configuration **X** throughout the whole process, and the intramolecular potential energy does not vary during this process. By construction, the total PMF is

$$W(\mathbf{X}) = U_u(\mathbf{X}) + \Delta W^{(np)}(\mathbf{X}) + \Delta W^{(elec)}(\mathbf{X}) \qquad ...(7.12)$$

where the nonpolar solvation contribution is

$$\exp\{-\Delta W^{(np)}(\mathbf{X})/k_B T\} = \frac{\int d\mathbf{Y} \exp\left\{-\left[U_{vv}(\mathbf{Y})+U_{uv}^{np}(\mathbf{X},\mathbf{Y})\right]/k_B T\right\}}{\int d\mathbf{Y} \exp\left\{-U_{vv}(\mathbf{Y})/k_B T\right\}} \qquad ...(7.13)$$

and the electostatic solvation contribution is

$$\exp\{-\Delta W^{(elec)}(\mathbf{X}) k_B T\} = \frac{\int d\mathbf{Y} \exp\{-[U_{uv}(\mathbf{Y}) + U_{uv}^{np}(\mathbf{X},\mathbf{Y}) + U_{uv}^{elec}(\mathbf{X},\mathbf{Y})]/k_B T\}}{\int d\mathbf{Y} \exp\{-[U_{vv}(\mathbf{Y}) + U_{uv}^{np}(\mathbf{X},\mathbf{Y})]/k_B T\}} \qquad ...(7.14)$$

Combining Eqs. (7.12)- (7.14) yields Eq. (7.7) directly. Although such a free energy decomposition is path-dependent it provides a useful and rigorous framework for understanding the nature of solvation and for constructing suitable approximations to the nonpolar and electrostatic free energy contributions.

In the following sections, we describe an inplicit solvent model based on this free energy decomposition that is widely used in biophysics. It consists in representing the nonpolar free energy contributions on the basis of the solvent-accessible surface area(SASA), a concept introduced by Lee and Richards and the electrostatic free energy contribution on the basis of the Poisson-Boltzmann (PB) equation of macroscopic electrostatics an idea that goes back to Born, Debye and Huckel, Kirkwood, and Onsager. The combination of these two approximations forms the SASA/PB implicit solvent model. In the next section we analyze the microscopic significance of the nonpolar and electrostatic free energy contributions and describe the SASA/PB implicit solvent model.

Nonpolar Free Energy

To clarify the significance of $\Delta W^{(np)}$, let us first consider the special case of a nonpolar molecule solvated in liquid water. We assume that the electrstatic free energy contribution is negligible. Typically, the solute-solvent van der Waals dispersion interactions are relatively weak and the nonpolar free energy contribution is dominated by the reversible work needed to displace the solvent molecules to accommodate the short-range harsh repulsive solute-solvent interaction. For this reason, $\Delta W^{(np)}$ is often refered to as the "free energy of cavity formation. : The reversible thermodynamic work corresponding to this process is positive and unfavorable. It gives rise to two aspects of the hydrophobic effect: hydrophobic solvation and hydropobic interaction. The former phenomenon in responsible for the poor solubility of nonpolar molecules in water; the latter accounts for the propensity of nonpolar molecules to cluster and form aggregates in water.

Modern understanding of the hydrophobic effect attributes it primarily to a decrease in the number of hydrogen bonds that can be achieved by the water molecules when they are near a nonpolar surface. This view is confirmed by computer simulations of nonpolar solutes in water. To a first approximation, the magnitude of the free energy associated with the nonpolar contribution can thus be considered to be proportional to the number of solvent molecules in the first solvation shell. This idea leads to a conveninet and attractive approximation that is used extensively in biophysical applications. it consists in assuming that the nonpolar free energy contribution is directly related to the SASA.

$$\Delta W^{(np)}(\mathbf{X}) = \gamma_v \mathcal{A}_{tot}(\mathbf{X}) \qquad ...(7.15)$$

where γ_v has the dimension of a surface tension and $\mathcal{A}_{tot}(\mathbf{X})$ is the configuration-depnedent SASA (note that both polar and nonpolar chemical groups must be included in the Sasa for a correct estimate of $\Delta W^{(np)}$. As pointed out by Tanford there should be a close relationship between the macroscopic oil-water surface tension, interfacial free energies, and the magnitude of the hydrophobic effect. However, in practical applications, the surface tension γ_v is usually adjusted empirically to reproduce the solvation free energy of alkane molecules in water. Its value is typically around 20-30 cal/(mol.Å^2) whereas the macroscopic oil—water surface tension is around 70 cal/(mol. Å^2). The difference between the optimal parameter γ_v for alkanes and the true macroscopic surface tension for oil/water interfaces reflects the influence of the microscopic length scale and the crudeness of the SASA model. A simple statistical mechanical approach describing the free energy of inserting hard sphers in water, called scaled particle theory (SPT), provides an important conceptual basis for understanding some of the limitations of SASA models. It is clear that the SASA doe not provide an ultimate representation of the nonpolar contribution to the solvation free energy. Other theories based on cavity distribution in liquied water. and long-range perturbaoon of water structure near large obstacles are currently being explored. A quantitative description of the hydrophobic effect remains a central problem in theoretical chemical physics and biophysics.

Electrostatic Free Energy

The electrostatic free energy contribution in Eq. (7.14) may be expressed as a thermodynamic integration corresponding to a reversible process between two states of the system: no solute-solvent electrosatatic interactions ($\lambda = 0$) and full electrostatic solute-solvent interactions ($\lambda = 1$). The electrostatic free energy has a particularly simple form if the the modynamic parameter λ corresponds to a scaling of the solute charges, *i. e.*, $U_{uv}^{(elec)}(\mathbf{X}, \mathbf{Y}; \lambda) = \lambda U_{uv}^{(elect)}(\mathbf{X}, \mathbf{Y})$, and the coupling is linear,

$$\Delta W^{elec}(\mathbf{X}) = \int_0^1 d\lambda \left\langle U_{uv}^{(elec)} \right\rangle_{(\lambda)} \quad ...(7.16)$$

For this reason, the quantity $\Delta W^{(elec)}(\mathbf{X})$ = is often called the "charging free energy." If one assumes that the solvent responds linearly to the charge of the solute, then $\left\langle U_{uv}^{(elect)} \right\rangle_{(\lambda)}$ is proportional to λ and the charging free energy can be written as

$$\Delta W^{elec}(\mathbf{X}) = \int_0^1 d\lambda \sum_i q_i \Phi_{rf}(X_i; \lambda) \approx \frac{1}{2} \sum_i q_i \Phi_{rf}(\mathbf{x}_i; \lambda = 1) \quad ...(7.17)$$

where $\Phi_{rf}(\mathbf{x}_i; \lambda = 1)$ is the solvent field acting on the ith solute atomic charge located at position X_i in reaction to the presence of all the solute charges (in "reaction field" is thus the elctrosataic potential exerted on the solute by the solvent that it has polarized. The assumption of linear response implies that $\Delta W^{elec} = (1/2)\left\langle U_{uv}^{(elec)} \right\rangle$, a relationship, that is often observed in calculations based on simulations with explicit solvent. The factor 1/2 is a characteristic signature of linear solvent response.

The dominant effects giving to the charging free energy are often modeled on the basis of calssical continuum electrostatics. This approximation, in which the polar solvent is represented as a structureless continuum dielectric medium, was originally pioneered by Born in 1920 to calculate the hydration free energy of spherical ions. It was later extended by Kirkwood and Onsager for the treatment of arbitrary charge distributions inside a spherical cavity. Nowadays, the treatment of solutes of arbitrary shape is possible with the use of powerful computers and numerical methods. In many cases, this is an excellent approximation. The classical electrostatics approach is remarkably successful in reproducing the electrstatic contribution to the solvation free energy of small solutes or amino acids, as shown by comparisons to free energy simulations with explicit solvent.

CONTINUUM ELECTROSTATICS

Poisson Equation

The continuum electrostatic approxsimation is based on the assumption that the solvent polarization density of the solvent at a position r in space is linearly related to the total local electric fied at that position. The Poisson equation for macroscopic continuum media follows from those assumptions about the local and linear electrostatic response of the solvent:

$$\nabla \cdot [\varepsilon(\mathbf{r}) \nabla \phi(\mathbf{r})] = -4\pi\rho_u(\mathbf{r}) \quad ...(7.18)$$

where $\phi(\mathbf{r})$, $\rho_u(\mathbf{r})$, and $\varepsilon(\mathbf{r})$ are the electrostatic potential, the charge density of the solute, and the postion-dependent dielectric constatnt at the point r, respectively. The Poisson equation can be solved numerically by mapping the system onto a discrete greid and using a finite difference relaxation algorithm. Several programs are available for computing the electrostatic potential using this approach, *e.g.*, DelPhi, UHBD and the PBEQ module incorporated in the simulation program CHARMM. Alternatively, one can use an approach based on finite elements distributed at the dielectric boundary (the boundary element methdo). Significant improvements can be obtained with this approach by using efficient algorithms for generating the mesh at the dielectric boundaries. FAMBE is one program that is available to compute the electrostatic potential using this method. Finally, a different (but physically equivalent) approach to incorporate the influence of a polar solvent, in which the solvent is modeled by a discrete lattice of dipoles that reorient under the influence of applied electric fields, has been proposed and developed by Warshel and coworkers.

It is generally assumed that the dielectric constant is uniform everywhere except in the vicinity of the solute/solvent boundary. If all the solute degrees of freedom are treated explicitly and the influence of induced electronic polarization is neglected, the position dependent dielectric constant $\varepsilon(\mathbf{r})$ varies sharply from, 1, in the interior of the solute, to ε_v in the bulk solvent region outside the solute. Such a form for $\varepsilon(\mathbf{r})$ follows rigorously from an analysis based on a statistical mechanical integral equation under the assumption that there are only short-range direct correlations in the solvent. To estimate the electrostatic contribution to the solvation free energy, the reaction field Φ_{rf} used in Eq. (7.17) is obtained as the electrostatic potential calculated from Eq. (7.18) with the nonuniform dielectric constant $\varepsilon(\mathbf{r})$, minus the electrostatic potential calculated with a uniform dielectric constant of 1.

Results obtained using macroscopic continuum electrostatic for biomolecular solutes depend sensitively on atomic partial charges assigned to the nuclei and the location of the dielectric boundary between the solute and the solvent. The dielectric boundary can be constructed on the basis of the molecular surface or the solvent-accessible surface (constructed as a surface formed by overlapping spheres). The parametrization of an accurate continuum electrostatic model thus requires the development of optimal sets of atomic radii for the solutes of interest. Various parametrization schemes aimed at reproducing the solvation free energy of a collection of molecules have been suggested. from a fundamental point of view, the dielectric boundary is closely related to the nearest density peak in the solute-solvent distribution function. As a consequence, the optimal radius of an atom is not a property of that atom alone but is an effective empirical parameter that depends on its charge, on its neighbors in the solute, and also on the nature of the molecules forming the bulk solvent. In contrast to the radii, the partial charges of the solute are generally taken from one of the standard biomolecular force fields without modification and are not considered as free parameters.

Continuum electrostatic approaches based on the Poisson equation have beeb used to address a wide variety of problems in biology. One particularly useful application is in the determination of the protonation state of titratable groups in proteins. For further drtails readers are referred to the reviews of Honig and Nicholls and Sharp and Honig.

Analytic Gradients

In most practical applications of continuum electrostatics, the solute is considered to be in a fixed conformation. However, this procedures has obvious limitations, because it ignors the importance of conformational flexibility. To proceed further requires knowledge of the "electrostatic solvation forces" associated with the continuum electrostatics description of the solvent. *i, e.,* the analytic first derivative of the solvation free energy with respect to the atomic coordinates of the solute. The computation of analytic gradients of the free energy of solvation with respect to nuclear coordinates is important for efficient geometric optimization based on energy minimization, conformational searches, and dynamics. Analytic gradients for finite-difference solutions to the Poisson equation have been presented by Gilson et al. and Im et al. Boundary element methods can also be used very effectively for computing analytic gradients (48). Nonetheleless, when repeated evaluation of the solvation energy is requested, the solution to the classical electrostatic problem and the calculation of analytic gradients may be too expensive computationally. For this purpose, approximations to the exact continuum electrostatics based on semianlytical functions have been developed. This is possible, in principle, because the free energy can be expressed as a superposition of pariwise additive terms (which depends on the geometry of the solute/solvent dielectric interface) in virtue of the linearity of continuum electrostatics. The general strategy of semianalytical approaches is to design a suitable closed-form pairwise deshielding function for the charge-charge coupling. One of the most popular approximation is the generalized born (GB), although alternative formulations such as the field integrated electrostatic approach a (FIESTA) [50], the inducible multipole solvation model (IMS) the analytical continuum electrostatics approach (ACE) and the solvation models (SM*x*) of Carmer and Truhlar are also based on this general idea. Semianalytical approximation such as GB represent a very promising approach for implicitly incorporating the influence of the solvent in biomolecular simulations. Extensions and improvements to the original form of the GB deshielding function have been proposed and parameterized. The results have been compared with those from numerical continuum electrostatic calculations and explicit solvent simulations. The GB approximation has been applied to various problems *e.g.*, protein and nucleic acid stability, conformational searches, macromolecular association, and ligand binding.

Ionic Strength

The concentration of salt in physiological systems is on the order of 150 *m*M, which corresponds to approximately 350 water molecules for each cation-anion pair. For this reason, investigations of salt effects in biological systems using detailed atomic models and molecular dynamic simulations become rapidly prohibitive, and mean-fields treatments based on continuum electrostatics are advantageous. Such approximations, which were pioneered by Debye and Huckel are valid at moderately low ionic concentration when core-core interaction between the mobile ions can be neglected. Briefly, the spatial density throughout the solvent is assumed to depend only on the local electrostatic potential $\rho_i(\mathbf{r}) = \bar{\rho}_i \exp\{-q_i\phi(\mathbf{r})/k_B T\}$, where i refers to a specific ion type (*e.g.*, counterion or co-ion) and $\bar{\rho}_i$ is the number density in the bulk solution. The total ion charge density (summed over the different ion types) is then inserted explicitly in the Poisson equation with the solute charge $\rho_u(\mathbf{r})$ resulting in the nonlinear form of the poisson -Boltzmann (PB) equation. Linearization with respect to the potential ϕ yeilds the familiar Debye—Huckel approximation,

$$\nabla \cdot [\varepsilon(\mathbf{r})\nabla\phi(\mathbf{r})] - \bar{\kappa}^2(\mathbf{r})\phi(\mathbf{r}) = -4\pi\rho^{(u)}(r) \qquad (7.19)$$

where $\bar{\kappa}^2(\mathbf{r})$ is the space-dependent screening factor, which varies from zero in the interior of the solute to $4\pi\sum_i q_i^2\bar{\rho}_i/k_B T$ in the bulk solvent. The spatial dependence of $\bar{\kappa}^2(\mathbf{r})$ is often assumed to be similar to that of $\varepsilon(\mathbf{r})$, though that is not necessary. The PB equation (linear and nonlinear) is a particularly simple and powerful approach to address question about the influence of saft on complex biological systems. In particular, it has been used to examine the salt dependence of the conformational stability of nucleic acids and protein-DNA association.

Transmembrane Potential

The electrostatic free energy of a macromolecule embedded in a membrane in the presence of a membrane potential V can be expressed as the sum of three separate terms involving the capacitance C of the system, the reaction field $\Phi_{rf}(\mathbf{r})$, and the membrane potential field $\Phi_{mp}(\mathbf{r})$,

$$\Delta W^{elec} = \frac{1}{2}CV^2 + \frac{1}{2}\sum_i q_i\Phi_{rf}(X_i) + \left[\sum_i q_i\Phi_{mp}(X_i)\right]V \qquad ...(7.20)$$

where q_i and $\mathbf{x}_i$ are the charge and position, respectively, of solute i, Generally, the capacitive energy contribution is negligible. The function $\Phi_{mp}(\mathbf{x}_i)$ corrresponds to the fraction of the electrostatic transmembrane potential interacting with a charge of the solute. It is calculated by solving a modified version of the linear PB equation,

$$\nabla \cdot [\varepsilon(\mathbf{r})\nabla\Phi_{mp}(\mathbf{r}) - \bar{\kappa}^2(\mathbf{r})[\Phi_{mp}(\mathbf{r}) - \Theta(\mathbf{r})] = 0 \qquad ...(7.21)$$

where the function $\Theta(\mathbf{r})$ is equal to 1 on the side of the membrane that is contact with the bulk solution set to the reference potential V, and zero otherwise. The Θ function in Eq. (7.21) ensures that the mobile ions are in equilibrium with the bath with which they are in contacts. In the case of perfectly planar system, the electric field across the membrane is constant and $\Phi_{mp}(\mathbf{x})$ is a linear function corresponding roughly to a fraction of the membrane thickness (for this reason, it is often referred to as the "*electric distance*" If the shape of the protein/solution interface is irregular, the interaction of the solute charges with the membrane potential is more complicated than the simple linear field.

Simple considerations show that the membrane potential cannot be treated with computer simulations, and continuum electrostatic methods may constitute the only practical approach to address such questions. The capacitance of a typical lipid membrane is on the order of μF/cm^2, which corresponds to a thickness of approximately 25 Å and a dielectric constant of 2 for the hydrophobic orre of a bilayer. In the presence of a membrane potential the bulk solution remins electrically neutral and a small charge imbalance is distributed in the neighborhood of the interfaces. The membrane potential arises from a strikingly small accumulation of net charge relative to the bulk ion density. Typical physiological conditions correspond to a membrane potential on the order of 100 m V and a salt concentration of 150 mM. In this situation, the net charge per unit area is CV = 10^{-7} C/cm^2, which corresponds to only one atomic unit charge per (130 Å)^2 of surface. For molecular dynamics simulations, a minimal salt solution at a concentration of 150 mM with membrane system of cross-sectional area (130 Å)^2 containing about 100 ion pairs would require nearly 50,000 water molecules and 500 phospholipid molecules, for a total of more than 200, 000 atoms, which is computationally prohibitive. At the present time, the modifed PB, Eq. (7.21) , with membrane potential may provide the only practical way to address question about the membrane potential and its influence on the configurational free energy of intrinsic protein. The approach has been implemented in the PBEQ module of the biomolecular simulation program CHARMM and has been used to calculate the influence of the transmembrane potential on the insertion of an α helix into a membrane.

MECHANICAL INTEGRAL EQUATIONS

The average solvent structure caused by granularity, packing, and hydrogen bonding gives rise to important effects that are ignored by continuum electrostatic approaches. Statistical mechanical theories based on distribution functions and integral equations are sophisticated approaches that can provide a rigorous framework for incorporating such effects into a description of solvation. A complete review of integral equations would be beyond the scope of this chapter; therefore we provide only a brief overview of this vast field.

One important class of integral equation theories is based on the reference interaction site model (RISM) proposed by Chandler. These RISM theories have been used to study the conformation of small peptides in liquid water. However, the approach is not appropriate for large molecular solutes such as proteins and nucleic acids. Because RISM is based on a reduction to site-site, solute-solvent radially symmetrical distribution functions, there is a loss of information about the three-dimensional spatial organization of the solvent density around a macromolecular solute of irregular shape. To circumvent this limitation, of RISM-like theories for three-dimensional space (3d-RISM) have been proposed

$$c_\alpha(\mathbf{r}) = \exp\left[-U_\alpha(\mathbf{r})/k_B T + h_\alpha(\mathbf{r}) - c_\alpha(\mathbf{r})\right] - h_\alpha(\mathbf{r}) + c_\alpha(\mathbf{r}) - 1 \quad \text{...(7.22)}$$

and

$$\bar{\rho} h_\alpha(\mathbf{r}) = \int d\mathbf{r}' \sum_\gamma c_\gamma(\mathbf{r}')\chi_{\gamma\alpha}(\mathbf{r}-\mathbf{r}') \quad \text{...(7.23)}$$

where $U_\alpha(\mathbf{r})$ is the solute-solvent interaction on the solvent site α, $h_\alpha(\mathbf{r})$ is the solute solvent site correlation function $h_\alpha(\mathbf{r}) \equiv [\rho_\alpha(\mathbf{r})\sqrt{\rho} - 1]$, $c_\alpha(\mathbf{r})$ is the solute-solvent site direct correlation function, and $\chi_{\gamma\alpha}(r - r)$ is the density susceptibility of the uniform unperturbed liquid. The solvent susceptibility (an input in this approach) is related to the equilibrium site-site density susceptibility (an input this approach is related to the equilibrium site-site density susceptibility of the uniform unperturbed liquid. Numerical solutions of the 3d-RISM equation indicate that this approach is able to incorporate important features of hydration such as hydrogen bonding and packing of the solvent molecules in the first solvation shell. Recent advances allow the accurate estimate of the solvation free energy for nonpolar as well as polar bimolecules.

Other statistical mechanical theories are also currently being explored. An extension to the mean spherical approximation integral equation in three dimension (3d-MSA) describing the distribution function of a liquid of spherical molecules with an embedded dipole around a polar solute as well as an integral equation describing the structure of water molecules in terms of stickly interaction points were formulated and solved numerically. A theory based on an expansion in terms of two-and three-body correlation functions has been proposed to describe the hydration structure around nucleic acids and proteins. A theory for inhomogeneous fluids in the neighborhood of large nonpolar solutes was proposed to describe the hydrophobic effect.

Solvent Boundary Potentials

A description in which all atomic and structural details of the solvent molecules are ignored may not always be desirable. In some cases, it may be advantageous to use a mixed scheme that combines an implicit solvent model with a limited number of explicit solvent molecules. An intermediate approach, illustrated schematically in Figure 7.3 consists in including a small

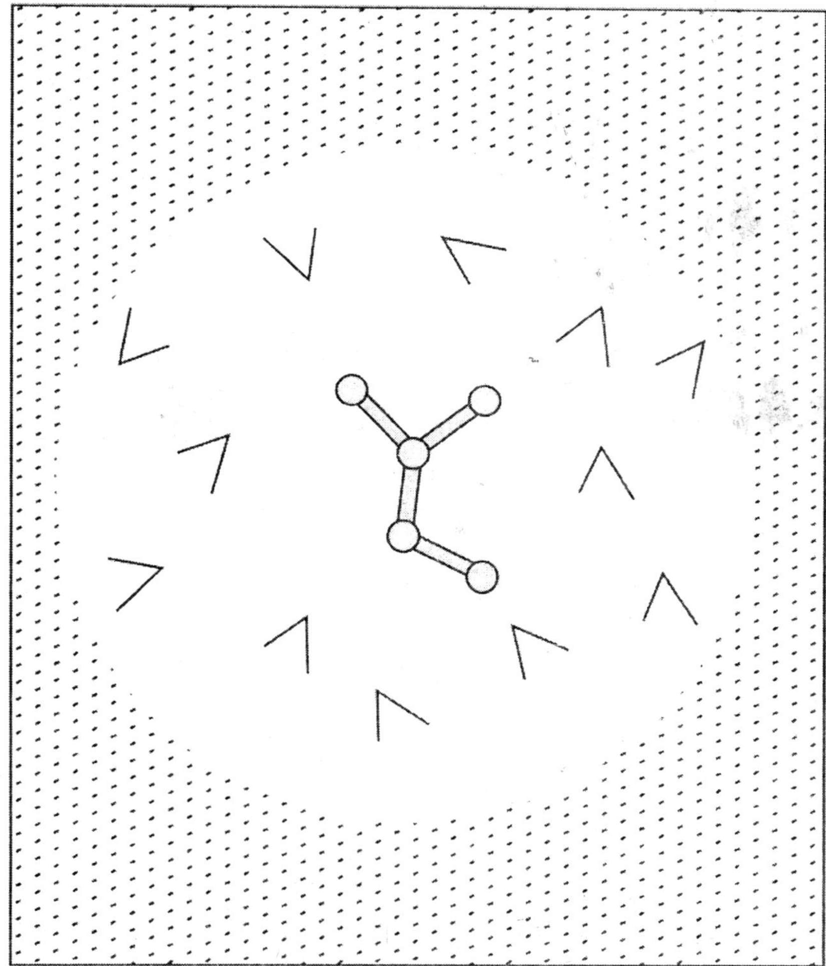

Fig. 7.3 Schematic representation of a mixed explicit-implicit solvent treatment. A small number of water molecules are included explicitly in the vacinity of the solute while the influence of the remaining bulk is taken into account implicitly.

number of explicit solvent molecules in the vicinity of the solute while representing the influence of the remaining bulk with an effective solvent boundary potential. The first to design such a simulation method appropriate for liquids were Berkowitz and Mc Cammon. In their method, simulation method appropriate for liquids were Berkowitz and Mc Cammon. In their method, the many-body system was divided into three main spherical regions: a central reaction region, a buffer region, and a surrounding static reservoir region. The force arising form the reservior region were calculated from fixed atomic centres. Instead of using explicit fixed atomic centres in the bath region, Brooks and Karplus introduced a mean force field approximation (MFFA) to calculate a soft boundary potential representing the average influence of the reservoir region on the reaction region. In the MFFA treatment, the boundary potential was calculated by integrating all contributions to the average force arising form the resrrvior region. The MFFA approach was extended by Brunger et al. For the simulation of bulk water. A similar potential for water droplets of TIP4P was developed by Essex and Jorgensen. The average electrostatic reaction field was taken into account in the surface constrained all-atom solvent (SCAAS) treatment of King and Warshel and in the reaction field with exclusion (RFE) of Rullmann and vau Duijnen.

The problem was reformulated on the basis of a separation of the multidimensional solute-solvent configurational integral in terms of "inner" solvent molecules nearest to the solute, and the remaining '*outer*" bulk solvent molecules. Following this formulation, the solvent boundary potential was identified as the solvation free energy of an effective cluster comprising the solute and inner explicit solvent molecules embedded in a large hard sphere. The hard sphere corresponds to a configurational restriction on the outer bulk solvent molecules; its radius is variable, such that it includes the most distant inner solvent molecule. An approximate spherical solvent boundary potential (SSBP) based on this formulation has been implemented in the biomolecular simulation program CHARMM. Using computer simulations it was shown that SSBP yields solvation free energies that do not depend sensitively on the number of explicit water molecules

Solvent-Accessible Surface

In Section III we described an approximation to the nonpolar free energy contribution based on the concept of the solvent accessible surface area (SASA). In the SASA/PB implicit solvent model, the nonpolar free energy contribution is complemented by a macroscopic continuum electrostatic calculation based on the PB equation, thus yielding an approximation to the total free energy, $\Delta W = \Delta W^{(np)} + \Delta W^{(elec)}$. A different implicit solvent model, which also makes use fo the concept of SASA, is based on the assumption that the entrire solvation free energy of a solute can be expressed in terms of a linear sum of atomic contributions weighted by partital exposed surface area,

$$\Delta W(\mathbf{X}) = \sum_i \gamma_i \mathcal{A}_i(\mathbf{X}). \qquad 7.24$$

Here, $\mathcal{A}_i$ (X) is the *partial SASA* of atom i (which depends on the solute configuration (**X**), and γ_i is an atomic free energy per unit area associated with atom i. We refer to those models as "**SASA**" Because it is so simple, this approach is widely used in computations on biomolecules. Variations of the solvent-exposed area models are the shell model of Scheraga, the excluded-volime model of Colonna-Cesari and Sander, and the Gaussain model of Lazaridis and Karplus . Full SASA models have been used for investigating the thermal denaturation of proteins and to examine protein-protein association.

One important limiation of full SASA models is the difficulty of taking into account the dielectric shielding of electrostatic interactions between charged particles in a physically realistic way. The SASA model incorporates, in an averagy way, the free energy cost of taking a charged particle and buying it in the interior of the protein. In the continuum electrostatic description that corresponds to the self- interaction energy, *i.e.,* the interaction of a charge with its own reaction field. However, as two charged particles, move from the solvent to the nonpolar core fo the protein, their electrostatic interaction should also vary progressively from a fully to an incompletly shielded form. Thus full SASA approximations require further assumptions about the treatment of electrostatic interactions and dielectric shielding in paractical applications. For example,, in Full SASA models residues carrying a net charge are usually neutralized and a distance-dependent dielectric function is introduced to shield the Coulomb potential at large distances.

Knowledge-based Potentials

One of the greatest problems in predicating the three-dimensional fold of a protein is the need to search over a large number of possible configurations to find the global free energy minimum. For extensive configurational searches, it is necessary to use a free energy function W(**X**) that is as simple and inexpensive as possible. Knowledge-based potentials are the simplest free energy functions that can be designed for this purpose. Such potentials are constructed empirically from statistical analyses of known protein structures taken from structural databases. The general idea is that the number of residue pairs at a certain distance observed in the database follows the statistics of a thermal ensemble, in other words a Boltzmann principle. Equivalently, it is assumed that the observed probability of finding a pair of residues at a distance R in a protein structure is related to the Boltzmann factor of an effective distance-dependent free energy. The simplest potentials distinguish only two types of residues: nonpolar and polar. Usually no attempts are made to establish a realistic description of the microscopic interactions at the atomic level, though some comparisons have been made with explicit solvent simulations. For example, one of the simples t potentials, designed by Sippl, is attractive for pairs of nonpolar residues and repulsive for pairs of polar residues. Nevertheless,, the resulting structures that are obtained via conformational searches, usually with an additional restraint on the protein radius of gyration, are reasonable: The nonpolar residues tend to form a hydrophobic core in the center of the structure, whereas the polar residues tend to be located at the protein surface. A growing number of potentials are constructed on the basis of similar ideas. In 1996, Mirny and Shakhnovich re-examined the methods for deriving knowledge-based potentials for protein folding. Their potential is obtained by a global optimization procedure that simultaneously maximizes thermodynamic stability for all proteins in the database. This field is in rapid expansion, and it is beyond the scope of the present review to cover all possible developments.

CONCLUDING REMARKS

A statistical mechanucal formulation of implicit solvent representations provides a robust theoretical framework for understanding the influence of solvation biomolecular systems. A decomposition of the free energy in terms of nonpolar and electrostatic contributions, $\Delta W = W^{(np)} + \Delta W^{(elec)}$, is central to many approximate treatments. An attractive and widely used treatment consists in representing the $\Delta W^{(elec)}$ by a using the finite-difference PB, Eq. 7.19. These two approximations constitute the SASA/PB implicit solvent model. Although SASA/PB does not incorporate solvation effect with all atomic details, it nevertheless relies on a physically consistent picture of solvation. A relationship with first principles and statistical mechanics can be established, and the significance of the approximations at the imcroscopic level can be calrified. The results can be compared with computer simulations including explicit solvent molecules. Implicit solvent models based on the SASA/PB approximation have been used to address a wide ragne of questions concerning biomolecular systems, *e.g.*, to discriminate misfolded proteins, assess the conformational stability of nucleic acids, and examine proteinligand, protein-DNA and protein-membrane association.

It is possible to go beyond the SASA/PB approximation and develop better approximations to current implicit solvent representations with sophisticated statistical mechanical models based on distribution functions or integral equations. An alternative intermediate approach consists in including a small number of explicit solvent molecules near the soulte while the influence of the remain bulk solvent molecules taken into account implicitly. On the other hand, in some cases it is necassary to use a treatment that its markedly simpler than SASA/PB to carry out extensive conformational searches. In such situations, it possible to use empirical models that describe the entire solvation free energy on the basis of the SASA (see Section V.C) An even simpler class of approximations consists in using information based potentials constructed to mimic and reproduce the statistical trends observed in macromolecular structures. Although the microscopic basis of these approximations is not yet formally linked to a statistical mechnical formulation of implicit solvent, full SASA models and empirical information-based potentials may be very effective for particular problems.

CHAPTER

8 BIOCHEMICAL EVENTS IN SOLUTIONS

Most chemical reactions in nature as well in the laboratory take place in liquid solutions. Chemical reactions of molecules in living systems take place exclusively in the solution phase. One of the most important tactics in organic chemistry is to choose an appropriate solvent for their reactions or to get higher yields. The reaction yield and rate are controlled by the solvent through changes in the free energy difference between reactants and prodcuts and thus in th activation free energy. In living cells, chemical reactions re controlled bythe solvent with an additional complication-conformational fluctuations in biomolecules. In all those reactions, the solvent effect manifests itself not only through solutesolvent interactions and solvent reorganizations but alos through intramolecular processes that are always associated with changes in electronic structure. in this regard, theroretical investigations of chemical processes in solutions are inevitable coupled with studies of quantum and statistical mechanics. One of the most straightfoward realizations of such couplinges is the ab inito molecular dynamics (MD) approach originated by Car and parrinello (CP) and published by Lassoninen et al.

The method, which consists of solving the Kohn-Sham density fuctional (DF) equation by simulated annealing,has been provent to be a capable therory in many applictions. However, the treatment has serious limitations or difficulties in terms of system size and the handling of excited states. In a sense, the approaches based on the path intergral technique share similar difficulties with the CP theory for chemical processes in solution. combined with molecular simulations and the statistical mechanics of liquids, the method could have revealed many intersting physical aspects of quantum processes in soultion. But, again, it is largely limited to a simple and/ or small system such as harmonic oscillator or a solvated electron. In this respect, it is highly desirable to exploit the molecular orbital (MO) theory for the electronic structure, which has been proven to be the most powerful tool for exploring molecular processes. This method does not have the serious limitions characteristic of the CP and path interal approaches. However,theories that are based on a basis set expansion do have a serious limitation with respect to number of electrons. Even if one consideres the rapid development of computer technology, it will be viratually impossible to treat by the MO method a small system of a size typical of callsical molecular simulation, say in 1000 water molecules. A logical solution to such a problem would be to employ a hybrid approach in which a chemical species of interest is handled by quantum chemistry while the solvent is treated caassically. Roughly speaking, thre types of hybrid approaches have been proposed depending on the level of coarse graining with respect to solvent coordinates.

Methods based on continuum models smear all solvent coordinates and represent solvent characterisitcs with a single parameter, namely, a dielectric constant. Molecular simulatiosn, on the other hand, take all solvent coordinates into account explicitly and sample many solvent configurations in order to realize meanigful statistics for the physical quantities of concern.

The approach based on the statistical mechanics of molecular liquid coarse-gains solvent coordinates in a level of the (density) pair correlation function. In what follow, we briefly outline these three methods.

Continuum Model

The *continuum model*, in which solvent is regraded as a continuum dielectric, has been used to study solvent effects for a long time. Because the electrostatic interaction in a polar system dominates over other forces such as van der Waals interactions, solvation energies can be approximated by a reaction field due to polarization of the dielectric continuum as solvent. Other contributions such as dispersion interactions, which must be explicitly considered for nonpolar solvent systems, have usually been treated with empirical quantity such as maceoscopic surface tension of solvent. A variety of methodologies have been implemented for the reaction field.

The basic equation for the disolvence-electric countium model is the Poisson-Laplace equation,by which the electrostatic field in a cavity with an arbitrary shape and size is calucated, although some methods do not satisfy the equation. Because the solute's electronic structure and the reaction field depend on each other, a nonlinear equation (modified Schrodinger equation) has to be solved in an iterative manner, which is defined through the shape and size of the cavity and description of the solute's electronic distribution. If one takes a dipole moment approximation for the solute's electronic distribution and spherical cavity , the interaction can be derived rathe easily and an analytical expression of the Fock operator is obtained. However, such an expression in not feasible for an arbitary electronic distribution in an arbitrary cavitry fitted to the molecular shape. In this case the Fock operator is very complicated and has to be prepard by a numerical procedure. Numerous attempts have been made to develop hybrid methodologies along these lines.

An obvious advantage of the method is its handiness, while its disadvantage is an artifact introduced at the boundary between the solute and solvent. You may obtain agreement between experiments and theory as colse as you7 desire by introducing many adjustable parameters are introduced, the more the physical significance of the parameter is obscured.

Molecular Simulations

Molecular simulation techniques, namely Monte Carlo and molecular dynamics methods, in which the liquid is regarded as an assembly of interacting particles, are the most popular approach. Various properties related to the macroscopic and microscopic quanties are computed by these method.

A typical hybrid approach is the QM/MM(quantum mechanical-molecular mechanical) simulation method. In this the solute molecule is treated quantum mechanically, whereas surrounding solvent molecules are approximated by molecular mechanical potentials. This idea is also used in bioligical systems by regarding a part of the system, *e, g.*, the activation site region of an enzyme, as a quantum "*solute,*" which is embedded in the rest of the molecule, which is represented by molecular mechanics.

The actual procedure used in this method is very simple: The total energy of the liqued system (or part of a protein) at an instantaneous configuration, generaged by a Monte Carlo or molecular dynamics procedure, is evaluated, and the modified Schrodinger equations are solved repeatedly until sufficent sampling is accumulated. Since millions of electronic structure calculations are needed for suffcient sampling, the ab intito inito MO method is usually too slow to be paractical in the simulation of chemical or biological system in soultion. Hence a semiempirical theory for electronic structure has been used in these type of simulations. The QM/MM methods have their own disadvantages, the obvious one being the computational load added to the already complex calculation of the electronic structure.

Integral Equation Method

The intergal equation method is free of the disadvantages of the continuum model and simulation techniques mentioned in the foregoing,and it gives a microscopic picture of the solvent effect within a reasonable computational time. Since details of the RISM-SCF/MCSCF method are discussed in the following we here briefly sketch the rererence interaction site model (RISM) therory. The statistical mechanicdes of liquids and liquid mixtures has its own long history, but the major breakthrough toward the theory in chemistry was made by Chandler and Anderssen in 1971 with the reference interaction site model (RISM) This theory can be regraded as a natural extesnionof the Ornstein-Zernike (OZ) equation for simple atomic liquids to a mixture of atoms with chemcial bonds represented by intramolecular correlation functions.

Introducing this correlation function enables us to take into account the geometry of molecules. However, it cannot handle electrostatics in its original form, thogh the charge distribution in a molecule plays an essential role in determining the chemical specificty of th emolecular system. The next important development in the theory was made in 1981 with the extended RISM theory. The extended RISM theory takes into account not only the geometry but alos the charge distribution of molecule, which completes the chemical characterization of a species for the statisticl mechanics of a molecular liquid. A general expression of the RISM requation for a system consisting of several moleculr species can be written as

$$\rho \mathbf{h} \rho = \omega c * \omega + \omega * c * \rho \mathbf{h} \rho \qquad \text{...(8.1)}$$

where the asterisk indicates convolution integrals and matric products, and ρ denotes a diagonal matric consisting to the density of molecular species. **c** and **h** are the direct and total correlation matrices with matrix elements of $c_{\alpha\beta}(\mathbf{r})$ and $h_{\alpha\beta}(\mathbf{r})$, respectively. These function, $c_{\alpha\beta}$ and $h_{\alpha\beta}$, represent the intermolecular correlation functions between the sites (atoms or atomic groups) α and β ω is the intramolecular correlation function, which embodies information about a molecular geometry. In the case of rigid molecules, ω is expressed by

$$\omega_{\alpha\beta}(r) = \rho_\alpha \delta_{\alpha\beta} \delta(\mathbf{r}) + (1 - \delta_{\alpha\beta}) \frac{1}{4\pi L_{\alpha\beta}^2} \delta\left(r - L_{\alpha\beta}\right) \qquad \text{...(8.2)}$$

where $L_{\alpha\beta}$ indicates the rigid constraint (bond length) representing the chemical bound between sites α and β.

The general equation can be further reduced to the case of infinite dilution limit, a binary mixture, ionic soultion, and so on. These equations are supplimented by closure relations such as the percus–Y evick (PY) and hypernetted chain (HNC) approximations.

$$c_{\alpha\beta}(r) = \exp\left[-\beta u_{\alpha\beta}(r)\right] \{1 + h_{\alpha\beta}(r) - c_{\alpha\beta}(r)\} - 1 \text{ (PY)} \qquad \text{...(8.3)}$$

and

$$c_{\alpha\beta}(r) = \exp\left[-\beta u_{\alpha\beta}(r)\right] h_{\alpha\beta}(r) - c_{\alpha\beta}(r)\} - \{h_{\alpha\beta}(r) - c_{\alpha\beta}(r)\} - 1 \text{ (HNC)} \qquad \text{...(8.4)}$$

where $u_{\alpha\beta}(r)$ is the intermolecular interaction potential and $\beta = 1/k_B T$. All the thermodynamic functions can be calculated form the correlation functions. The excess chemical potential or the solvation energy of a molecule has particular importance and is calculated from the correlation functions.

$$\Delta\mu = -\frac{\rho}{\beta}\sum_{\alpha s}\int dr\left[c_{\alpha s}(r)-\frac{1}{2}h_{\alpha s}^{2}(r)+\frac{1}{2}h_{\alpha s}(r)c_{\alpha s}(r)\right] \quad ...(8.5)$$

Essentially, the RISM and extended RISM theories can provide information equivalent to that obtained form simultion techniques, namely, thermodynamic properties, microscopic liquid structure, and so on. But is is noteworthy that the computational cost is dramatically brduced by this analytical treatmet, whic can be combined with the computationally expensive ab inito MO therory. Another aspect of such treatment is the transparent logica that enables phenomena to be understood in terms of statistical mechanics. Many application have been based on the RISM and extended RISM theories.

RISM-SCF/MCSCF METHOD

We recently proposed a new method referred to a s RISM-SCF/MCSCF based on the ab inito electronic structure theory and the integral equation theory of molecular liquids (RISM). Ten-no et al. Proposed the original RISM-SCF method in 1993. The basic idea of the method is to replace the reaction field in the continuum models with a microscopic expression in terms of the site-site radial distribution function between solute and solvent, which can be calculated from the RISM theory. Exploiging the microscopic reaction fields, the Fock operator of a molecule in solution can be expressed by

$$F_i^{solv} = F_i - f_i \sum_{\lambda \in solute} b_\lambda V_\lambda \quad ...(8.6)$$

where the first term on the right-hand side is the operator for an isolated molecular and the second term represents the solvation effect. b_λ is a population operator of solute atoms, and V_λ represents the electrostaric potential of the reaction field at solute atom λ produced by solvent molecules. The potentail V_λ takes a microscopic expression in terms of the site-site radial distribution functions between solute and solvent:

$$V_{\lambda \in solute} = \rho \sum_{\alpha \in solvent} \int \frac{q_\alpha}{r} h_{\lambda\alpha}(r)\, d\mathbf{r} \quad ...(8.7)$$

where q_α is the partial charge on site α, ρ is the bulk density of the solvent, and $g_{\lambda\alpha}$ is the stite-site radial distribution function (RDF) or pair correlation function (PCF).

In the RISM-SCF theory, the statistical solvent distribution around the solute is determined by the electronic structure of the solute, whereas the electronic structure of the solute is influenced by the surrounding solvent distribution. Therefore, the ab into *MO* calcualtion and the RISM equation must be solved in a self-consistent manner. It is noted that "SCF" (self consistent fiedl) applies not only toth eelectronic structure calculation but to the whole system *e.g.*, a self-consistent treatment of electronic manner. It is noted that "SFCF" (self-consistent fied) applies not only to the electronic structur calculation but to the whole system *e. g.* a self-consistent treatment of electronic structure and solvent distribution. The MO part of the method can be readily extended to the more sophisticated levels beyond-Hartree-Fock (HF), such as configuration interaction (CI) and couplet cluster (CC).

The solvated Fock operator canbe naturally derived from the variational principles defining the Helmholtz free energy of the system ($\mathcal{A}$) by

$$\mathcal{A} = \left\langle \psi \mid \hat{H} + \Delta\hat{\mu} \mid \psi \right\rangle + E_{nuc} \quad ...(8.8)$$

Here, E_{nuc} is the nuclear repulsion energy and

$$\left\langle \Psi \mid \hat{\Delta\mu} \mid \psi \right\rangle = \Delta\mu \qquad ...(8.9)$$

is the modifed version of the solvation free energy originally defined by Singer and Chandler Eq. 8.5. This is a functional of the total correlation function $h_{\alpha s}(r)$, the direct correlation function $c_{\alpha s}(r)$, and the solute wave function $|\psi\rangle$. Energy of the solute molecule E_{solute} is defined as follows:

$$E_{solute} = \left\langle \Psi \mid \hat{H} \mid \psi \right\rangle + E_{nuc} \qquad ...(8.10)$$

Note that this is also a functional of $h_{\alpha s}(r)$, $c_{\alpha s}(r)$, and $|\psi\rangle$. Imposing constraints concerning the orthonormality of the configuration state function (C) and one-particle orbitals (ϕ_i) on the equation, one can derive the Fock operator from $\mathcal{A}$ based on the Variational principle:

$$\delta(\mathcal{A}\,[\mathbf{c, h, t, v, C}] - [\text{ constraints to orthonormality}]) = 0 \qquad ...(8.11)$$

and

$$F_{ij}^{solv} = F_{ij} - \gamma_{ij} \sum_{\lambda \in solute} b_\lambda \frac{\partial}{\partial q\lambda}\left(-\frac{\rho}{\beta} \sum_{\alpha s} \int d\mathbf{r} \exp[-\beta u_{\alpha s}(r) + h_{\alpha s}(r) - c_{\alpha s}(r)] \right) \qquad ...(8.12)$$

If classical Coderulombc interactions are assumed among point charges for electrostatic interactions between solute and solvent Fock operator is reduced to Eq. (8.6). The significance of this definition of the Fock operator from a variational principle is that it enables us to express the analytical first derviative of the free energy with respect to the nuclear coordinate of the solute molecule $R_{\alpha,}$

$$\frac{\partial \mathcal{A}}{\partial R_a} = \frac{\partial E_{nuc}}{\partial R_a} - \frac{1}{2(2\pi)^3 \beta} \sum_{\alpha, \lambda, s, s'} \int dk\, \hat{c}_{\alpha s}(k) \frac{\partial \hat{\omega}_{\alpha\gamma}(k)}{\partial R_a} \hat{\chi} ss'(k) \qquad ...(8.13)$$

$$\sum_{ij} \gamma_{ij} h_{ij}^{\alpha} + \frac{1}{2} \sum_{i,j,k,l} \Gamma_{ijkl} \left(\phi_i \phi_j \mid \phi_k \phi_l \right)^a - \mathbf{V}' \mathbf{q}^{a} - \sum_i \sum_j \varepsilon_{ij} S_{ij}^{a}$$

The second term of the right-hand side of Eq. (8.13) corresponds to the change of the solutesolvent distribution function due to the modification of the intramolecular correlation function ω Other notations used here have the usual meanings. It has been well recognized that the energy gradient technique in the ab inito electronic structure theory is a powerful tool for investigating the mechanism of chemical processes in solution: carrying out the egeometric optimization of reactant, transition state, and product in the solvated molecular system; constructing the free energy surfaces along the proper reaction coordinates; computing the vibrational frequencies and modes; and so on.

In analyzing the compuitational results, the following quantities are very important:

$$E_{recog} = E_{solute} - E_{solates} \qquad ...(8.14)$$

where $E_{isolate}$ is the total energy of the solute molecule in an isolated condition and E_{solute} is the energy of the solute molecule defined above. The quantity E_{recorg} represents the reorganization energy associated with the relaxation or distortion of the electronic cloud and molecular geometry in solution.

Now we have the tools in hand to tackle various problems in solvated molecules. In the following sections, we present our recent efforts to explore such phenomena by means of the RISM-SCF/MVDVG method.

MOLECULAR POLARIZATION

The molecular and liquid properties of water have been subjects of intensive reserch in the field of molecular science. Most theoretical approaches, including molecular simulation and integral equation methods, have relied on the effective potential, which was determined empirically or semiempirically with the aid of ab inito MO calculations for isolated molecules. The potential parameters so determined from the ab initio MO in vaccum should have been readjusted so as to reproduce experimental observables in solutions. An obvious problem in such a way of determining molecular parameters is that it requires the reevalution of the parameters whenever the theromodynamic conditions such as temperature and presurre are changed, because the effective potentials are state properties.

One of the most efficient ways to treat this problem is to combine the ab initio MO method and the RISM theory, and this has been achieved by a slight modification of the original RISM-SCF method. Effective atomic charges in liquid water are determined such that the electronic structure and liquid properties become self-consistent, and along the Route of convergence the polarization effct can be natuarally incorporated.

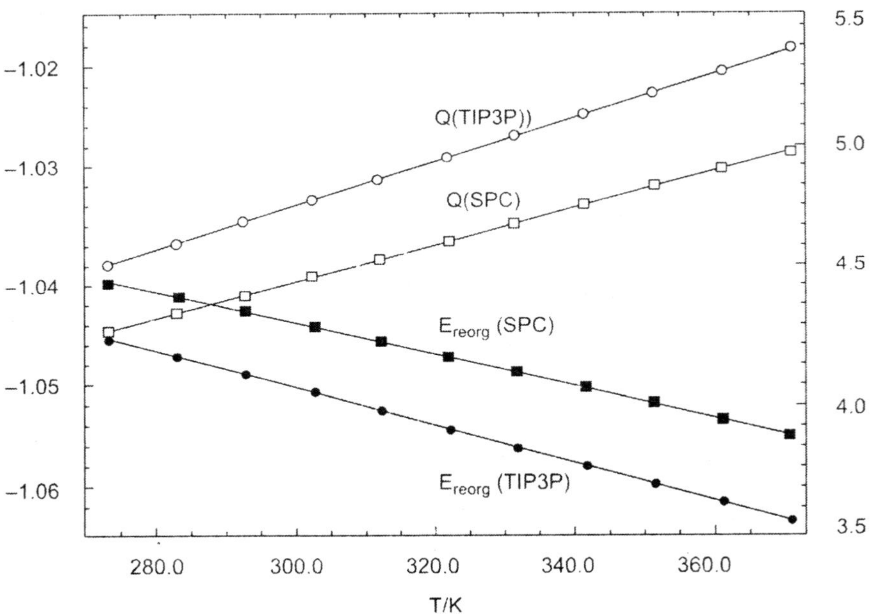

Fig. 8.1. The Temperature dependence of the recorganization energy (E_{reorg}) and effective charges on oxygen atom based on ( ) SPC and (o, •) TIP3P models.

The temperature dependence of the effective charges and dipole moment of water are plotted in Figure 8.1. The parameters associated with the short-range part of the interaction and geometry are borrowed from two typical models of water, SPC nad TIP3P. In both models, the magnitudes of the effective charges and dipole moment monotonically decrease with increasing temperature. The results can be explained in terms of the increased in molecular motion, especially rotational motion, with increasing termperature. As the motion of a molecule (molecule A) increases, the average electrostatic field produced by the surrounding water molecules becomes less anisotropic, which decreases the polarity of molecule A. Conversely, the reaction field from the water (molecule A) become more isotropic, which decreased the polarity of other molecules.

The pair correlation function of water has marked feature that distinguishes water from other liquids (Fig. 8.2) One of the important features characterizing the liquid water structure is a pek around r = 1.8 Å observed in the oxygen-hydrogen (O–O) pair, which is a direct minifestation of the hydrogen bond between a pair of water molecules. Another feature is the position of the second peak in the oxygen-oxygen (O–O) PCF, which is caused by the tetrachedral icelike coordination. Since the icelike structure becomes less prononced as tempeature increases becuse of the thermal disruption of the hydrogen bonded network, those features in PCF become less prominent.

Autoionization of Water*

A water molecule has amphoteric character. This means it can act as both an acid and a base. The autoionization equilibrium process in water,

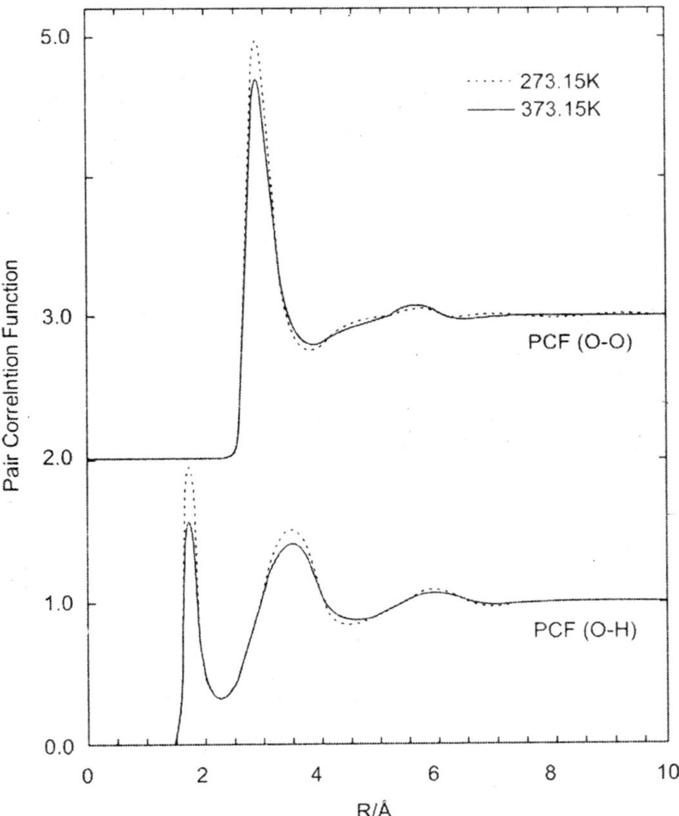

Fig. 8.2 Parir corrlation functuons of O–O and O–H at (...) 273.15 and (—) 375.15 K computed with the parameters of the SPC water model.

$$H_2O + H_2O \rightleftharpoons H_3O^+ + OH^- \qquad ...(8.15)$$

is one of the most inportant and fundamental reactions in a variety of fields in chemistry, biology, and biochemistry. The ionic product (K_w) and its logarithm defined by

$$K_w = [H_3O^+]\, OH^-], PK_w = -\log K_W \qquad ...(8.16)$$

are measures of the autoionization. The quanitity can be related to the free energy change (ΔG^{az}) associated with the reaction of Eq. (8.15) by the standard thermodynamic relation

$$\Delta G^{aq} = 2.303 \, RT \, pK_w \qquad ...(8.17)$$

It is experimentally known that the pK_w values shows significant temperature dependence, *i.e.*, it decreases with increasing temperature. However, there is no easy explanation for this phenomenon even from the phenomenological point of view. The free energy change consists of various contributions, including changes in the electronic energy and solvation free energy of the molecular species taking part inthe reaction, which are related to each other. Therefore, a theory that accounts for both the electronic and liquid structurs of water with a microsopic description of the reaction is required.

The free energy change associated with the reaction in Eq. (8.15) can be written in terms of the energyh change associated with the recation in vacuo (ΔG^{vac}) and the free energy chage of the reacting species due to solvation as

$$\Delta G^{ad} = \Delta G^{vac} + \delta G(H_3O^+) + \delta G(OH^-) - 2\delta G\,(H_2O) \qquad ...(8.19)$$

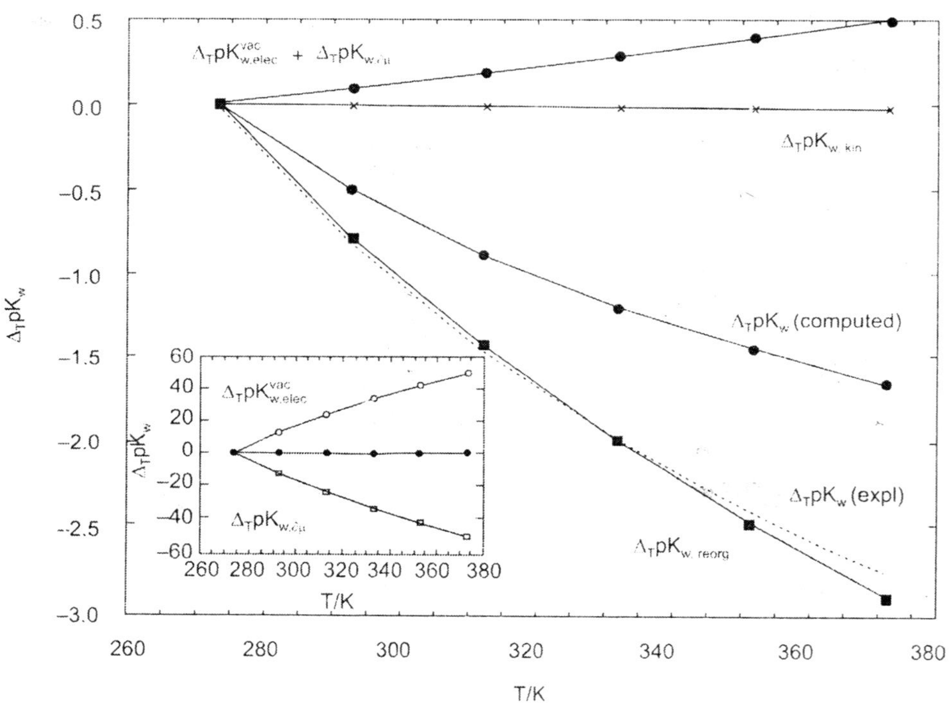

Fig. 8.3. Temperature dependence of calculated pK_w. Dashed line indicates experimental values.

where $\delta G(H_3O^+)$, $\delta G(OH^-)$, and $\delta G(H_2O)$ are, respectively, the free energy changes of H_3O^+, HO^-, and H_2O upon solvation. It is also posible to decomposed ΔG^{aq} into intra-and intermolecular contributions as

$$\Delta G^{aq} = \Delta E^{vac}_{elec.} + \Delta\delta G_{kin} + \Delta\delta G_{reorg} + \Delta\delta\mu \qquad ...(8.19)$$

where ΔG^{vac}_{elec}, and $\Delta\delta G_{reorg}$ are electronic energy in vacuo, kinetic free energy, and electronic reorganization energy, respectively, which are intramolecular contributions. $\Delta\delta\mu$ is the the solvation free energy change. (We use Δ for changes of quantities associated with the chemical reaction and δ for changes due to solvation.)

The value of pK_w at temperature T relative to that at T = 273.15 K, given by

$$\Delta_{T\,p}K_w(T) = pK_w(T) - pK_w\,(273.15) \qquad ...(8.20)$$

is further decomposed into four contributions corresponding to the free energy componetns:

$$\Delta_T\,pK_w(T) = \Delta_T\,pK^{vac}_{w,elec} + \Delta_T\,pK_{w,\,kin}(T) + \Delta_T\,pK_{w,\,reorg}(T) + \Delta_T\,pK_{w,\,\delta\mu}(T) \qquad ...(8.21)$$

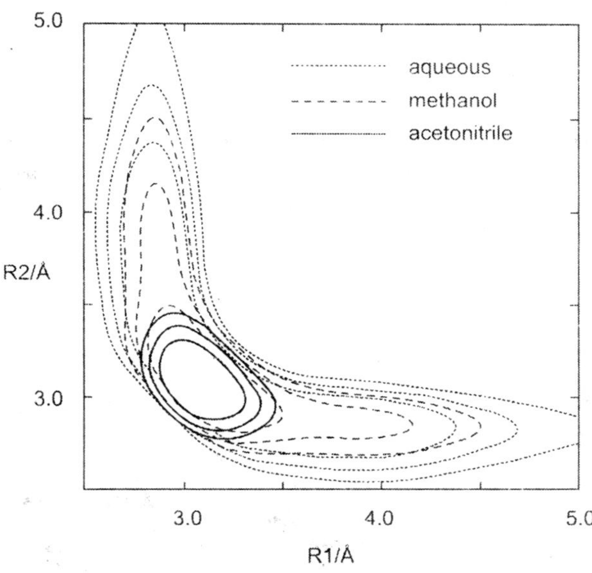

Fig. 8.4. Free energy surfaces of triiodide ion in various solutions. Contours correspond to isoenergy lines of $1k_BT$, $2k_BT$, and $3k_BT$, respectively.

The resultant $\Delta_T\,pK_w(T)$ values and their components are plotted in Fig. 8.3 As shown in the figure, contributions form $\Delta_T\,pK_{w,\,\delta\mu}$ and $\Delta_T\,pK^{vac}_{w,elec}$ are very large, but they compensate for each other. The final temperature dependence of pK_w is determined by an interplay of several contributions with different physical origins. it is also interesting that the temperature dependence is dominated by $\Delta_T\,pK_{w,\,reorg}$ after the compensation for the largest contributions. The theoretical results for temperature dependence of the ionic product show fairly good agreement with experiments ans also demonstrate the importance of polarization effects.

Solvatochromism

The molecular properties of the triiodide ion in polar liquids have been studied by many techniques, motivated in part by the expected strong coupling between the solute I_3—electronic structure and the environment. Interestingly, Raman and spectra in several solvents show a weak band corresponding to the antisymmetrical stretch mode, which is expected to be symmetry-forbidden, whereas the infrared spectra of many triiodide complexes with cations were also reported to show a band corresponding to the symmetrical stretching mode, which again should be summetryforbidden, Ab initio MO calculations for the solvated triiodide ihad been impractcable because of the difficulties in dealing with the cahrcter of the solvation, which is strongly coupled with solute electronic structure.

Computed free energy surfaces of the triiodide ion in its ground state in acetonitrile, methanol, and aqueous solution are presented in Figure 8.4 in which the two I–I bond lengths R1 and R2) are taken as coordinates. Note that the free energy surface in solution does not correspond to the potential energy surface but governs the relative population of different structures in solution. It would be misleading to estimate vibrational frequencies from the curvatures of the surfaces. Contours in the Figures represent isoenergy lines of 1 k_BT at room temperature, 298.15 K, showing how large a population of triodide ions exists in the different structures. The free energy profiles in solution strongly depend on the solvent. The profiles in acetonitrile solution are very localized and similar to those in the gas phase, consistent with Resonance Raman experimental results in which the symmtry-forbidden band does not appear. The free energy surface in aqueous solution is markedly different and indicates a dramatically enhanced probability of structures with lower symmetry. The obnservation of nominallyh symmetry-forbidden bands in vibrationl spectra is attributed to these species.

Conformational Equilibrium

The acidity or basicity of organic acids and bases such as craboxylic acid and amines is governed by many factors: solvent, substitution, conformation, and so forth. Among those factors, the effefct of conformational chage is of special interest in terms of its significance in biological systems. In bimolecular systems such as protein, the acidity or basicity of related functional groups depends sensitively upon the molecular conformations; due to such sensitivity, the property is sometimes exploited to detect the conformational chagne of protein.

A prototype of such phenomena can be seen in even the simplest carboxylic acid, acetic acid (CH_3CHOOH). Acidity is determined by the energy or free energy difference between the dissoicated and nondissociated forms, whose energetics usually depend significantly on their conformation, *e.g.*, the *syn*/*anti* conformational change of the carboxyl ate group in the compound substantially affects the acide-base equilibrium.

Potential and potential of mean force curves along the torsional angle θ(H—O—C—C) is illustrated in Figure 8.5. In the gas phase the *syn*-acetic acide ($\theta = \pm 180^o$) is more stable than the *anti* conformer by 6.9 kcal/mol, and the barrier heigh of rotation between these conformers is estimated as 13.2 kcal/mol. In aqueous solution, the calculated free energy difference is significantly reduced to 1.7 kcal/mol. The rotational barrier also becomes lower than that in the gas phase, 10.3 kcal/mol. The reducation of the free energy gap indicates that the pK_a difference between the two conformers is drastically changed from 5.1 to 1.2 on tansferring form the gas phase to aqueous solution at room temperature.

Stabilization of the *syn* conformer in the gas phase is explained rather intuitively in terms of the extra stabilization due to increased interactions between the H atom in the OH group and the O atom in C = O group. The extra stablization in the *anti* conformer in aquous solution arise from the solvation ergy, especially at the carbonyl oxygen site.

The chagne in the electronic redistribution on transferring the molecule from the gas phase to aqueous solution is another interesting issue. Analysis of the computed Mulliken charge population demonstrates a substantial chagne on the hydrogen and oxygen in the OH group: for the *anti*-conformer, the partial chages are altered from 0.335 (H) and –0.428 (O) to 0.377 (H) and –0.434 (O) upon transfering from the gas to the aqueous soultion phases, and for the *syn* conformer, from 0.373 (H) and –0.463 (O) to 0.395 (H) and – 0.476 (O), respectively. One can notic that these values are evaluated at the optimzed geomerty in aqueous solution. Geometrical chagnes from the gas to aqueous phases should also contribute to the modification of the electron density).

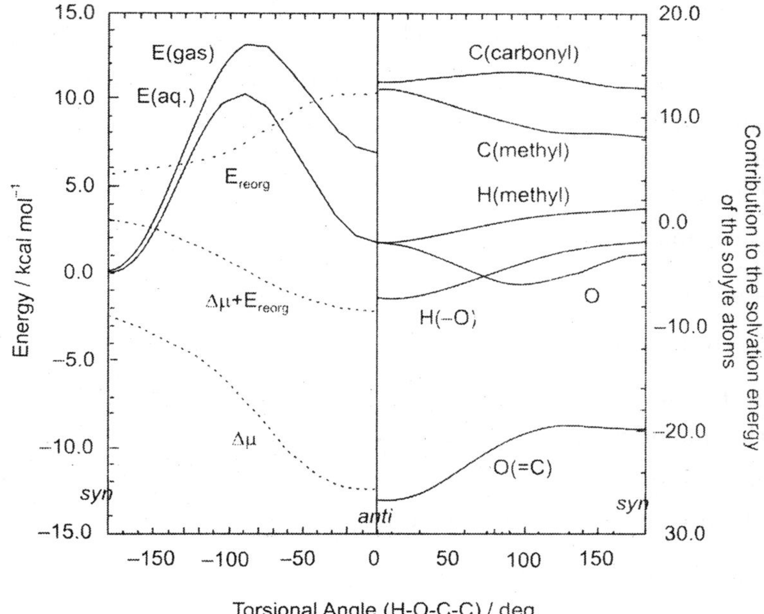

Fig. 8.5. Calculated potentail and potential of mean force of acid along the torsional angle of θ(H—O—C—C). The left-hand side show the total energies and the components,and the right hand side shows the decomposed Δμ into the site.

Acid-Base Equilibrium

In this section, we review three studies on the coupled substitution and solvent effects on basicity and acidity.

Basicity of the Methylamines in Aqueous Solution

Basicity and acidity are fundamental and familiar concepts in chemistry and biochemistry. Quantum chemistry has provided a theoretical understanding of the phenomena as far as the gas phase in concerned. However, it is konwn that in solution reactivity is seriously affected by solvents. One example of such as well-known phenomenon is that the basicity of the methylamines increases monotonically with successive methyl substitutions in the gas phase,

$$NH_3 < (CH_3)\,NH_2 < (CH_3)_2\,NH < (CH_3)_3\,N$$

while the order reverses at the trimethylamine in an aqueous environment

$$NH_3 < (CH_3)\,NH_2 < (CH_3)_2\,NH > (CH_3)_3N$$

The monotonic increase in basicity in the gas phase has been explained in terms of the "negative induction" or the polarization effect due to the methyl groups. Essentially two important factors are considered responsible for the solvent effect on the proton affinity: the solvation free energy and the energy chagne associated with the electron reorganization upon solvation. The solvation free energy in turn consists of the solute-solvent interaction energyand the free energy cnahge associated with the solvent reorganization.

Let us define the respective basicity by $-\Delta G^g$ in the gas phase and $-\Delta G^s$ in aqueous solution. For discussions concerning the relative strength in basicity of a series of methyl-amines, only the relative magnitudes of these quantities are needed. Thus the free energy changes associated with the protonation of the methy lamines relative to those of ammonia are drfined as

$$\Delta\Delta^{i}_{298}\,[(CH_3)_n\,NH_{3-n}] \;=\; \Delta G^{i}_{298}[(CH_3)_n NH_{3-n} - \Delta G^{i}_{298}[NH_3],\;\; n = 1, 2, 3 \qquad ...(8.22)$$

Computed values are plotted in Figure 8.6 against the number of methlyl groups. Note that these components include the electronic contributions (net contribution for isolated molecule and E_{reorg}), solvation energy ($\Delta\mu$) described above, and kinetic contribitions evaluated from the elementry statistical mechanics of ideal systems. The contribution form solute itself, $\Delta\Delta^{s}_{298}$ (solute), exhibits similr monotonic behavior with the gas phase result ($\Delta\Delta^{g}_{298}$), which is in good agreement with the experimental data. The difference between $\Delta\Delta^{s}_{298}$ (solute) and $\Delta\Delta^{g}_{298}$ is due essentially to the electron reorganization energy. The solvation free energy ($\Delta\Delta^{s}_{298}$) (solvent) shows the monotonic increase with succęssive metlyl substiution. The sim of the two contribitions produces an inversion in the overall free energy cnhage $\Delta\Delta^{s}_{298}$, which is in qualitative accord with the experimental result.

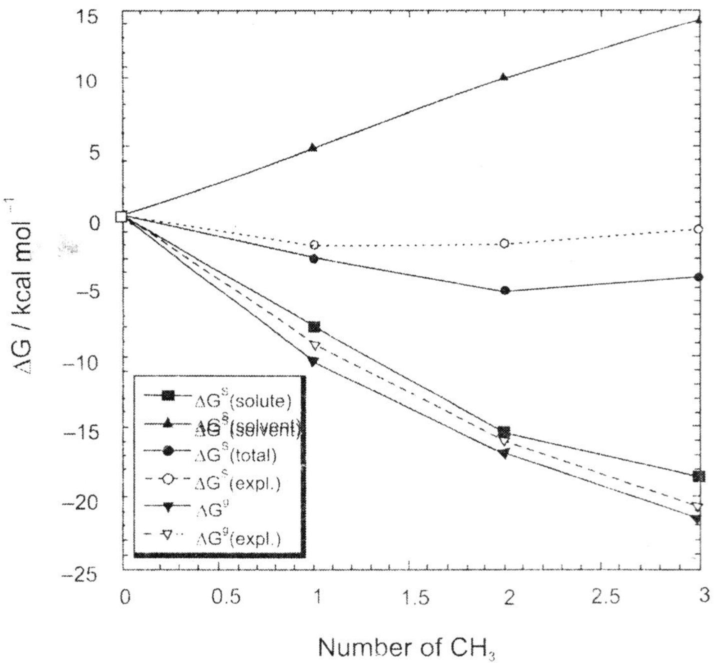

Fig. 8.6. Free energy changes of metlylamines in aqueous solution upon protonationreferred to NH_3.

Acidities of Haloacetic Acids

Another example is the acidities of a series of carboxylic acids. it is known that the substitution effect on these compounds also depends on the environment. The behavior of the halo-substituted acetic acids is one of the prototype problems for the solvent effect on acidity: The order in strenght of the haloacetic acids in the gas phase is

$$CH_3\ COOH < CH_2FCOOH < CH_2CICOOH < CH_2BrCOOH$$

whereas in aqueous solution it is drastically altered [27] to

$$CH_3\ COOH < CH_2FCOOH < CH_2CICOOH < CH_2BrCOOH$$

The observed acidities in the gas phase are interpreted in terms of the negative induction effect of the halo substituents; however, the microscopic picture of the solvent effects in addition to such induction effects of the solute have not been clarified.

Procedures to compute acidities are essentially similar to those for the basicities discussed in the previous section. The acidities in the gas phase and in solution can be calculated as the free energy changes ΔG^g and ΔG^s upon proton release of the isolated and solvated molecules, respectively. To discuss the relative strenths of acidity in the gas and aqueous solution phases, we only need the magnitued of $-\Delta G^g$ and ΔG^s for halocetic acids relative to those for acetic acids. Thus the free energy calculations for acetic acid, haloacetic acids, and each conjugate base are carried out in the gas phase and in aqueous solution.

In Figure 8.7, ΔG^s and (solvent), are plotted. ΔG^s (solute) contributes to the increase in the net free energy upon proton release or to the acidity, which is similar to ΔG^g in the gas phase. ΔG^g (solvent) is positive relative to aceitc acid and contributes to hinder the proton release and to decrease the acidity. The net increase in ΔG^g (solvent) form the fluoro dervative to the choloro dervivative is cause by the greater destabilization of the negative ions compared

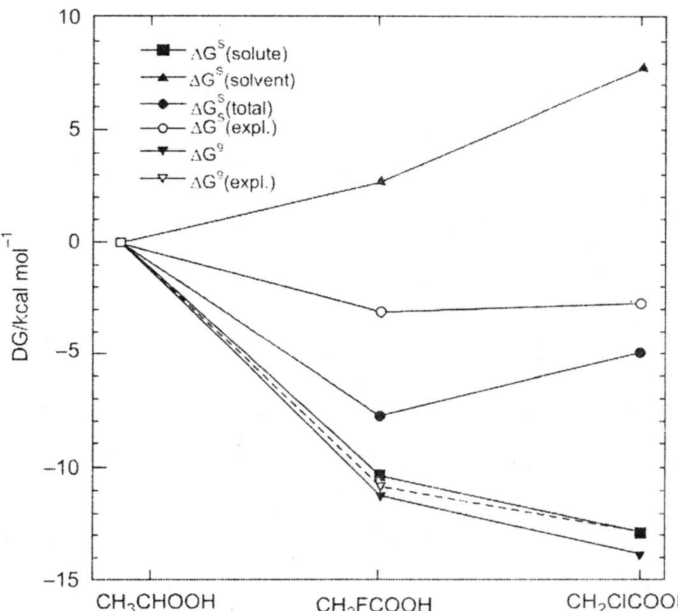

Fig. 8.7. Free energy chagnes of halo-substituted carboxy1 acid in aqueous solution upon deprotonation referred to acetic acid.

to that of the neutral molecules. The essentail difference between the ionic and neutral species lies in the elecrtrostatic contribution in the solvationfree energy compared to the decrease in the contribution from the change in the soulte electronic structure. Agreement with respect to the relative order in acidity is obntained in aqueous solution as well as in the gas phase.

Hydrogen Halides in Aqueous Solution

It is known that the order of acidity of hydrogen jalides (HX, were X = F, CI, Br, I) in the gas phase can be sucessfully predicated by quantum chemical considerations, namely, F< CI < Br < I. However, in aqueous solution, whereas hydrogen chloride, bromide, and iodide completely dissociate in aqueous solution, hydrogen fluoride shows a small dissociation constant. This phenomenon is explained by studying free energy changes associated with the chemical quilibrium $HX + H_2O \rightleftharpoons X^- + H_3O^+$ in the solution phase for a series of hydrogen halides. In this study, ths species in the equilibrium reaction, HX, H–, H_2O, and H_3 O +, are regarded as "*solute*" in the infinitely dilute solution. Thus the free energy difference in aqueous solution can be obtained in terms of the free energy difference associated with the reaction in vacuo and solvation free energy.

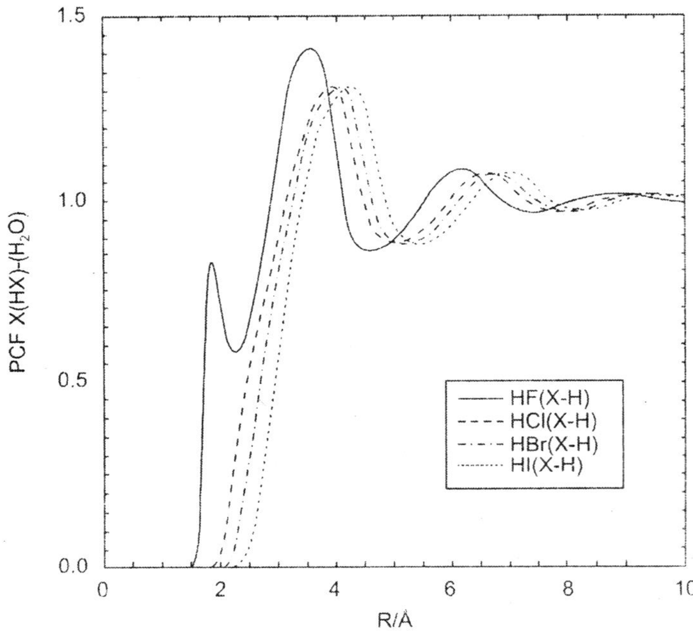

Fig. 8.8. Pair correlation function between X (in HX) and H (in solvent water) for a series of hydrogen halides.

Figure 8.8 shows the PCF between the halogen site in HX and the hydrogen site in solvent water. The fluoride shows a distinct peak at 1.82 Å. There is no corresponding peak in the oether X–H correlation functions. From other PCFs and geometrical considerations, we can conclude that hydrogen bonds between solute hydrogen and solvent oxygen are strong and are found in all the hydrogen halides and that only hydrogen fluoride forms a distict F–H (solvent water) hydrogen bond. The liquid structure around HF is expected to be markebly different form those around the other hydrogen halides.

The free energy difference is mainly governed by the subtle balance of the two energetic components, the formation energies of hydrogen halides and the solvation energies of the halid anions. The characterstic behaviour of hydrogen fluoride as a weak acid is explained in terms of the enhanced stability of the nondissociated form of the molecule in aqueous solution due to the extra hydrogen bonding with solvent. From these modern theoretical considerations, one can say that Pauling's heuristic agruments seems to be correct in a qualitative sense.

Tautomerization

Solvent effect on chemical equilibria and reactions have been an important issue in physical organic chemistrty. Several empirical relationships have been proposed to characterize systematically the various trypes of properties in protic and aprotic solvents. One of the simplest models is the continuum reaction field characterized by the dielectric constant, of the solvent, which is still widely used. Taft and coworkers presented more sophisticated solvent parameters that can take solute-solvent hydrogen bonding and polarity into account. Althogh this parameter has been successfully applied to rationalize experimentally observed solvent effects, it seems still far from satisfactory to interpret solvent effects on the bais of microscopic information of the solute-solvet interaction and solvation free energy.

Among many examples of the solvent effects on chemical equilibria and reactions, the solvent effect on tautomerization has been one of the most extensively studied. Experimentally, it is known that such equilibrium constants depend sensitively on the solvent polarity in solution. Tautomerization in formamide has been studies to obtain the micro-scopic solvation structure around the solute and to find a relationship between the empirical relation (Raft's parameter) and microscopic information given by ab initio theory.

The empirical parameters derived by Taft and coworkerws are well known as well known as measure of the hydrogen-bonding abilities of solvents. For aprotic solvents, β parameters indicate the strength of hydrogen bonding. On the other hand, the height of the first peak in PCFs around solute hydrogen, which are obtained from the RISM-SCF calculation, also represent the strength of hydrogen bonding. Taft's β parameters are actually well correlated to the calculated well depth of the hydrogen bonding corrsponding to the logarithm of the height of the first peak in the PCFs. The figures give a microscopic explantion for the origin of the empirical solvent parameters representing the ability of a solvent to form hydrogen bonds.

Figure 8.10 presents a correlation between the solvation energy computed by the RISM SCF method and the Onsager-Kirkwood parameter, $(e - 1)/(2e + 1)$, which is a typical parameter empiricallyderived form the macroscopic parameter ε. If the parameters are good, all the solvation energies must lie on a striaght line. One can see that the irregularity of the solvation free energy in acetonitrile is remakable for both tautomes. Considering the situation that the present theory reproduces well the experimentally derived hydrogen bonding strengths, the irregularity observe here clearly demonstrates the break down of the continuum solvation model.

The S_N2 Reaction

Chemical reactions are undoubtedly the most important issue in theroretical chemistry, where electronic structure plays an essential role. However, as will be demonstrated in this section, solvent effects also often play a crucial role in the mechanism of a chemical reaction in solution.

The Menshutkin-type S_N2 reaction in aqueous solution,

$$HN_3 + CH_3CI \rightarrow NH^+_3\,CH_3 + CI^-$$

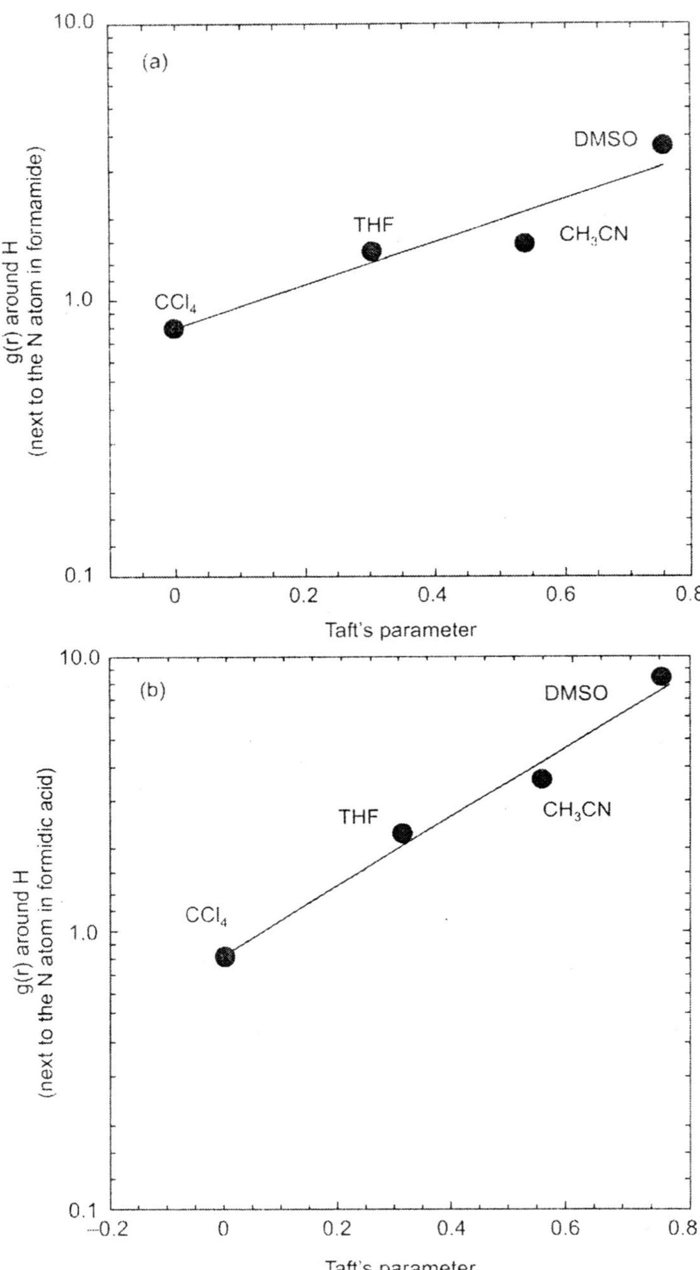

Fig. 8.9. Logarithmic plots of the heights of the first peak in the PCF against Taft's β parameters: (*a*) formamide; (*b*) formamidic acid.

is a prototype reaction in quantum chemistry that requires treatmetn of the solvent effect and substantial computational studies have been reported based on the dielectric continuum model and QM/MM method.

A free energy profile along the reaction path is constructed by taking the diffrence of the C—CI and C—N distances as the reaction coordinate. Although this reaction is found by the

Hartree-Fock method to be endothermic in the gas phase by 106.3 kcal/mol, it becomes exothermic in aqueous solution by computed by the MP2 method (not shown) is slightly higher, 20.9 kcl/mol, but the global feactue of the energy surfaces in not afferected by taking the electron correlatin into account.

It is interesting that the molecular structure in the transition state is also subject to a solvent effect. Compared to the gas phase, the solute molecular geometry at the transition state shift toward the reactant side in aquous solutionn; the C—N and C—CI distances are almost the same (2.258Å) in aqueous solution, whereas those in the gas phase are 1.890 and 2.490 Å, respectively, which is consistent with Hammond's postulate.

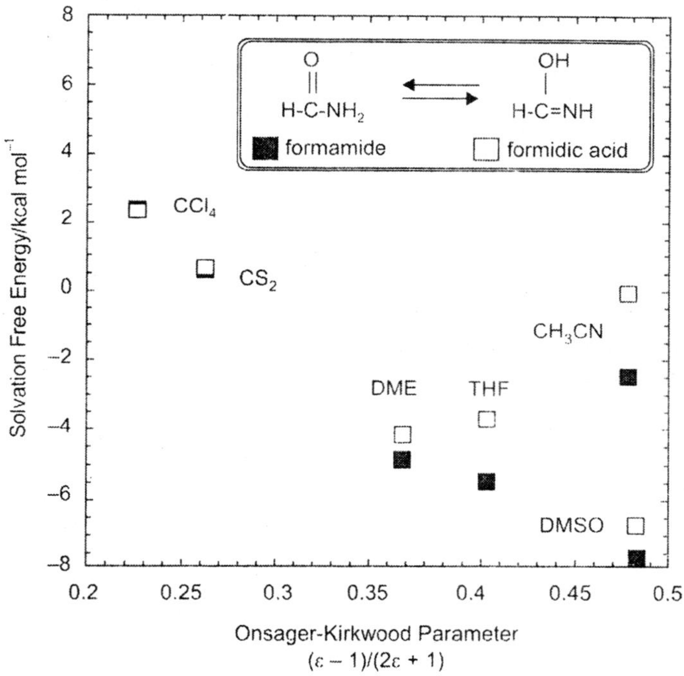

Fig. 8.10. Solvation free energy versus the Onsager-Kirkwood parameter $(\varepsilon - 1)/(2\varepsilon + 1)$.

The great stabilization of the product in aquesous solution is explained form changes in FCFs along the reaction path, which are shown in Figure 8.12. The PCF of Cl—H obviosuly illustrates the progress of hydrogen bonding along the reaction coordinate: for the reactant no peak corresponding to the CI—H interaction is observed, but a distinct peak around 2.2 Å gradually appears with slight inward shift as the reaction proceeds. In constrast, the hydrogen bond between the ammonia N and water H is observed to break. The first peak around 2.0 Å corresponds to the hydrogen bonding of a water dydrogen to the lone pair electron in nitrogen. The lone pair electron participates in a ew chemical bond with C of CH_3 Cl, and and the peak disappers as the reaction proceeds. The formation of Cl—H hydrogen bonds (solvation) and breaking of N—H bounds are key features in the understanding of solvent effects on the reaction mechanism.

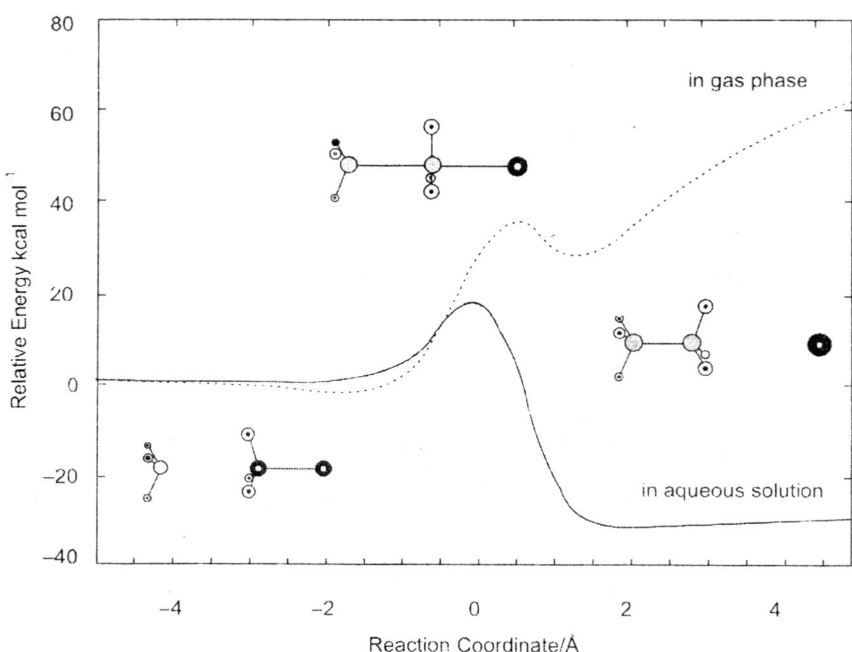

Fig. 8.11. Potential energy and potential of mean force of the Menshutkin reaction. The dashed line is for reaction in the gas phase, and the solid line for reaction in aqueous solution.

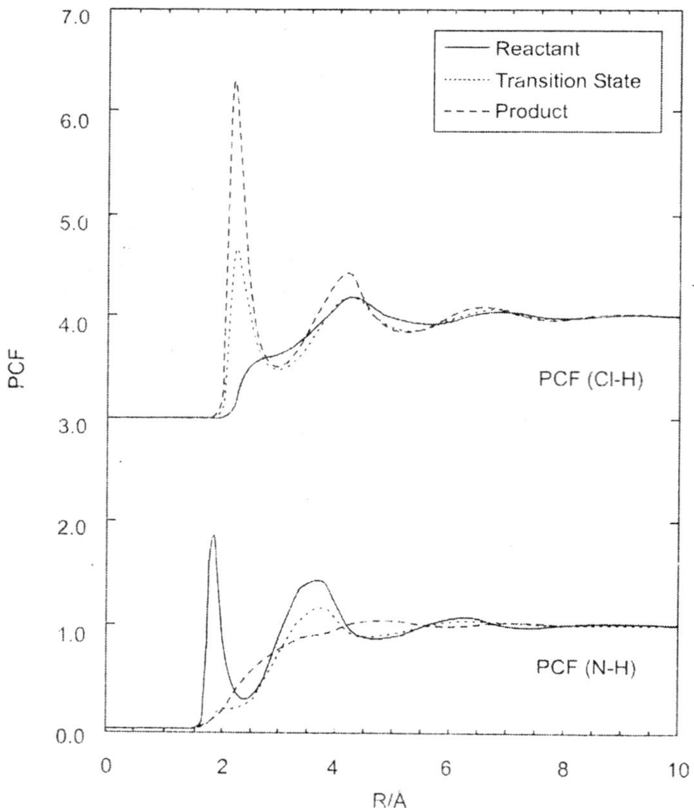

Fig. 8.12. Pair correlation functions of Cl—H and N—H along the reaction path.

CONCLUDING REMARKS

In this chapter, we have reviewed the RISM-SCF/MCSCF method, which combines ectronic structure and liquid state theories to deal with the chemistry of solutions. The ability of the method to treat solvent effects on chemical processes has been demonstrated. Electronic structure plays a primary role in determining the structure of molecule. However, changes in the electronic energy associated with a chemical process are comparable,in many cases, with those due to solvation in solution. This subtle balance between changes in electronic energy and in the solvation free energy sometimes causes drastic changes in the stability of chemical species, as we have seen in several examples.

The solvent effect even reverses the equilibrium between a reactant and a product As has been repeatedly stated throughtout the cahpter, the theory provides good qualitative descriptions to solution chemistry in most cases, but quantitative agreement between theoretical and experimental results is only moderate. It is always possible to improve the results numberwise by tuning the molecular parameters and/or by introducing empirical parameters in to either or both elements of the theory, MO and RISM. This direction, in fact, has been pursued by Gao and coworkers, who have shown almost perfect quantitative agreement between the theory and experiments by replacing the ab initio method by the semiempirical approach for the MO part and by adjusting the Lennard-Jones parameters of atoms.

The effort should be greatly appreciated, because is could have demonstrated the capability of the combined MO and RISM method to account for experimental data at least at the same quantitative level as the continuum model descriptions. However, it will not be an ultimate goal of the combined quantum and statistical mechanics theory. The real strength of the therory lies in the fact that *it does not require in principle any adjustable or empirical parameter* to describe complicated solution chemistry. The theory could be or should be natuarlly improved by theoretical development of either or both elements of the method and by coupling them, not by adjusting or introducing empirical parameters.

Since considerable efforts have been continuosuly devoted to improvement in both theoretical fieds in chemical physics, there is no doubt that the RISM–SCF that MCSCF approach will find greater application in the future. There are several directions conceivable to extend further the horizon of the theory. One such direction is to seek an experimental method to prove electron distributions in a molecule in solution: the partial atomic charges are an *effective* representation of the electron distribution. As has been described in the preceding sections, the RISM-SCF/MCSCF method provides information o the electron distribution that is more detailed than information on the dipole moment. Therefore, if we could find a menas to observe the electron distiubution, it will provide more detailed information on molecular structure in solution. It will also provide a reliable tool for testing theory experimentally.

One possible candidate among experimental methods to observe the electron distribution may be the MR chemical shift,because the chemical shift is a mainfestation of changes in the screening of nuclear magnetic fields due to electron clouds. it is highly desirable to establish a theory to bridge the NMR chemical shift and the RISM-SCF/MCSCF method. In the first example of application of the theory in this chapter, we made a point with respect to the polarizability of molecules and showed how the problem could have been handle by theRISM-SCF/MCSCF theroy. However, the current level of our method has serious limitation in this respect.

The method can handle the polarizability of molecules in neat liquids or that of a single molecule in solution in a reasonable manner. But in order to be able to treat the polarizability of both solute and solvent molecules in soultion, considerable generalization of the RISM side of the theory is required. When solvent molecules are situated within the influence from the solute molecules, and the solvent can no longer be "neat". Therefore, the polartiable model developed for neatliquids is not valid. In such a case, solvent-solvent PCF should be treated under the soulte field, which is typical on in-homogeneuos liquids. The density functional theory in classical theory may be the best choice for extending the theory to such a problem [34].

Nonequilibrium processes including chemical reactions present a most challenging problem in theoretical chemistry. There are two aspects to chemical reactions: the reactivity or chemical equilibrium and the reaction dynamics. The chemical equlibrium of molecules is a synonym for the free energy difference between reactant and product. Two important factors determing the chemical equlibrium in solution are the changes in electronic structure and the solvation free energy. Those quantities can be evaluated by the coupled quantum and extended RISM equations, or RISM-SCF theory. Exploration of the reaction dynamics is much more demanding. Two elements of reaction dynamics in solution must be considered: the determination of reaction paths and the time evolution along the reacation path.

The reaction path can be determined most naively by calculating the free energy map of reacting species. The RISM-SCF procedure can be employed for such calculations. If the rate-determining from the free energy diffreence of the two states based on transition state theory. On the other hand, for such a reaction in which the dynamics of solvent reorganization determines the reaction rate, the time evolution along the reaction path may be described by coupling RISM and the generalized Langeving equation (GLE) in the same spirit as the Karmesrs theory; The time evolution along a reaction path can be ewed as a stochastic barrier crossing driven by thermal fluctuations and damped by friction. Our treatment features the microscopic treatment of solvent structure on the level of the density pair correlation functions, which distinguisher it from earlier attempts that used phenomenological solvent models.

One of the prerequisites for developing such a treat,treatment is a theory to describe liquid dynamics on the molecular level. We recently proposed a new theory based on the interaction site model in which liquid dynamics is decoupled into the collective modes of density fluctuation: the acoustic and optical modes corresponding, respectively, to transationl and rotational motion of molecules [35]. From this point of view, transport coefficients such as the friction coefficient can be realized as a response of the collective modes of the solvent to perturbations due to solute. It is the first step in developing the theory of reaction dynamics to describe stoachasic barrier crossing in terms of the collective fluctuations of solvent to reacting species alosng a properly defined reaction corrrdinate.

CHAPTER

9 BIOCHEMICAL AND BIOPHYSICAL RHYTHMS

Biochemical and biophysical rhythms are ubiquitous characteristics of living organisms, from rapid oscillations of membrance potential in nerve cell to slow cycles of ovulation in mammals. One of the first biochemical oscillations to be discovered was the periodic conversion of sugar to alcohol "(glycolysis)" in anaerobic yeast cultures. The oscillation can be observed as periodic changes in fluorescence from an essential intermediate, NADH; In the laboratories of Brittion Chance and Benno Hess, these oscillations were shown to arise from a curious property of the enzyme phosphofructokinase (PFK), which catalyzes the phosphorylation of fructose-6, phosphate to fructose-1, 6-bisphosphate using ATP as the phosphate donor; . Although PFK consumes ATP, the glycolytic pathway produces more ATP than it consumes . To Properly regulate ATP Production, ATP inhibits PFK and ADP activates PFK. Hence, if the cell is energy "*rich*" (ATP high, ADP low), then PFK activity is inhibited, and the flux of sugars into the glycolytic pathway is shut down. As ATP level drops and ADP level increases, PFK is activated and glycolysis recommences. In principle, this negative freedback loop should stabilize the energy supply of the cell. However, because ATP is both a substrate of PFK and an inhibitor of the enzyme (like-wise, ADP is both a product of PFK and an activator of the enzyme), the steady-state flux through the glycolytic pathway can be unstable, and the regulatory system generates sustined oscillations in all intermediates of the pathyway.

Another classic example of rhythmic behaviour in biology is periodic growth and division of well-nourished cells. This phenomenon will be studied thoroughly in the next chapter, but for now, to illustrate some basic ideas, we focus on the periodic accumulation and degradation of cyclins during the division cycle of yeast cells; Kim Nasmyth and others have shown that these oscillations are intimately connected to dynamical interactions between CLN-type cyclins and CLB-type cyclins; CLNs (in combination with a kinase subunit called CDC28) activate their own synthesis (product activation of "*autocatalysis*") and inhibit the degradation of CLBs. AS CLBs accumulate, they inhibit the synthesis of CLNs, causing CLN-dependent kinase activity to drop and CLB degradation to increase. The mutual interplay of CLN and CLB generate the periodic appearance of their associated kinase activities, which drive the crucial events of the budding yeast cell cycle (bud emergence, DNA sysnthesis , mitosis, and cell division.)

The third example that we shall use in this cahpter concerns periodic changes in physiological properties (physical activity, body temperature, reproduction, etc.) entrained to the 24th cycle of light and darkness so prevalent to life on earth. These rhythms are not driven solely by the externsal timekeeper, because they persist under constant conditions of illumination and temperature; Under constant conditions, the organism exhibits its own "endogenous" rhythm, which is colse to, but not exactly, 24 *h* (hence *"circadian" or "nearly daily"* . The basic molecular mechanism of circadian rhythms has been uncovered has been uncovered only recnetly, by resarch in the laboratories of Michael Rosbash, Michael Young Jay Dunlap, and others. Central to the mechanism is a protein called PER, which, after being synthesized in the cytoplasm, moves into the nucleus to inhibit the transcription of its own mRNA; Hence, the regulatory mechanism for PER synthesis contains a time-delayed negative-freeback loop, which, as we shall see, is another common theme in biochemical oscillatiors.

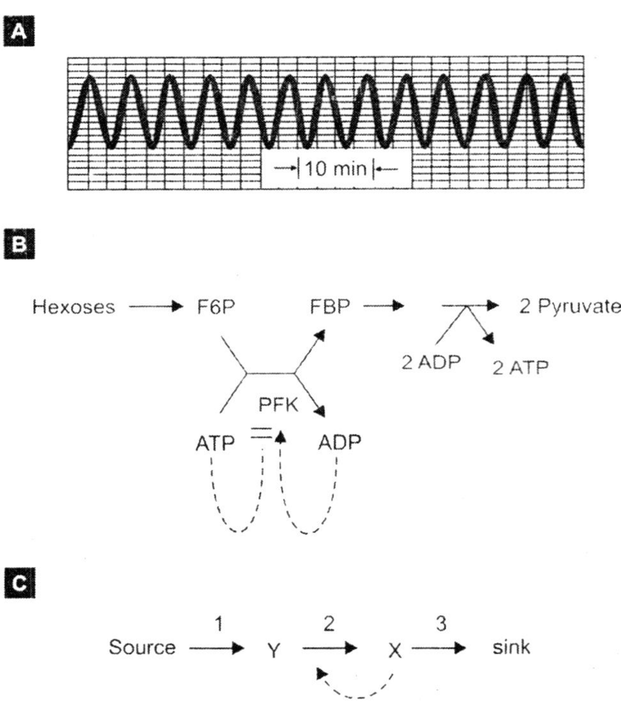

Fig. 9.1. (A) Sustained oscillations in NADH fluorescence in yeast cells *Saccharomyces cerevisiae.* (B) The control properties of the enzyme phospho fructokinase (PFK) are thought to be responsible for the generation of oscillations in the glycolytic paths-way. PFK catalyzes the conversion of fructose-1, 6 b_1 phosphate (FBP), using ATP as phosphate-group donor. PFK activity is allosterically modulated by ATP (inhibitor) and ADP (activator). FBP is steadily supplied to PFK by a sugar source ("Hexoses") and FBP is steadily utilized in the production of metabolites ("Pyruvate") ADP and ATP are also recycled by other metabolic processes. (C) Simplified mechanism. Reaction 1 is a steady supply of substrate Y for rection 2, whose product X is removed by reaction 3. X activates the enzyme (PFK) catalyzing reaction 2. Roughly speaking, Y = F6P + ATP, = FBP + ADP, "source" = hexoses, and "sink" = pyruvate.

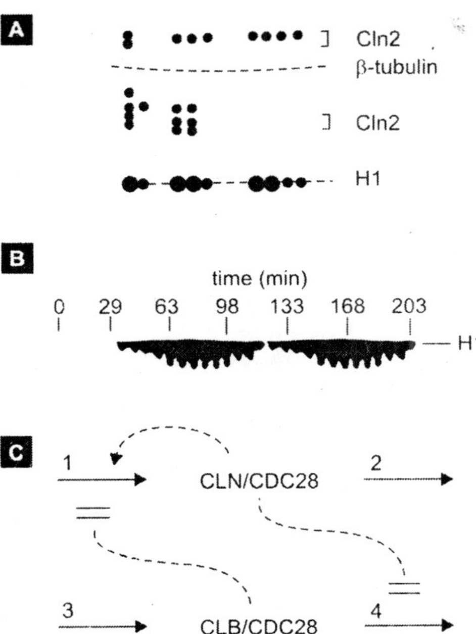

Fig. 9.2. Cyclin fluctuations during the cell cycle in budding Yeast. (A) Samples are taken at increasing time points (left to right) from a synchronous culture of budding yeast cells and analyzed for CLN2 / CDC28 protein content (top row) and CLN2/CDC28 kinase activity (last raw, lableled" H1") β tubulin serves as a control, nonoscillating, protein. (b) In a similar experiment, a synchronous culture of yeast cells is analyzed for CLB2/CDC28 activity (labeled" H1" ; each column represents a 7 min increment). (C) mechanism of cyclin oscillations. Budding yeast cells contain two classes of cyclins: CLN and CLB. These cyclins combine with a kinase partner ((CDC28) to make active dimers. CDC28 subunits are always in excess, so dimer activity is limited by cyclin availability (*i. e.* cyclin synthesis and degradation, arrows in the diagram). CLN/CDC28 activates the trnascription factor that promotes CLN synthesis and inhibits the proteolytic enzymes tha degrade CLB. In return, CLB/CDC28 inhibits the transcription factor that promotes CLN synthesis.

FEEDBACK

In general, a biochemical reaction network is a schematic diagram of "boxes and arrows." Each box is a chemical species. Solid arrows represent chemical reactions producing or consuming a molecule. Dashed arrows represent control of a chemical reaction rate by a species that may not be a reactant or product of the reaction. Regulatory signals may be activatory (barded end) or inhibitory (blunt end).

To each reaction in a biochemical network is associated a rate law (with accompanying kinetic parameteres). Hence, the network implies a set of rate equations of the form

$$\frac{dx}{dt} = v_{in1} + v_{in2} + \ldots - v_{out1} - v_{out2} - \ldots = f(x;p) \tag{9.1}$$

where t = time, x = concentration of species X, and v_{in1}, v_{out1}, ... are the rates of the various reactions that produce and consume X. We lump together all these rate laws into a nonlinear function $f(x; p)$, where p is a vector of kinetic parameters. In general, we can think of x as a vector of concentrations of all the time-varying components in the reaction network, $x_i = [X_i]$, and f as a vector-valued function. In component form,

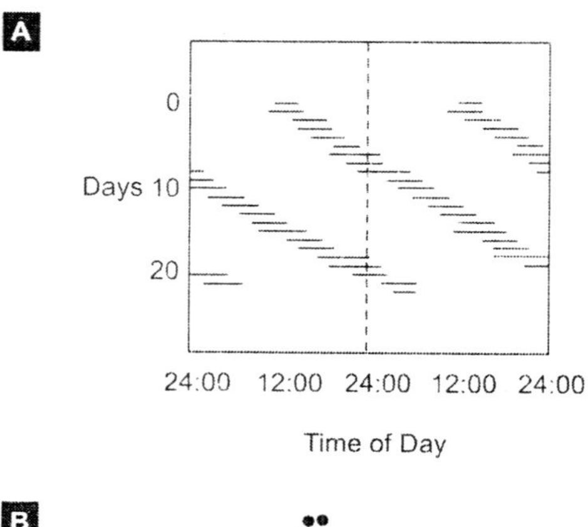

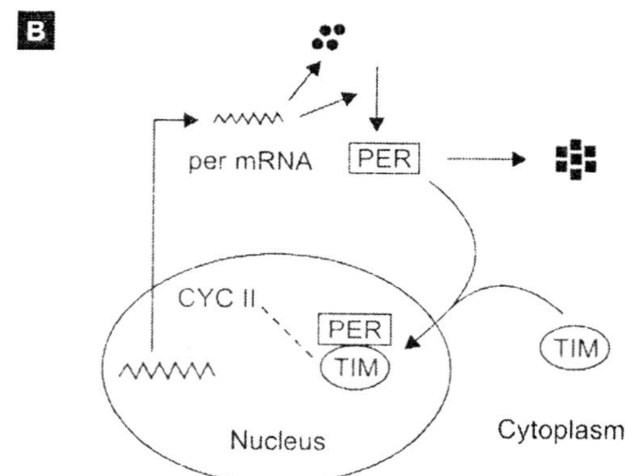

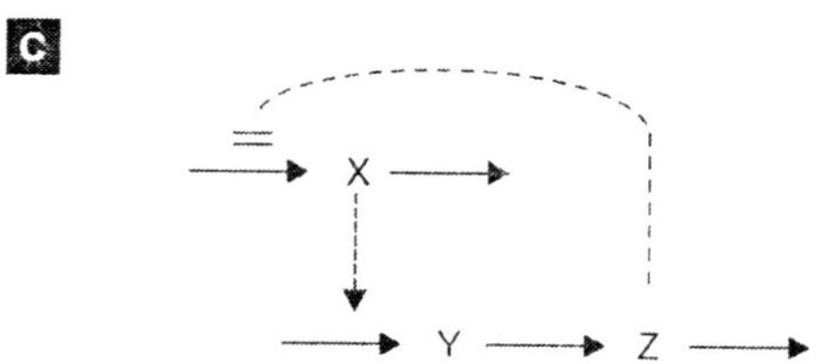

Fig. 9.3. (A) Endogenous circadian rhythm of activity. The black bars represent the sleep episodes of a human being isolated form all external temporal cues (variable light, temperature, etc.). The sleep episodes are plotted twice on each line to emphasize that the endogenous period of sleepiness is longer than 24 *h*. (B) A protein called PER is known to play a crucial role in circadian rhythms in fruit files and mice. The *per* gene is transcribed in the nucleus, with the help of two transcription factors, CLK and CYC. Then *per m* RAN is transported to the cytoplasm, where it codes for PER protein. PER protein is processed in the cytoplasm by phos-phorylation and by binding to other proteins, such as TIM and CRY Properly processed PER then moves into the nucleus, where it disrupts the binding of CLK and CYC, turning off the transcription of Per. (C) A schematic representation of the negative-feedback loop in panle (B). This is Goodwin's classical mechanism for periodic protein expression, driven by feedback repression of transcription.

Table 9.1. Biochemical and Cellular Rhythms.

Rhythm	*Period*
Membrane potential oscillatios	10 ms-10s
Cardiac rhythms	1 s
Smooth muscle contraction	Seconds- hours
Calcium oscillations	seconds-minutes
Protoplasmic streaming	1 min
Glycolytic oscillations	1 min-1 h
CAMP oscillations	10 min
Insulin secretion (pancreas)	minutes
Gonadotropic hormone secretion	hours
Cell cycle	30 min–24 h
Cricadian rhythms	24b
Ovarian cycle	weeks-months

the reaction network, x; = $[x_i]$ and f as a vector-valued function. In component form,

$$\frac{dx_i}{dt} = f_i\left(x_i, x_2, \ldots, x_n; p_1, p_2, \ldots p_r\right) \quad i = 1, \ldots n. \tag{9.2}$$

A "*steady-state*" solution $(x^*{}_1{}^*, x^*{}_2, \ldots, x^*{}_n)$ of the kinetic equations satisfies the algebraic equations

$$f_i(x_1, x_2, \ldots\ldots, x_n; p_1, p_2, \ldots, p_r) = 0, i = 1, \ldots, n. \tag{9.3}$$

Notice that steady-state concentrations depend on the parameter values $x_i^* = x_i^* (p_1, \ldots p_r)$. Moreover, the algebraic equations are generally nonlinear, so there may be multiple steady-state solutions. Of course, since we are dealing with chemical reaction systems, we are interested only in solutions that satisfy $x_i = \geq 0$ for all i.

In the theory of biochemical oscillations, based on rate equations of this sort, a crucial role is played by elements of the Jacobian matrix $\mathbf{J} = [a_{ij}]$, where $a_{ij} = \partial f_i / \partial x_i$ evaluated at a steady-state solution. The matrix $\mathbf{J}$ is a square matrix $(n \times n)$ dependent on parameter values and the steady state under consideration. The diagonal elements of the jacobian matrix of a chemical reaction system are usually negative, $a_{ii} < 0$, because v_{out} terms in are always proportional to the concentration of the substance being destroyed, so $\partial f_i / \partial x_i$ always has one or more negative terms. Positive contributions to a_{ii} are rare because autocatalysis (reactions that produce X_i at a rate that increases with $[X_i]$ is rare.

Common to biochemical networks are complex feedback loops, whereby the products of one reaction affect the rates of other reactions. A steady-state feedback loop can be defined as a set of nonzero elements of the Jacobian matrix that connect in a loop: $a_{ij}\, a_{jk}\, a_{kl} \ldots a_{mi} \neq 0.$

Representative feedback loops are illustrated in Figure 9.4

(A) Autocatalysis $(a_{ii} > 0)$, though rare, plays a major role in biochemical oscillations, as will become clear. We have already seen examples of autocatalysis in the mechanism of glycolysis and yeast division; figure 9.1 and Figure 9.2.

(B) Autocatalysis also occurs when a chemical declecrates the rate of its own destruction.

(C) Indirect autocatalysis occurs through a positive feedback loop (a_{ij} and $a_{ji} > 0$), whereby X_i activates the production of X_i, and X_i retruns the favour.

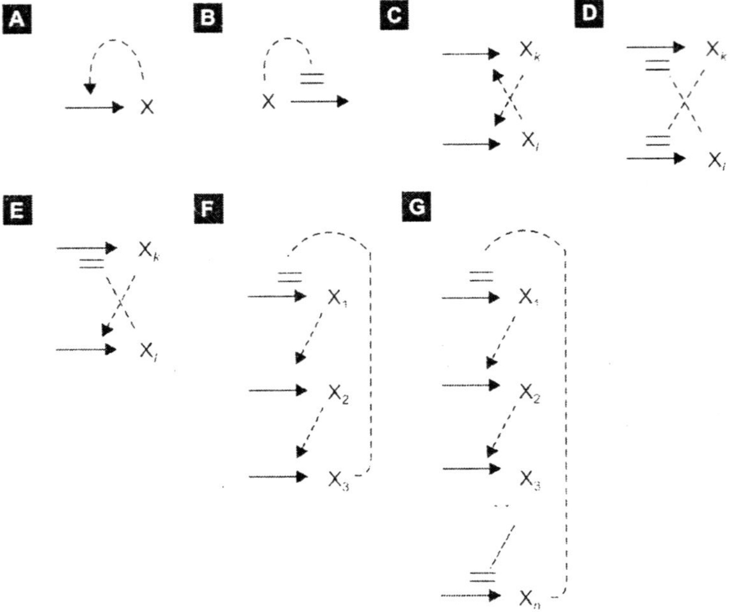

Fig. 9.4. Representative feedback loops.

(D) A two-component positive-feedback loop is also created by a pair of antagomstic species ($a_{ij} <$ 0 and $a_{ji} < 0$).

(E) A two-componet negaitve-feedback loop ($a_{ij}aji > 0$) is created when X_i activates the production of X_i, and X_i, inhibits the production of X_i.

(F) Longer negative feedback loops are common.

(G) Long feedback loops are either positive or negative, depending on the sign of the product $a_{1n}\, a_{n},\, _{n-1} \ldots a_{43}\, a_{32}\, a_{21}$.

Before we can speak definitively about the effects of feedback in biochemical reaction networks, we need to know how to characterize the rates of enzyme-catalyzed reactions that are subject to regulation by "*Distant*" effectors, *i.e.,* chemical species other than are reactants and products of the enzyme.

REGULATORY ENZYMES

Regulatory enzymes are usually multisubunit proteins with binding sties for reactants; and products (on the ctalytic subunits) and for activators and inhibitors (on the regulatory units). Although threre are more sophisticated and accurate ways to characterize the binding of small molecules to multisubunit proteins, we shall limit ourselves to straightforward generalization of the Michaelis-Metnen equation. For simplicity, consider a two-subunit enzyme (EE) that converts substrate (S) into product (P). The binding constant and turnover number of each subunit depend on whether the other (identical) subunit is bound to S or not. The mechanism can be written

$EE + S \leftrightarrow EES$	dissociation constant = $k_{-1}/K_1,$
$EES \rightarrow EE + P$	Rate constant = $k_3,$
$EES + S \leftrightarrow SEES$	dissociation constant = $k^{-2}/k_2,$
$SEES \rightarrow EES + P$	rate constant = $k_4.$

We can make rapid-equilibrium approximation on the enzyme substrate complexes:

$$\frac{[\mathrm{EES}]}{[\mathrm{EE}]_T} = \frac{[\mathrm{S}]\mathrm{K}_{m1}}{1+([\mathrm{S}]/\mathrm{K}_{m1})+(\mathrm{S})^2/(\mathrm{K}_{m1}\mathrm{K}_{m2})}$$

$$\frac{[\mathrm{SEES}]}{[\mathrm{EE}]_T} = \frac{[\mathrm{S}]^2/\mathrm{K}_{m1}K_{m2}}{1+([\mathrm{S}/\mathrm{K}_{m1}])+(\mathrm{S})^2/(\mathrm{K}_{m1}\mathrm{K}_{m2})}$$

where $[\mathrm{EE}]_T$ = [EE] + [EES] + SEES], $Km1 = (k_{-1} + k_3)/k_1$, and $km_2 = (k_{-2} + k_4)/k_2$. Here K_{m1} is the Michaelis constant for the interaction of the enzyme with the first substrate to bind; K_{m1} is inversely proportional to the affinity of the enzyme for the first substrate; Km_2 is inverserly proportional to the affinity of the enzyme for the second substrate to bind.

Now it is easy to write an equation for the rate of the reaction:

$$v = \frac{d[p]}{dt} = -\frac{d[\mathrm{S}]}{dt} = \frac{[\mathrm{EE}]_T\left(\frac{\mathrm{S}}{\mathrm{K}_{m1}}\right)(k_3+k_4)\left(\frac{[\mathrm{S}]}{K_{m2}}\right)}{1+\left(\frac{[\mathrm{S}]}{\mathrm{K}_{m1}}\right)+\left(\frac{[S]^2}{\mathrm{K}_{m1}\mathrm{K}_{m2}}\right)} \quad ...(9.4)$$

We can distinguish two interesting limiting cases of (9.4).

Hill equation: $K_{m1} \to \infty$, $K_{m2} \to 0$, such that $K_{m1}K_{m2} = K^2_m$ constant,

$$v = \frac{\mathrm{V}_{\max}([\mathrm{S}]/\mathrm{K}_m)^2}{1+([\mathrm{S}]/\mathrm{K}_m)^2}, \mathrm{V}_{\max} = k_4[\mathrm{EE}]_T; \quad (9.5)$$

Substrate inhibition: $k_3 \to \infty$, $K_{m1} \to \infty$, $k_{m2} \to 0$ such that $K_{m1}K_{m2} = K^2_m$ constant, $k_3Km_2 \to \infty$, and $k_3\,[\mathrm{EE}]_T\; K_m/K_{m1} = V_{max}$ = constant,

$$v = \frac{\mathrm{V}_{\max}[\mathrm{S}]/\mathrm{K}_m}{1+([\mathrm{S}]/\mathrm{K}_m)^2}. \quad (9.6)$$

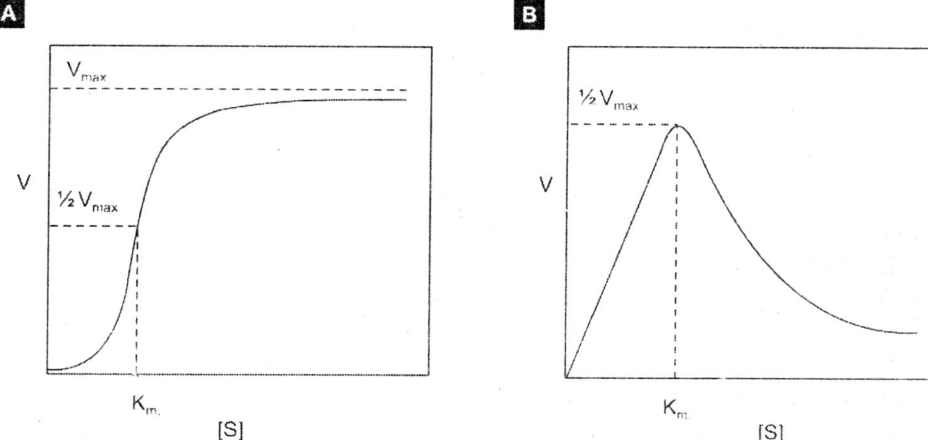

Fig. 9.5 Rate laws for activation and inhibition of multisubunit enzymes by cooperative binding.

Equation (9.5) and (9.6) are plotted in Figure 9.5

Rate laws like (9.4) are said to express "cooperative" kinetics. The "empty" enzyme (EE) has low affinity for substrate (K_{m1} large), but as the enzyme pick up its first substrate molecule, the two subunits change their conformation to a high affinity form (K_{m2} small, such that $K_m = (K_{m1} K_{m2})^{1/2}$ is physiologically significant).

If the subunits do not exhibit cooperative binding, then $k_1 = 2k_2$, $2k_{-1} = k_{-2}$, $k_4 = 2k_3$, and hence

$$v = \frac{V_{mx}[S]}{K_m + [S]}. \quad ...(9.7)$$

That is, we regain the Michaelis-Menten rate law, with Vmax = $2k_3$ and $k_m = 2(k_{-1} + k_3)/k1$.

Cooperative binding of ligands to a multisubunit enzyme is not limited to substrates. Other small molecules may bind to the enzyme and alter its catalytic properties (eigher its affinity for substrates or its rate of converting bound substrates into products). Such enzymes are called "*allosteric*" because in addition to substrate-binding sites, thye have " *other sites*" for binding regulatory molecules that either activate or inhibit the enzyme. Allosteric proteins play crucial roles in the regulation of metabolic pathways, membrane transport, gene expression, etc. Relatively simple algebraic expressions for allosteric effects can be dervied by the reasoning behind.

For instance, consider a tetrameric enzyme, with two catalytic subunits (EE) and two regulatory subnits (RR), which bind substrate (S) and ligand (L), respectively. In this case, the holonzyme may exist in nine different forms. If each assoication dissociation reaction is in rapid equilibrium at given ocncentrations of substrate and ligand, then

$$[E_T] = [EE_{oo}]\left(1 + \frac{[S]}{K_{m1}} + \frac{[S]^2}{K_{m1}K_{m2}}\right)\left(1 + \frac{[L]}{K_{n1}} + \frac{[L]^2}{K_{n1}K_{n2}}\right)$$

where $[E]_T$ = total concentration of enzyme in all nine forms. [EEoo] = concentration of holoenzyme unbound to substrate or ligand, [S] = substrate concentration, [L] = ligand concentration, *Km's* = Michaelis constants for substrate binding (*e.g.*, $K_{m1} = k_{-1}/k_1$), and $K_n s$ = Michaelis constants for ligand binding (defind similarly). If, for example, the only form of the enzyme with significant catlytic activity is

$$\begin{bmatrix} \text{SEES} \\ \text{LRRL} \end{bmatrix},$$

than the rate of the reaction is

$$v = V_{max}\frac{[S]^2/(K_{m1}K_{m2})}{1 + [S]/K_{m1} + [S]^2/(K_{m1}K_{m2})} \cdot \frac{[L]^2/(K_{n1}K_{n2})}{1 + [L]/K_{n1} + [L]^2/(K_{n1}K_{n2})} \quad ...(9.8)$$

In this case, the ligand is an allosteric *activator* or the enzyme. For an allosteric *inhibitor*, the enzyme is most active when no ligand is bound to the regulatory subunits; hence

$$v = V_{max}\frac{[S]^2/(K_{m1}K_{m2})}{1 + [S]/K_{m1} + [S]^2/(K_{m1}K_{m2})} \cdot \frac{1}{1 + [L]/K_{n1} + [L]^2/(K_{n1}K_{n2})} \quad (9.9)$$

A comprehensive and accurate kinetic theory of allosteric enzymes is much more complicated than what has been presented, but (9.8) and (9.9) will serve our purposes.

Finally, suppose EE and S above are not "multisubunit enzyme" and "substrate" but "dimeric transcription factor" and "ligand." The active form of the transcription factor promotes the expression of some gene, and the presence of ligand alters the distribution of transcription factor among its various forms: EE, EES, and SEES. Depending on which form of transcription for is most active the rate of gene expression can be activated and/or inhibited by ligand:

ligand effect	***active from relative***	***rate of gene expression***
inhibition	EE	$\dfrac{1}{1+([S]/K_{m1})+([S]^2/K_{m1}K_{m2})}$
activation	SEES	$\dfrac{[S]^2/K_{m1}K_{m2}}{1+([S]/K_{m1})+([S]^2/K_{m1}K_{m2})}$
mixed	EES	$\dfrac{[S]/K_{m1}}{1+([S]/K_{m1})+([S]^2/K_{m1}K_{m2})}$

AUTOCATALYSIS

First, we consider two minimal requirements for oscillations in chemical reaction systems.

1. If the "*network*" is absurdly simple, with only one time-varying component, then oscillations are, generally speaking, impossible. For x (t) to be periodic, dx/dt must take on both positive and negative values at some values of x, which is impossible if $(x;p)$ is a continuous single-valued function of x). Thus, to understand biochemical oscillations, we must start with two-component networks, described by a pair of ODEs

$$\frac{dx}{dt}=f(x,y)$$

$$\frac{dy}{dt}=g(x,y) \qquad \text{...(9.10)}$$

   where x and y are (nonegative) concentrations of the two components, and we have suppressed the dependence of f and g on parameters, for the time beign.

2. Bendixon's negative criterion states that if $\partial f/\partial x+\partial g/\partial y$ is of constant sign in some region R of the x, y plane, then there can be no periodic solution of solution of system in R since ∂x ∂y, and $\partial g/\partial y$are usually negative (refer to our earlier discussion of the Jacobian matrix), Bendixson's criterion requires that a two-component chemical reaction system have an autocatalytic step ($\partial f/\partial x>0$ or $\partial g/\partial y>0$) in order to exhibit sustained oscillations.

In general, we can expect system to have one or more steady-state solutions (x^*, y^*), satisfying $f(x^*, y^*) = 0$ and $g(x^*, y^*) = 0$. The stability of such steady states is determined by the eigenvalues of the Jacob in matrix evaluated at the stready state

$$\mathbf{J}=\begin{bmatrix} f_x(x^*,y^*) & f_y(x^*,y^*) \\ g_x(x^*,y^*) & g_y(x^*,y^*) \end{bmatrix}=\begin{bmatrix} a_{11} & a_{12} \\ a_{21} & a_{22} \end{bmatrix}.$$

The eigenvalues of J are the roots of the characteristic equation $-\lambda^2-(a_{11}+a_{22})\lambda+(a_{11}+a_{22}-a_{12}a_{21})=0$ namely

$$\lambda \pm = \frac{1}{2}\left[a_{11} + a_{22} \pm \sqrt{(a_{11} + a_{22})^2 - 4(a_{11}a_{22} - a_{12}a_{21})}\right].$$

For the steady state to be stable, Re (λ) must be negative for both eigenvalues. If $a_{11}\, a_{22} - a_{12}\, a_{21}$) = det (**J**) < 0, the **J** has one positive and one negative eigenvalue, and the steady state is a saddle point. **IF** det (**J**) > 0 and $a_{11} + a_{22}$ = tr (**j**) < 0, then the steady state is an unstable node or focus.

In general, the trace and determination of **J** depend continuously on the kinetic parameters in *f* and *g*. If by varying one of these parameters (call it *p*1) we can carry tr (**J**) from negative to positive values, with det (***j***) > 0, then the steady state loses stability at tr (**J**) = 0 (when $p_1 = p_{\text{crit}}$, Say). At the bifurcation point, tr (**J**) = $a_{11} + a_{22} = 0$, the eigenvalues are pure imaginary numbers, $\lambda \pm = \pm i\,\omega$, $\omega = \sqrt{(a_{11} + a_{22})^2 - 4(a_{11}a_{22} - a_{12}a_{21})}$. Close to the bifurcation point *i, e,. *$p_1 \approx P_{\text{crit}}$, small—amplitude limit-cycle solutions surround the steady state, and the period of oscillation is colse to $2\pi/\omega$. We say that periodic solutions arise by a ***Hopf bifurcation*** at $p_1\, p_{crit}$.

Biochemical oscillations usually arise by this mechanism, so we will consider first the requirements for Hopf bifurcation in a two-component network. In chemical reaction systems the diagonal elements of the jacobian matrix are usually negative numbers, reflecting the various steps by which species X is transformed into something else. If both a_{11} and a_{22} are always negative, then tr (**J**) never changes sign, and a Hopf bifurcation cannot occur. At least one of them must be positive for some values of the kinetic parameters. For a diagonal element to be positive, species X must be "autocatalytic;" *i,e.,* with increasing [X], the rates of production of X increase faster than the rates of destruction. IF a_{11} and a_{22} are of opposite sign, then a_{12} and a_{21} must also be of opposite sign in order for det (**J**) to be positive. Thus we have two characteristic sign patterns for Jacobian matrices that typically produce Hopf bifurcations in chemical reaction systems with two time-dependent components:

$$J = \begin{bmatrix} + & + \\ - & - \end{bmatrix} \text{ and } J = \begin{bmatrix} + & - \\ + & - \end{bmatrix} \qquad ...(9.11)$$

Oscillatory Mechanism

The simplest oscillatory mechasnism is probably the linear pathway inFigure 9.1C. Species Y is converted into X by an enzyme that is activated by its product. Hence, the production of X is autocatalytic; the reaction speeds up as [X] increases, until the substrate Y is depleted so much that the reaction ceases.

Using a rate law for the auosteric enzyme we can write a pair of ODEs describing this mechanism.

$$\frac{d[X]}{dt} = v_2[Y]\frac{\epsilon^2 + ([X]/K_n)^2}{1 + ([X]/K_n)^2} - k_3[X],$$

$$\frac{d[Y]}{dt} = k_1 - v_2[Y]\frac{\epsilon^2 + ([X]/K_n)^2}{1 + ([X]/K_n)^2}. \qquad ...(9.12)$$

The rational function

$$\frac{\in^2 + ([X]/K_n)^2}{1 + ([X]/K_n)^2}$$

represents activation of the allosteric enzyme by binding of product X to the eenzyme's regulatory sites. The constant $\in_2$ is the activity of the enzyme with no product bound relative to its activity when its regulatory sites are saturated by product molecules. We assume that product binding to regulatory sites is highly cooperative, so that we can neglect the fraction of enzyme with only one product molecule bound. We shall also assume that $\in \ll 1$.

It is convent to define "dimensionless" variables $x = [X]/K_n$, $y = [Y]K_n$, and $t' = k_3 t$, and a new variable $z = x + y$, and write system as

$$\frac{dx}{dt'} v = (z - x) \frac{\in^2 + x^2}{1 + x^2} - x,$$

$$\frac{dz}{dt'} = k - x, \qquad \text{...(9.13)}$$

where $v = v_2/k_3$ and $k_1/(k_3 k_n)$. The steady state for this model satisfies $x^* = k$ and $v\,(z^* - k) = k\,(1 + k^2)/(\in^2 + k^2)$ For the Jacobian matrix of system at the steady state, it is easy to show that det $(\mathbf{J}) = v\,(\in^2 + k^2)/(1 + k^2) > 0$ and

$$\text{tr}\,(\mathbf{J}) = -1 - v\frac{\in^2 + k^2}{1 + k^2} + \frac{2k^2(1 - \in^2)}{(\in^2 + k^2)(1 + k^2)}$$

$$= \frac{(1+v)k^4 - (1 - 3\in^2 - 2v\in^2)k^2 + \in^2(1 + v\in^2)}{(\in^2 + k^2)(1 + k^2)} \qquad \text{...(9.14)}$$

It should be obvious from that $tr\,(\mathbf{J}) < 0$ and the steady state is stable if k is either close to 0 or very large. On the other hand, $tr\,(\mathbf{J}) > 0$ and the steady state is unstable if k takes on intermediary values. For $\in$ small, the steady state is unstable when $k^2 - \in^2 > 0$ and when $-(1 + v)k^4 + k^2 > 0$, *i.e.*, when

$$\in\ < k < (1 - v)^{-1/2}, \qquad \text{...(9.15)}$$

which may, alternatively, be expressed as conditions on the input rate of Y:

$$\in < \frac{k_1}{k_3 k_n} < \sqrt{\frac{k_3}{k_3 + v_2}}. \qquad \text{...(9.16)}$$

Between these limits the system executes stable limit cycle oscillations.

As we have often seen in this book, for two-component dynamical systems it is informative to plot nullclines in the phase plane. For system

x- nullcline: $z = x + \dfrac{x(1 + x^2)}{v(\in^2 + x^2)}$,

z–nullcline: $x = k$.

These nullclines are plotted in Figure 9.6 The x- nullcline has extrema at the roots of

$$(1 + v)\,x^4 - (1 - 3\in 2 - 2v\in^2)\,x^2 + \in^2(1 + v\in 2) = 0$$

Notice that the steady state loses stability excatly when the steady state passes through the extrema of the x-nullcline.

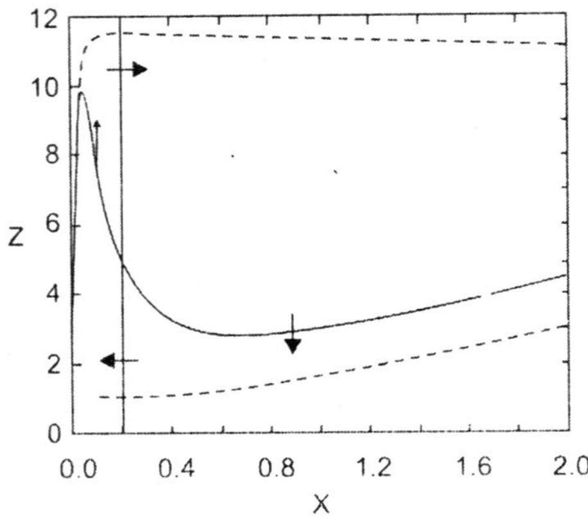

Fig. 9.6. Phase plane portrait of a typical substrate-depletion oscillator, (9.13). Parameter values: $v = 1$ $k = 0.2$, $\epsilon = 0.05$. The solid lines are the x-nullcline (marked byshort vertical arrows) and the y-nullcline (horizontal arrows). The nullclines intersect at an unstable steady state (o). The dashed line is the stable limit cycle solution (periodic orbit) of the dynamical system.

Activator-Inhibitor Oscillator

Looking back to we see that there are two sign patterns consistent with Hopf bifurcation in a two-component biochemical reaction system. The first is illustrated by the substrate-depletion oscillator inFigure 9.1, and the second by the activator inhibitor model in Figure 9.2. The ODEs describing Figure 9.2 are

$$\frac{d[X]}{dt} = v_1 \frac{\epsilon^2 + ([X]/K_m)^2}{1 + ([X]/K_m)^2} \cdot \frac{1}{1 + ([Y]/K_n)} - k_2 [X],$$

$$\frac{d[Y]}{dt} = k_3 - \frac{k_4 [Y]}{1 + ([X]/K_j)^2}.$$

where X = CLN/CDC28 and Y = CLB/CDC 28. In terms of dimensionless variables ($x = [X]/K_m$, $y = [Y]/K_n$' $t' = v_1 t / K_m$' these equations becomes

$$\frac{dx}{dt'} = \frac{\epsilon^2 + x^2}{1 + x^2} \cdot \frac{1}{1+y} - ax,$$

$$\tau \frac{dy}{dt'} = b - \frac{y}{1 + cx^2}, \qquad \text{...(9.17)}$$

where $a = k_2 K_m / v_1$, $b = k_3/(k_4 K_n)$, $c = (K_m/K_j)^2$, and $\tau = v_1/(k_4 K_m)$.

Rather than analyze the stability of the steady state state algebraically, which is difficult, we go directly to phase plane portraits of Figure 9.7. It is easy to find parameter values that produce either limit cycle oscillations or bistability.

Intuitively, the origin of the oscillations is clear. When Y is rare, X increases autocatalytically. Abundant X stimulates accumulation of Y (by inhibiting Y's degradation), which feeds back to inhibit the productiono of X . After X discappears, Y is also destroyed, and then X can make a comeback. Mechanisms like this one, whose jacobian matrix has sign pattern

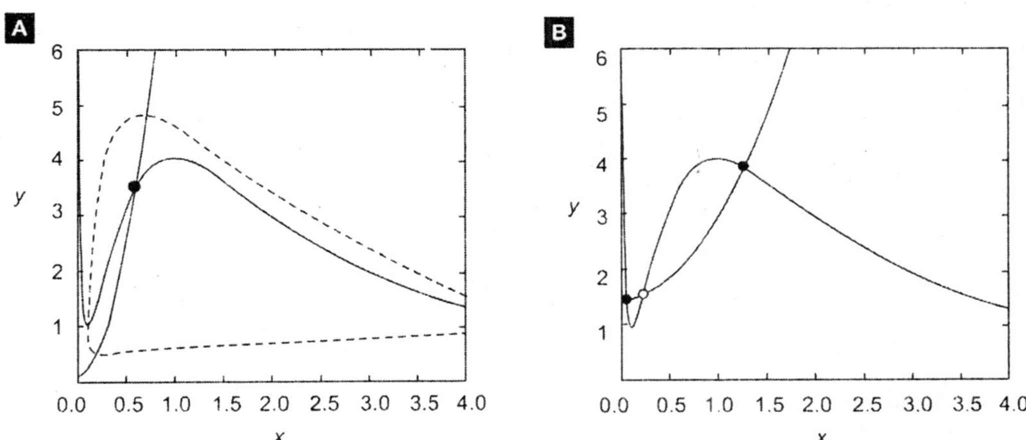

Fig. 9.7. Phase plane portrait of a typical activator-inhibitor system, (9.17). The solid lines are nullclines, intersecting at stable (•) or unstable (o) steady states. The dashed line is a closed orbit of the dynamical system. (A) Oscillation, for $a = 0.1$ b = 0.1, c = 100, $\epsilon = 0.1$, $\tau = 5$. (B) Bistability, for $a = 0.1$, $b = 1.5$, $c = 1$, $\epsilon = 0.1$, $\tau = 5$.

has sign pattern

$$\begin{bmatrix} + & - \\ + & - \end{bmatrix}$$

are called activator-inhibitor models.

Two-component mechanisms with autocatalysis easily generate oscillations and bistability, as we have just seen. They also exhibit a rich structure of bifurcations to more complicated behaviour.

THREE-COMPONENT NETWORKS

In the previous section we have seen that two-component reaction systems can oscillate if they have autocatalysis (a_{11} or a_{22} positive) and negative feedback (a_{12} and a_{21} opposite sign). In this section we examine networks of three components (x, y, z) with Jacobian matrices of the form

$$J\pm = \begin{bmatrix} -\alpha & 0 & \pm\phi \\ c_1 & -\beta & 0 \\ 0 & c_2 & -\lambda \end{bmatrix}.$$

where α, β, γ, c_1, c_2, ϕ are all positive constants. Since the diagonal elements of the Jacobian are all negative, the system lacks autocatalysis. The Jacobian J_+ describes a system with a positive feedback loop, and J_- one with a negative feedback loop.

Positive Feed Back

First, let us see whether Hopf bifurcations are possible in a system with a pure positive feedback loop, *i.e.,* a Jacobian of the form J_+ at the steady state. Recall that the condition for a Hopf bifurcation is that the Jacobian matrix has a pair of purely imaginary eigenvalues $\lambda \pm = \pm i\omega$. The eigenvalues of J_+ are roots of the characteristic equation

$$G(\lambda) = \lambda^3 + (\alpha + \beta + \gamma)\lambda^2 + (\alpha\beta + \beta\gamma + \gamma\alpha)\lambda + \alpha\beta\gamma - c_1 c_2 \phi$$
$$= \lambda^3 + A\lambda^2 + B\lambda + C = 0.$$

The roots of this equation can be characterized by the Routh-Hurwitz theorem. Let $G(\lambda_i) = 0$ for $i = 1, 2, 3$, Then $Re(\lambda_i) < 0$ for $i = 1, 2, 3$, if and only if (*i*) $A > 0$, (*ii*) $C > 0$, and (*iii*) $AB > C$. Hence, in order for the steady state of the positive feedback loop to be unstable, we must insist that $C = \alpha\beta\gamma - c_1c_2\phi < 0$. In this case, J+ has at least one real positiveroot, call it $\lambda_1 > 0$. Then, $G(\lambda) = (\lambda - \lambda) H(\lambda)$, where $H(\lambda) = \lambda^2 + D\lambda + E$ and $D = A + \lambda_1 > 0$, $E = -C/\lambda_1 > 0$. From the quadratic formula, it follows that the two roots of $H(\lambda) = 0$ must have $Re(\lambda_i) < o$. Hence, it is impossible for a steady state with Jacobian matrix J+ to undergo a Hopf bifurcation.

It is possible for a positive feedback loop to have multiple steady state solutions, with two stable nodes separated by a saddle point. At the saddle point, $\lambda_1 < 0$ for $i = 2, 3$.

Negative Feedback

The Jacobian J_- determines the stability of the steady state in a three-variable system with a pure negative feedback loop. In this case,

$$G(\lambda) = \lambda^3 + (\alpha + \beta + \gamma)\lambda 2 + (\alpha\beta + \beta\gamma + \gamma\alpha)\gamma + \alpha\beta\gamma + c_1c_2\phi = 0,$$

and the Routh-Hurwitz theorem implies that the steady state is unstable if and only if

$$(\alpha + \beta + \gamma)(\alpha\beta + \beta\gamma + \gamma\alpha) < (\alpha\beta\gamma + c_1c_2\phi = 0.$$

Furtheromore, if equality holds then $G(\lambda) = (\lambda + A)(\lambda^2 + B)$, So $G(\lambda)$ has conjugate roots on the imaginary axis at $\lambda = \pm i\sqrt{\alpha\beta + \beta\gamma + \gamma\alpha}$.

The Goodwin Oscillator

The quintessential example of biochemical oscillator based on negative feedback alone was invented by Brian Goodwin see also Griffith (1968a). The kinetic equations describing this mecahanism are

$$\frac{d[X_1]}{dt} = \frac{v_o}{1 + ([X_3]/K_m)^p} - k_1[X_1]$$

$$\frac{d[X_2]}{dt} = v_1[X_1] - k_2[X_2].$$

$$\frac{d[X_3]}{dt} = v_2[X_2] - k_2[X_3].$$

Here $[X_1]$, $[X_2]$, and $[X_3]$ are concentrations of mRNA, potein, and product, respectively; v_0, v_1 *and* v_2 determine the rates of transcription translation, and catalysis; k_1, k_2, and k_3 are rate constants for degradation of each component; $1/K_m$ is the binding constant of end product to transcription factor; and p is a measure of the cooperativity of end product repressions.

Next we introduce dimensionless variables:

$$x_1 = \frac{v_1 v_2 [X_1]}{k_2 k_3 K_m}$$

$$x_2 = \frac{v_2[X_2]}{k_3 K_m}$$

$$x_3 = [X_3]/K_m.$$

where $\alpha = (v_o\, v_1\, v_2)/\ Km\ k_2\ k_3)$.

In terms of these new variables, the dynamical system becomes

$$\frac{dx_1}{dt'} = \frac{1}{1+x_3^p} - b_1 x_1, = \frac{dx_2}{dt'} = b_2(x_1 - x_2),$$

$$\frac{dx_3}{dt'} = b_3(x_2 - x_3),$$

where $b_i = k_i / \alpha$

Furthermore, to make the example easier, we shall assume that b_1 n= $b_2 = b_3$. In this case, the dynamical system has steady state at $x_1 = x_2 = x_3 = \xi$, where ξ is the unique real positive root of $1/(1 + \xi^p)\ b\xi$. The Jacobian matrix at this stady state is J–, with $\alpha = \beta = \gamma = b$, c_1 = = $c_2 = b$, and $\phi = (p\xi^{\,p-1}) / (1 + \xi^p)^2 = bp\ (1 - b\xi) > 0$. Hence, the characteristic equation is $(b + \lambda)^2 + b^2\phi = 0$, whose roots are

$$\lambda_1 = -b - b\sqrt[3]{p(1-b\xi)} < 0,$$

$$\lambda_{2,3} = -b + b\sqrt[3]{p(1-b\xi)}\ \left[\mathrm{Cost}(\pi/3) \pm i\sin(\pi/3)\right].$$

The steady state of Goodwin's model is unstable if Re $(\lambda_{\,2,3}) > 0$, *i, e.*, if $-b + (b/2)\sqrt[3]{p(1-b\zeta)} > 0$. This condition is equivalent to $p\ (1- b\xi) > 8$, or $b\xi < (p - 8)/p$. Hence, if p (the cooperativity of end product repression) is greater than 8, then we can choose k samll enough to destabilize the steady–state solution of Goodwin's equation. At the critical value of k, when Re $(\lambda\ 2,3) = 0$, the steady state undergoes a Hopf bifucrcation, spinning off small-amplitude periodic soultion with peroid colse to $2\pi/\mathrm{IM}\ (\lambda_{\,2,3}) = 2\ \pi/\left(b\sqrt{3}\right)$.

In Exercise 8 you are asked to generalize this derivation to negative feedback loops with an aribitrary number n of components. You will find that the steady state is unstable when $b\xi < (p - p_{min})$ / where p min. = $\sec^n(\pi/n)$. Notice that $p_{min} \rightarrow 1^+$ as $n\ \infty$ *i, e,* the minimum cooperativity of endproduct repression required for oscillations becomes the minimum coopertativity of endproduct repression required for oscillations becomes small as the length of the feedback loop increases.

The analysis of Hopf bifurcations in Goodwin's model uncovers a number of problems with his negative-feeback mechanism for biochemical oscillations (Griffith 1968a). In a three variable systme (m RNA, protein, end product,), the cooperativity off feedback must be very high, $p >$ 8. Also, it is necessary, in this case, for the degradation rate constants of the three components to be nearly equal. If not, p_{min} increases dramatically; *e. g.*, if one of the k_i s is tenfold larger than the other two, then $p_{\mathrm{min}} = 24$. The value of p_{min} can be reduced by lengthening the loop, but one must still ensure that the k_i's are nearly equal.

Bliss, Painter and Marr (Bliss et al. 1982) fixed these problems by a slight modification of Goodwin's equations:

$$\frac{dx_1}{dt} = \frac{a}{1+x_3} - b_1 x_1,$$

$$\frac{dx_2}{dt} = b_1 x_1 - b_2 x_2,$$

$$\frac{dx3}{dt} = b_2 x_2 - \frac{cx_3}{\mathrm{K}+x_3}.$$

Notice that the feedback step is no longer cooperative (p = 1), and the uptake of end product is now a Michaelis-Menten function. The steady state of this system is $x^*_1 = a\ (b\ (1 + \xi))\ x^*_2 = a/\ (b_2\ (1 + \xi), x^*_3 = \xi$ where ξ is the unique real positive root of $a/\ (1 + \xi) = c\xi/\ (K + \xi)$. The stability of this steady state is determined by the roots of the characteristic equation $(b_1 + \lambda)\ (b_2 + \lambda)\ (\beta + \lambda) + b_1 + b_2\ \phi = 0$, where $\beta = cK/\ (K + \xi)^2$ and $= a/(1 + \xi)^2$.

The characteristic equation is hard to solve in this completely general case. In order to get a start on it, we make some simplifying assumptions. First, suppose that K = 1, so $\xi = a/c$. Next, suppose $b_1 = b_2 < c$, and choose $a = c\left(\sqrt{c/b_1}-1\right)$. so that $\beta = b_1$ as well. In this case, $\phi = b_1\ a/c$, and the characteristic equation becomes

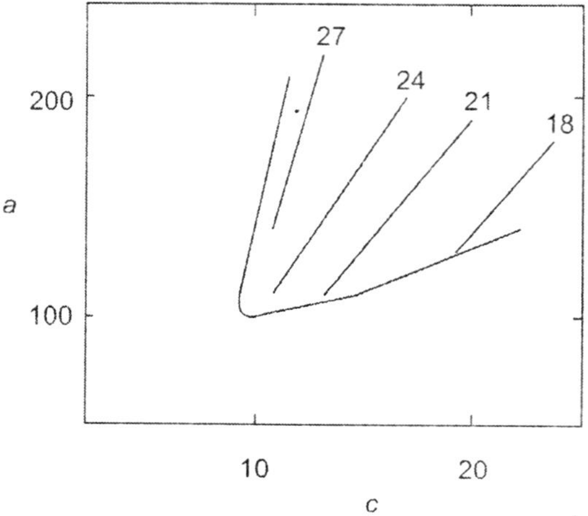

Fig. 9.8. Locus of Hopf bifurcations in the Bliss Painter-Marr equations given in system, for $b_1 = b_2 = 0.2$ We also plot loci of constant period within the region of limit cycle oscillations.

$(\lambda + b)^3 + b^3_1\ a/c = 0$. The solutions of this characteristic equation are $\lambda_1 = -\ b_1\left(1+\sqrt[3]{a/c}\right)$, $\lambda_{2,3} = -b_1 + b_1\sqrt[3]{a/c}\ \left[\cos(\pi/3) \pm i\sin(\pi/3)\right]$. The dynamical system has a Hopf bifurcation when Re $(\lambda_{2,3}) = 0$, *i, e,*. when $a = 8c$. Hence, at the Hopf bifurcation, c, 81 b_1 and $a = 8c = 648b_1$. If we set $b_1 = 0.2$, then the Hopf bifurcation occurs at $c = 16.2$, *an* 129.6 At this Hipf bifurcation the period of oscillation is colse to $2\pi/1$ m $(r)\ \left(b_1\sqrt{3}\right) = 18$. Starting at this point, we can trace out the locus of Hopf bifurcation numerically as a and c vary at fixed b_1.

TIME DELAYED FEEDBACK

In Goodwin's equations and Bliss-Painter-Marr's modiffied version, we assumed implicitly that there are no time delays in the processes of transcription, translation, or end product repression. However, there are surely some delays in transcription and translation associated with mRNA and protein processing in the nucleus and cytoplasm, respectively. And there are also bound to be delays in the feedback term, because the end product must move into the necleus, bind with transciption factors, and interact with the "*upstream*" regulatory sites of the gene to affect its rate of transcription. If we lump all these delays together, we can write a delayed differential equation for negative feedback:

$$\frac{d[X]}{dt} = \frac{a}{1+(Z/K_m)^p} - b[X]. \quad ...(9.18)$$

where Z (t) is a functional of the past history of [X] (t) For a discrete time lag,

$$Z(t) + [X](t - \tau), \quad ...(9.19)$$

with τ = constant. For a distributed time lag,

$$Z(t) = \int_{-\infty}^{t} [X](S)G_c^n(t-s)ds, \text{ with } G_c^n(S) = \frac{c^{n+1}}{n!}s^n e^{-cs}. \quad ...(9.20)$$

The kernel $G_c^n(s)$ is plotted you are asked to show that $G_c^n(s)$ has a maximum at $s = n/c$. As n and c increase, with n/c fixed, the kernel approaches a delta function, and the distributed time lag approaches the discrete time lage with $\tau = n/c$. In the same exercise, you are asked to prove that d/ds $G_c^n(s) = c\left[G_c^{n-1}(s) - G_c^n(s)\right]$, a fact that we shall presently put to good use.

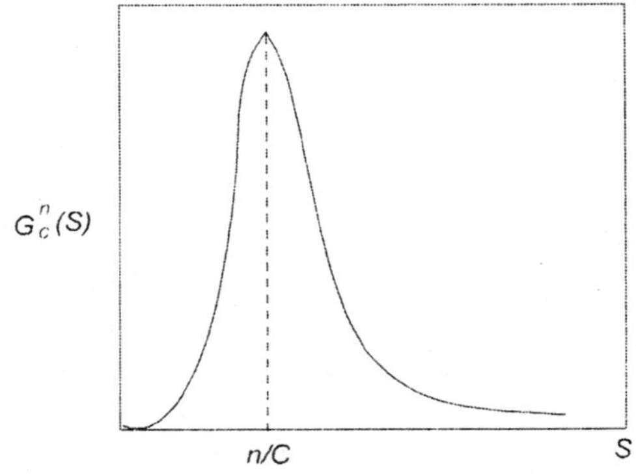

Fig. 9.9. The shape of the kernel $G_c^n(s)$ for distributed time lage (9.20).

Linear Chain Trick

Let us introduce a family of Z'_i's,

$$Z_i(t) = \int_{-\infty}^{t} [X](s)\, G_c^j(t-s)\,ds,\ j=0,1,\ldots\ldots n,$$

Notice that $Z_n(t)$ is just the functional Z (t) in (9.20). Differentiating the definite intergral, we find that

$$\frac{dZ_i}{dt} = [X](t)\, G_c^j(0) + \int_{-\infty}^{t} [X](s) \frac{d}{dt} G_c^j(t-s)\,ds,\ j=0,1,\ldots\ldots,n.$$

Since $G_c^j(0) = c$ and $G_c^j(0) = 0$ for $j = 1\ldots.n$, we see that

$$\frac{dZ_0}{dt} = c[X](t) + \int_{-\infty c}^{t} [X](s)\left[-G_c^0(t-s)\right]ds,$$

$$\frac{dZ_j}{dt} = \int_{-\infty}^{t} [X](s)\left[C - G_c^{j-1}(t-s) - cG_c^j(t-s)\right]ds,\ j=1,\ldots n.$$

Hence, (9.18) with Z defined in (9.20) can be written equivalently as a set of ODEs:

$$\frac{d[X]}{dt} = \frac{a}{1+(Z_n/K_m)^p} - b[X],$$

$$\frac{dZ_0}{dt} = c[X - Z_0],$$

$$\frac{dZ_j}{dt} = c(Z_{j-1} - Z_j),\ j = 1,2,\ldots,n.$$

That is, the distributed time-dealy model, with kernel $G_c^n(s)$, is identical to a Goodwin negative feedback loop of length $n + 2$. To see this, scale [X], $Z_{0,\ldots\ldots,}$ Z_n by K_m and t by K_m/a to put these equations in to the classical, dimensionless form of Goodwin's equations, (9.18). For simplicity, let $c = b$. Then according to the results of Exercise 8, the loop has a Hopf bifurcation at $(bK_m)/a = (p - p_{min}) / (\xi p)$, where $p_{min} = [\sec(\pi / (n+2))]^{n+2}$ and the dimensionless number ξ is the unique real positive root of $\xi^{p+1} + - a/(bK_m) = 0$.

As an example, suppose $n = 6$ and $p = 4$. In this case, P min = 1.884 and the Hopf bifurcation occurs at $0.529 = (bK_m/a)\xi = 1/(1+\xi^4)$, or $\xi = 0.9714$. Hence, the critical value of b is $b_{crit} = 0.5446a/k_m$, and oscillations occur for $b < bc_{rit}$. The period of oscillation close to the Hopf bifurcation is $2\pi/\left(b_{crit}\sqrt{3}\right) \approx 6.66 K_m/a$.

Time Lag

For the case of discrete time lag, we must solve the delay-differential equation

$$\frac{dx}{dt} = \frac{1}{1+x(t-\tau)^p} - bx. \qquad ...(9.21)$$

where x and t have been scaled to eliminate the parameters a and K_m from (9.18). (To do this, let $x = [X]/K_m$, dimensionless time $= at/K_m$, and dimensionless "b" $= b\ Km/a$.) Then (9.21) has a steady-state solution x^* satisfying $x^{p+1} + x - b^{-1} = 0$. To investigate the stability of the seeady state, we use Taylor's theorme to expand (9.21) in terms of small deviations from the steady state, $y, (t) = x(t) - x^*$,

$$\frac{dy}{dt} = -\phi y(t-\tau) - by(t) + \text{higher-order terms},$$

$$\phi = \frac{p\left(x^*\right)^{p-1}}{\left(1+\left(x^*\right)^p\right)^2} = pb\left(1-bx^*\right) = \frac{pb}{1+\left(x^*\right)^{-p}},$$

Looking for solutions of the form $y(t) = yoe^{\lambda t}$, we find that λ must satisfy the characteristic equation $\lambda + b = -\phi e^{-\lambda\tau}$. At a Hopf bifurcation, the eigenvalue λ must be purely imaginary; $\lambda = \pm i\omega$. Thus, for (9.21) to exhibit periodic solutions at a Hopf bifurcation, we must insist that

$$b = -\phi\cos(\omega\tau),\ \omega\ \phi\sin(\omega\tau) \qquad ...(9.22)$$

From these equation we can determine the oscillatory frequency (ω) and critical time delay (τ_{crit}) at the onest of limit cycle oscillations:

$$\omega\sqrt{\phi^2 - b^2} = b\sqrt{\left[p/\left(1+\left(x^*\right)^{-p}\right)\right]^2 - 1},$$

$$\tau_{crit} = \frac{\cos^{-1}\left(-\left[1+\left(x^*\right)^{-p}\right]/p\right)}{b\sqrt{\left[p/\left(1+\left(x^*\right)^{-p}\right)\right]^2 - 1}}. \qquad (9.23)$$

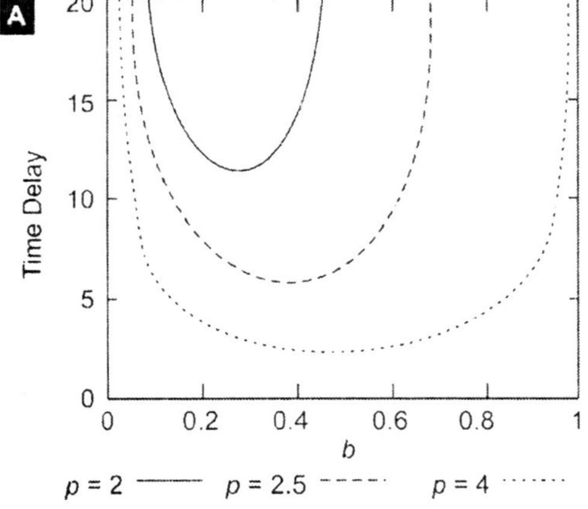

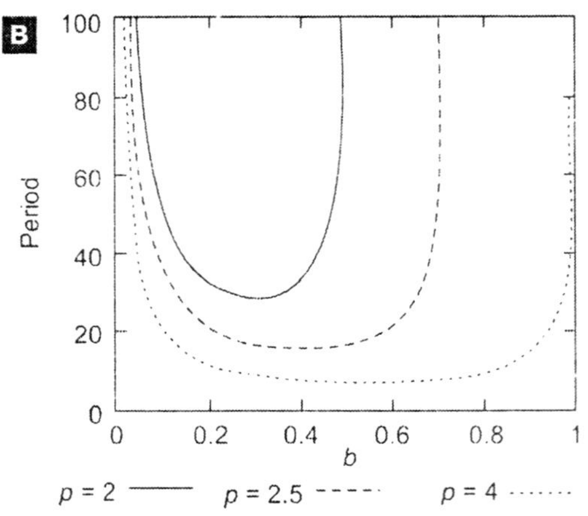

Fig. 9.10. Loci of Hopf bifurcations in 2.5 (dashed line) (9.21). We plot τ_{crit} and period $(2\pi/\omega)$ as functions of b, for three values of p : 2 (solid line), and 4 (dotted line). Oscillatory solutions lie within the U-shaped domains. Notice that for $p = 4$, $b = 0.5$, as calculted in the text, $\tau_{crit} = 2.418$ and period = 7.255.

Because the domain of $\cos^{-1}$ is [–1, 1], a necessary condition for Hopf bifurcation is $1 + (x^*)^{-p} < p$. For example, if $p = 4$, then x^* must be $> (1/3)^{1/4} \approx 0.760$, which implies that $b = (x^{p+1} + x)^{-1} < 3^{5/4}/4 \approx 0.987$. **IF** $b = 1/2$, then $x^* = 1$ and $\omega = \sqrt{3/2}$. Because $\cos(\omega\tau) = -b/\phi = -1/2$, $\tau = 4\pi = \sqrt{3/9}$. Hence, for $b = 1/2$ and $p = 4$, small- amplitude oscillations, with period $= 2\pi/\omega \approx 7.255$, bifurcate from the steady state as the time delay increases beyond 4.418. In Figure 9.10 we show how the charcteristics of these Hopf bifurcations (τ_{crit} and period) 2.418 depend on b, for fixed values of p.

CIRCADIAN RHYTHMS

Everyone is familar with his or her own 24-hour sleep-wake cycle. Many other aspects of human physicology also exhibit daily rhythms, including body temperature, urine production, hormone secretion, and skin cell division. Such rhythms are observed in all kinds of plants, animals, and fungi, as well as unicellular organism, and even cyanobacteria. Because these rhythms persist in the absence of external cues (light intensity, temperature, etc.), they reflect an endogenous oscillator within cells that runs at a period colse to 24 h.

Biologists have long been puzzled by the molecular basic of circadian rhythms. Although a fundamental breakthrough was made by Konopka and Benzer in 1971, with their discovery of the *per* gene in *Drosophila* (mutations of which alter the endogenous circadian rhythm of affected files), it was 25 years before the molecular details of the circadian oscillator began to become clear. We know now that PER protein inhibits transcription of the *Per gene,* through a complicated process involving phosphorylation by DBT kinase, binding to TIM subunits transport into the nucleus, and interaction with the transcription factors (CLK and CYC).

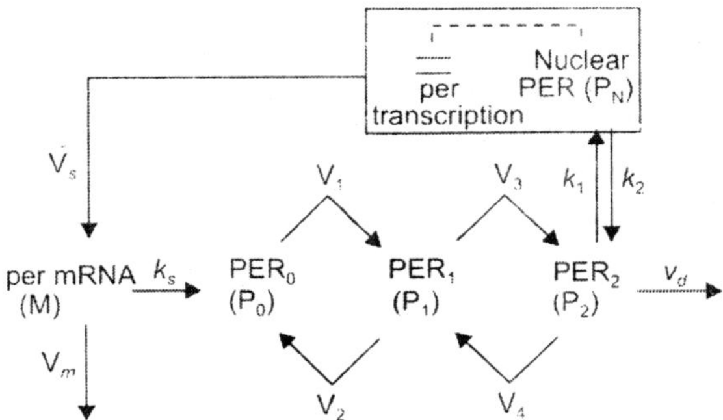

Fig. 9.11. Goldbeter's model of circadian rhythms. PER protein is synthesized in the cytoplasm, where it is successively phosphorylated, as indicated by subscripts. The doubly phosphorylated form enters the nucleus and represses transcription of the *per* gene.

It is clear to all that the control system is dominated by a time-delayed negative-feedback loop, quit in principle to Goodwin's original negative feedback oscillator. Numerous theoreticians have exploited the intersting nonliner dynamics of delayed negative feedback in order to model certain charcteristics of circadian rhythms. Ruoff and Rensing have explored the capabilities of Goodwin's Equations (9.18), with $p = 9$, to acount for temperature compensation, entrainment, and phase resetting.

Goldbeter proposed a more complicated model, based loosely on Goodwin's idea, supplemented with reversible phosphorylation steps and nuclear transport; The kinetic equations describing this mechanism are

$$\frac{dM}{dt}=\frac{v_s}{1+(P_N/K_1)^4}-\frac{v_m M}{K_{m1}+M},$$

$$\frac{dp_0}{dt}=k_s M-\frac{V_1 P_0}{K_1+P_0}+\frac{V_2 P_1}{K_2+P_1},$$

$$\frac{dp_1}{dt}=\frac{V_1 P_0}{K_1+P_0}-\frac{V_2 P_1}{K_2+P_1}-\frac{V_3 P_1}{K_3+P_1}+\frac{V_4 P_2}{K_4+P_2},$$

$$\frac{dp_2}{dt}=\frac{V_3 P_1}{K_3+P_1}-\frac{V_4 P_2}{K_4+P_2}-k_1 p_2+k_2 P_N-\frac{V_d/P_2}{K_4+P_2},$$

$$\frac{dp_1}{dt}=\frac{V_1 P_0}{K_1+P_0}-\frac{V_2 P_1}{K_2+P_1}-\frac{V_3 P_1}{K_3+P_1}+\frac{V_4 P_2}{K_4+P_2},$$

$$\frac{dp_N}{dt}=k_1 P_2-P_N.$$

The basal parameter values are:

$V_s = 0.76\ \mu\ M/h\ v_m = 0.65\ \mu\ M/h,\ v_d = 0.95\ \mu M/h,$

$k_s = 0.38\ h^{-1},\ k_1 = 0.9h^{-1},\ k_2 = 1.3h^{-1},$

$V_1 = 3.2\ \mu\ M/h\ V_2 = 1.58\ \mu\ M/h,\ V_3 = 5\mu M/h,\ V_4\ 2.5\ \mu\ M/h,$

$K_1 = K_2 = K_2 = K_4 = 2\ \mu\ M/K_1 = 1\ \mu\ M,\ Km_1 = 0.5\ \mu\ M, K_d = 0.2\ \mu\ M.$

Figure 9.12 A shows a numerical simulation of this system of ODEs, with a period close to $24h$. Figure 9.12 B shows how the period of oscillation depends on v_d, the degradation rate of PER.Goldbeter suggested that the short-period mutant of *per* (*per*^{s} has an autonomous period of 19 *h*) encodes a more stable form of PER, and the long period mutant (*per*^{L} has an autonomous period of 28 h) encodes a less stable form of PER. In subsequnt papers, Lelopu and Goldbeter have studied temperature compensation and phase resetting in this model.

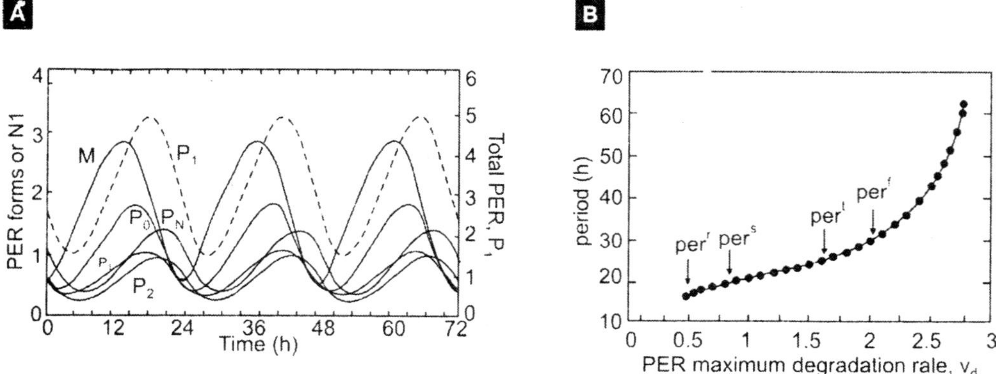

Fig. 9.12. Simulations of Goldbeter's model. (A) mRNA and protein concentrations as function of time. (B) Period of oscillation as a function of V_d.

Perhaps the simplest model of circadian rhythms is negative feedback with discrete time dealy described by (9.18) and (9.19), which has been explored in a recent paper by Lema, Golombek, and Echave. In this case, X is PER, a = maximum rate of PER synthesis, b = rate constant for PER degradation, K_m = PER concentration at half-maximal synthesis rate, and τ = time dealy (because the rate of synthesis of PER at the persent moment depends on its cytoplasmic concentration some time in the past). (Lema et al. chose $a = 1\ h^{-1}$, $b = 0.4\ h^{-1}$, $K_m = 0.04$, $\iota = 8$ h, and $p = 2.5$.). In section 9.5.2 we showed how to do stability analysis of the steady state of this dealy-differential equation, in order to find the points of Hopf bifurcation to periodic solutions. Letting x^* be the be the unieuq real positive root of $x^{p+1} + x - [a/(bK_m)] = 0$, we find that periodic solutions exist for $\tau > (K_m/a)\tau_{crit}$ With τ_{crit} (dimensionless) given by (9.23). The period of oscillation of the bifurcating solutions is close to $2\pi/\omega$, where $\omega\tau = \cos -1 (-[1 + (x^*)^{-p}]\,p)$. For the parameter values chosen by Lema et al, $a/(bK_m) = 62.5$, $x^* = 3.21, \omega = 0.0244$, and $\tau_{crit} = 58.1$. Hence, for $\tau > 2.33\ h$, their model oscillates with period close to $2\pi/\omega \approx \pi\tau \approx 7.32\ h$. To get a period close to 24 h they chose $\tau = 8\ h$. In their paper Lema et al. explored entratiment by phase-resetting in response to light pluses, assuming that light interacts with PER dynamics at either the synthesis (s) or degradation (b) step.

Tyson and coworkers have taken a different approach, noting that phosphorylation of PER by DBT induces rapid degradation of PER, but multimers of PER and TIM are not readily phosphorlated by DBT; see Figure 9.13. They describe this mechanism by three ODEs

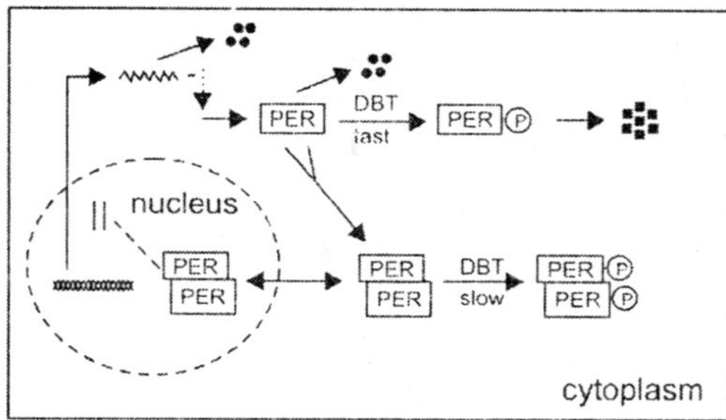

Fig. 9.13. An alternative model of circadian rhythms. The circadian protein PER is synthesized in the cytoplasm, where it can form homodimers, PER:PER. (Multimeric protein complexes with TIM are neglected in this simple model.) PER monomers aer rapidly phosphorylated by a protein kinase called DBT, and phosphorylated PER is rapidly degraded. PER dimers are only slowly phosphorylated by DBT.

$$\frac{dM}{dt} = \frac{v_m}{1+(P_2/A)^2} - k_m M,$$

$$\frac{dp_1}{dt} = v_p M - \frac{k_1 p_1}{J + P_1 + 2P_2} - k_3 P_1 - 2k_a P_1^2 + 2k_d P_2 + \frac{2k_2 P_2}{J + P_1 + 2P_2} + 2k_2 P_2.$$

$$\frac{dP_2}{dt} = k_a P_1^2 - k_d P_2 - \frac{2k_2 P_2}{J + P_1 + 2P_2} - 2k_3 P_2.$$

where

M = [*per m* RNA], P_1 = [PER monomer], P_2 = [PER dimer],

v_m, v_p = rate constants for synthesis of m RNA and protein,

k_m, k_3 = rate constants for non-specific degradation of mRNA and protein,

k_1, k_2 = rate constants for phosphorylation of monomes and dimer.

k_a, k_d = rate constants for association and dissociation of dimer,

J, A = Michaelis constants for the binding of PER to phosphatse and transcriptional regulation factors.

If the dimerization reaction is in rapid equilibrium, then $P_1 = qp_T$ and $P_2 = ((1 - q/2)\ P_T$, where $P_T = P_1 + 2P_2$ = [total protein], and

$$q \frac{2}{1+\sqrt{1+8KP_t},}$$

where $K = ka/kd$ = the equilibrium binding constant for dimer formation. In this case the system of three ODEs reduces to a pair of ODEs:

$$\frac{dM}{dt} = \frac{v_m}{1+(P_2/A)^2} - k_m M,$$

$$\frac{dp_T}{dt} = v_p M - \frac{k_1 P_1 + 2k_2 P_2}{J + P_T} - k_3 P_T.$$

with P_1 and P_2 given as functions of P_T by the equations above. Figure 9.14 A shows a phase plane portrait for this system, with limit cycle oscillations of period close to 24 h. Figure 9.14 B shows how these oscillations depend on k_1, the rate constant for phosphorylation of PER monomers byDBT, and K, the equilibrium binding constant for dimer formation.

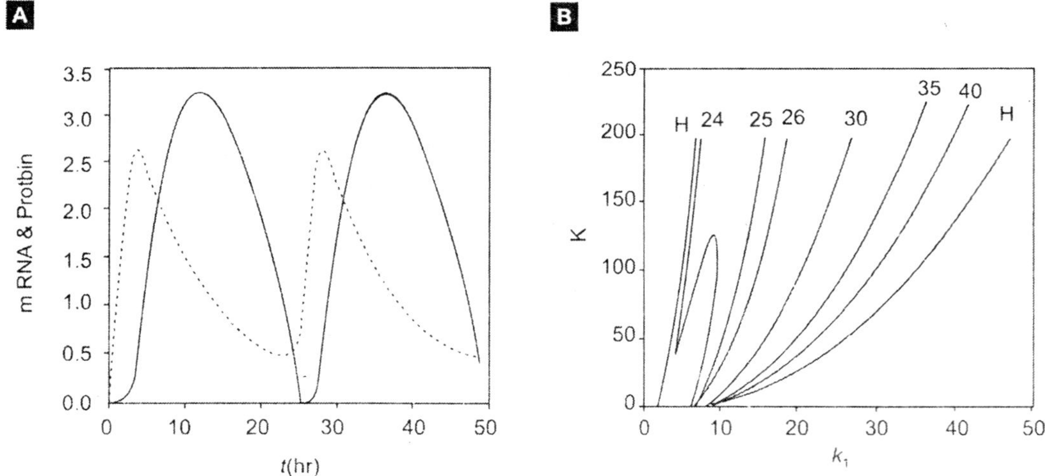

Fig. 9.14. Simulations of the model of Tyson et al. (1999). Basal parameter values: $v_m = 1$, $v_p = 0.5$ $k_m = k_3 = 0.1$, $k_1 = 10$, $k_2 = 0.03$ J= 0.05, K= 200, A = 0.1 (A) mRNA and protein concentrations as functions of time. (B) Locus of Hopf bifurcation (H) in dependence on k_1 and K. Also indicated are loci of limit cycles of constant period (24 h, 25h, ..., 40h).

CHAPTER

10 BUILDING BLOCKS

Secondary structure describes regular features of the main chain of protein molecules. Experimental investigations on polypeptides and small proteins suggest that secondary structure can form in insolation, implying the possibility of identifying rules for its computational prediction. Predicting secondary structure from the amino acid sequence alone is an important step toward our understanding of protein structure and function. It may provide a starting point for tertiary structure modeling, especially in the absence of a suitable homologues template structure, reducing the search space in simulation of protein folding. Advances in fold recognition (or threading) techniques have been where secondary structure predictions and other protein characteristics were combined to suggest resemblance to a known fold. The predictions can also be used in various aspects of molecular biology research to provide clues about the functional properties of proteins under analysis. The goal in secondary structure prediction approaches is to extract the maximum information from the primary sequence in the absence of a tertiary structure model. In this chapter we will deal with the description of secondary structure characteristics, assignment of secondary structure using known 3D coordinates, and predictions of secondary structure based on primary sequence.

GLOBULAR PROTEINS

The term *"secondary structure"* was first introduced by Linderstrom-Lang more than 40 years ago to describe regular structural patterns of polypeptide backbone independent of the types of amino acid side chains. There are three major types of secondary structure: α helix, β structure and β–turn.

The occurrence of different kinds of secondary structure can be see in the Ram-achandran plot. In figure 10.1, one can see the distribution of φ and ψ angles for a total of 9,156 amino acid residues from 4,413 protein chains, based on crystallographic data. There are two areas where the density of points is high: (1) around $\varphi = -60^o$ and $\psi = 60^o$, which corresponds to the α helix; (2) and around $\varphi = -90^o$ and $\psi = 120^o$, which corresponds to the β-structure.

Figure 10.2 shows the Ramachandran plots for valine (a) and leucine (c) residues calculated for five thousand randomly chosen amino acids. The highly occupied areas of these plots have a good correspondence with low energy conformation of amino acid residues. All other redieues have very similar plots, with the exception of glycine and proline. Glycine has a much wider low energy aea becuuse it does not have a C^{α} atom, Proline has its side chain covalently bound to bcakbone amine; hence its φ angle is limited to the range of $= -60^o \pm 20^o$.

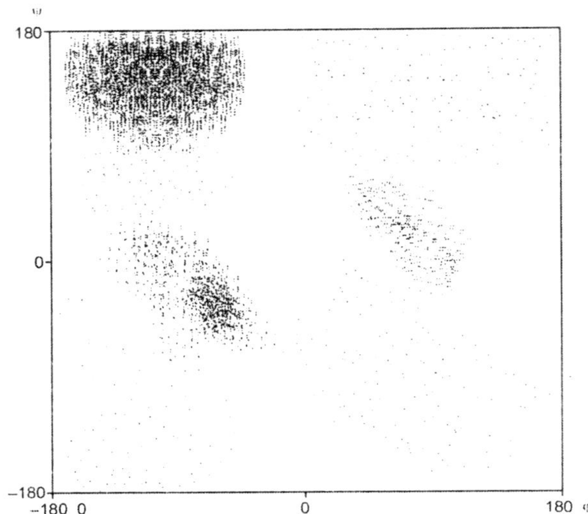

Fig. 10.1 Distribution of main chain dihedral angles for random sample of 9,156 amino acids from 4,413 protein chains.

φ and ψ angles associated with low-energy areas on energy plots are the angles observed in major types of secondary structure: α helices and β– structures, The largest area corresponding to β-structures can be seen as divided into four parts due to the limit of representation of continuous map. Two separate areas (α_R and α_L) correspond to right-handed and left-handed α

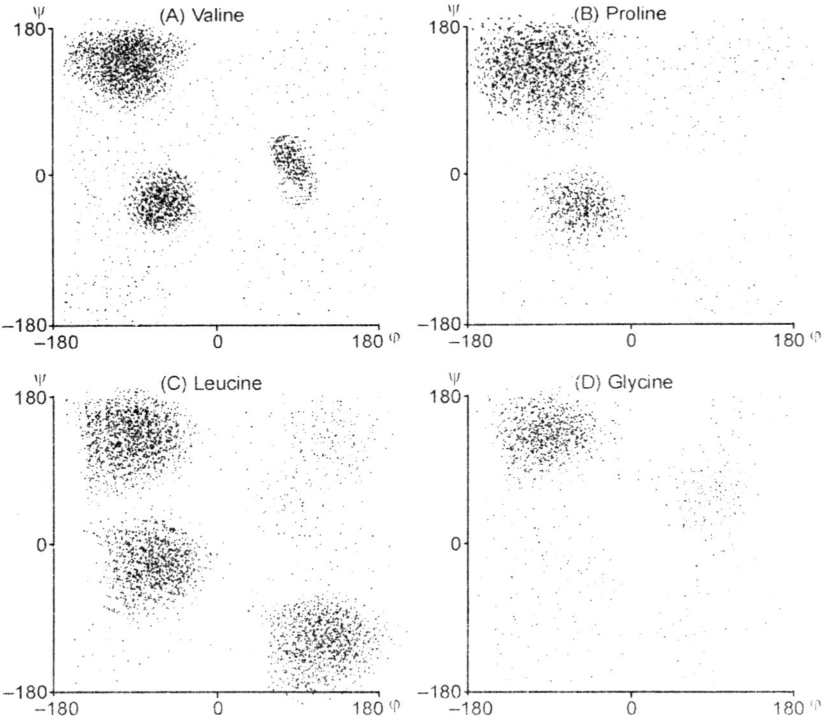

Fig. 10.2. Comparison of main chain dihedral angles distributions for (A) Valine' 9B) Proline; (C) Leucine' (D) Glycine.

Helices

Regular backbone conformations with the elements maintaining nearly identical orientations (the same φ and Ψ angles) can be described as helices, which can be described by the shift along the helices axis (d), the number of residues per turn (n), and the distance (r) from the specific residue location (*e.g.,* C^{α} atom) to the lelical axis.

Figure 10.4 shows the overall geometry and the schematic diagrams for 3_{10}-and α–helices. Protein helices are stabilized by hydrogen bonds between the amino and carhboxyl groups of the amino acid residue main chains: i, $i + 3$ (3_{10}-helix); i, $i + 4$ (α-helix); and i, $i + 5$ (α-helix).

The α-helix was first described by Pauling, who modeled a stable polypetide chain based on peptide unit geomerty. There are about 35 percent of amiono acids in the α-helical conformation, based on crystallographic data for 50 different proteins.

The average length of the α-helix is about 10-11 residues, which is apprximately 17Å, or three helical turns, The main chain angles in the α-helix are approximately $\varphi = \psi = -60^{o}$.

Some amino acid residues have small but distinguished preference to forming α–helical conformation. Ala, Glu, Leu, and Met are often found in α–helices, wheres Pro, Gly, Ser, Thr, and Val occur relatively rarely. Proline mainly occurs in the first turn of an α-helix because it can not donate a hydrogen bond in the middle of a helix, and it creates sterical problems in α-helical conformation.

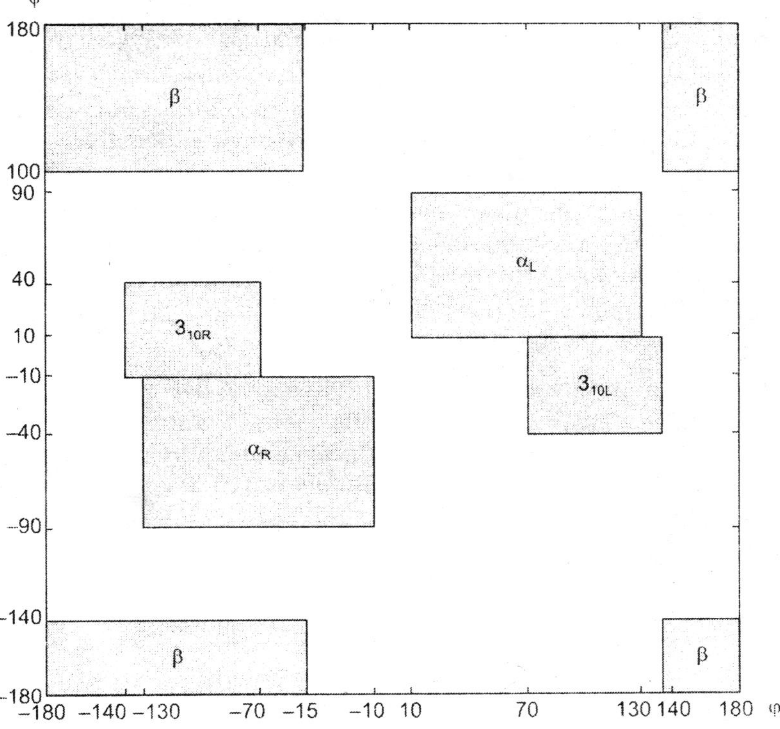

Fig. 10.3. Conformational states in proteins α_R and α_L right and left α–helices; 3_{10R} and 3_{10L}, right and left 3_{10} helix (or β–turn of type III); β,β-structural conformation (extended polyhpeptide chain).

In a α-regular helix, all dipoles formed by the N-H...O–C main chain groups point along the helical axis. it creates total nonzero dipole moment of the α–helix and partial charges of about one-half of the electron charge (positive charge at N-terminus of the α–helix and negative charge at the C-terminus).

In addition to local interactions within negihboring residues, the α-helix is stablized by the gain of hydrophobic energy when nonpolar side chains of amino acids are shielded from the solvent. According to Chothia (1976), when an α-helix is formed, the energy goes down by 2-3 kcal/mol per reside.

Most α-helices are immesed into protein interior from one side and form an exterior protein surface from the other side.Analysis has shown that nonpolar residues are usually located on one side of α-helix (forming a hydrophobic cluster) and polar and charged residues are on the other side. A typical cluster includes hydrophobic amino acids in positions i, $i \pm 1$, $i \pm 3$, $i \pm 4$,... In the folded protein, the α-helices are additionally stabilized by interacting with their hydrophobic clustes, thus forming hydrophobic core of the protein globule.

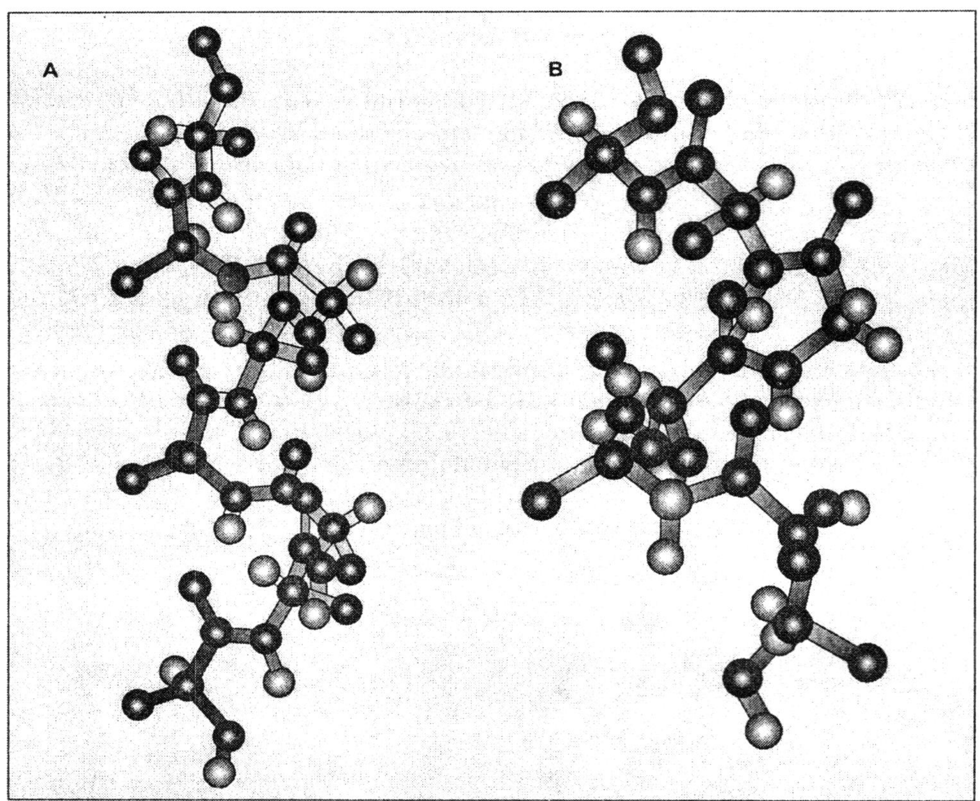

Fig. 10.4. Helices with internal hydrogen bonds in proteins. (A) 310-helix; (B) α-helix.

In addition to the α-helix there is a 3_{10}-helical conformation that is relatively common in proteins. The 3_{10}-helix contains three residues and 10 main chain atoms per turn φ–and ψ angles in the 3_{10}-helix equal approximately–60° and –30°, respectively. Long 3_{10}-helices are energetically unfavorable, compared to α–helixes; hence the majority of such helices contain no more than two turns. The 3_{10}-helix is often located at the C-terminus of an α-helix. Dipoles formed with hydrogen bonds in 3_{10} -helices are not parallel to the helical axis because the energy minimum is not reached. Furthermore, the side chains are placed into a sterically unfavorable conformation, where the residues i and i + 3 are not on the same line along the helical axis.

Several left α–helices have been found in proteins. The longest one cccurs in alcohol-dehydrogenase and has three hydrogen bonds. A left α-helix containing six amino acids has been found in thermolysine (Kabsch and Sander 1983)

β-Structure

The second common type of ordered polypetide conformation (after the α-helix) is the -β structure. About 36 percent of amino acid residues in globular proteins are in–β structural state.

The φ and ψ main angles of the β–structure are spread widely in the upperleft corner of the Ramachandran plot. φ = ψ =180° corresponds to the allowed conformation on (φ, ψ)-map and represents the full extended conformation of the polypeptide chain. When looking along the polypetide framework, one can see that the neighbouring side chain groups are pointing to the opposite directions. However, such fully extended conformation is favorable for polyglycine only. In the presence of other amino acids, the φ and ψ angles are slightly different.

The flat layer of β-stands (β-sheet) was first described by Pauling et al. (1951) as one of the structure with the maximum hydrogen bonding between the C = O and N-H groups of the main chain. There are two possible mutual arrangements of β–strands in the φ-sheet with respect to polypeptide chain directipn: parallel and antiparallel.

Most β-sheets in proteins are not flat but have a left twist, if one looks along a single β-strand. Every β-strand can be considered a very extended left helix, which turns 60° per two resideues. The twist in β–structure allows for conformational stabilization, providing energetically favorable contacts between the side chains of negihbouring 3-strands and the optimal orientation of the hydrogen bonds. In addition, such β-sheets are capable of dense packing between them as wall as with α-helices

Parallel β-structures usually occur inside a portein. They are often surrounded by α-helics protecting them from the solvent on both sides. Parallel β-structures usually do not form layers with fewer than five β-strands. The φ-and ψ-angles in parallel β-strands are more regular than in antiparallel β-strands. This indicates that parallel β-sheets are less stable than the antiparalle. In β-strands, one side is typically exposed to solvent and the other side is embedded inside the protein. This results in characteristic interchange of hydrophobic and polar amino acids in antiparalle β-sheets. The area in a β-strand involved in H-bonding with another β-strand ranges from one to nine residues in antiparallel β-sheets, comprise less than 20 percent of cases. Main chain angles are distroted in the mixe β-sheets because some β-stands re involved in both parallel and antiparallel interactions on the opposite sides.

Amino acids Val, Ile, Tyr, and Thr have a preference for the β-strictiral conformation, whereas Glu, Gln, Lys, Asp, pro, and Cys are rare found in it In general, β-structures ae favored by bulky amino acids, which usually have limited Conformationl flexibility due to branching at the C^{β} atom a large aromatic group.

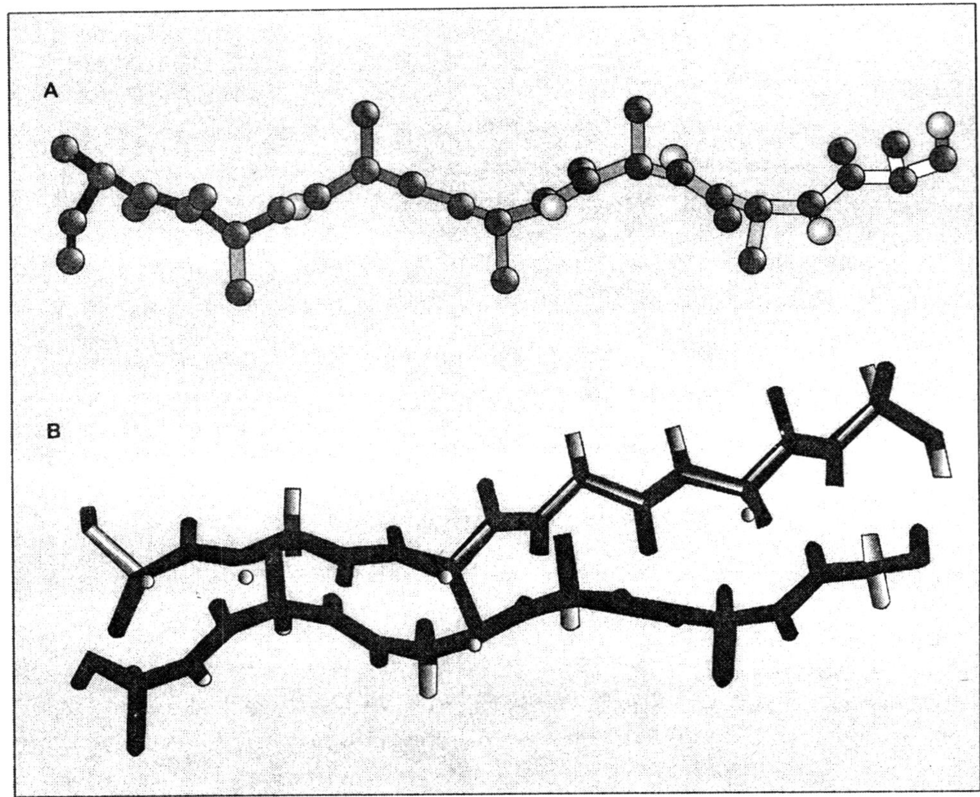

Fig. 10.5 . β-sttructual conformation. (A) β-strand geomerty; (B) interacting β-strands.

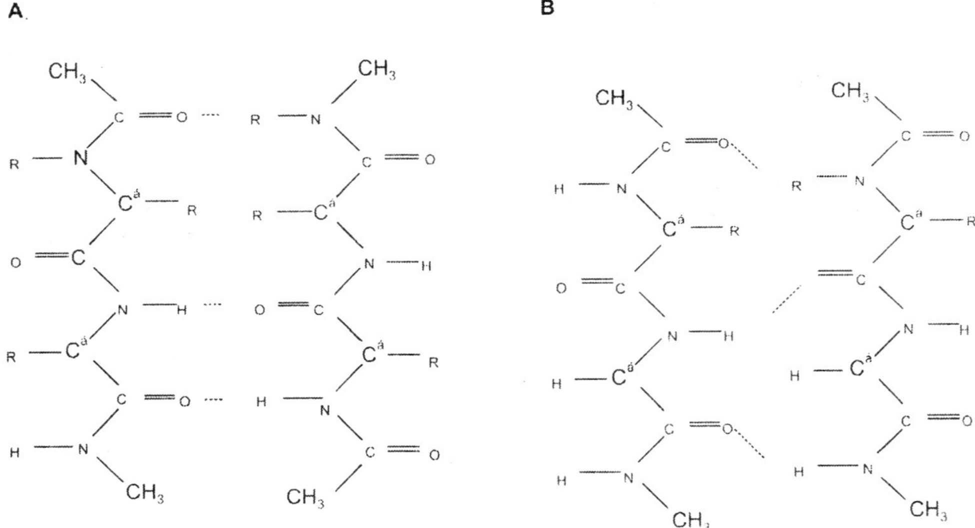

Fig. 10.6. Feactures of β-sttuctural conformaiton. IA) Antiparallel β-sheet; (b) Parallel β-steet.

β-turns

After the exclusion of α-helices and β-sttucturs, the rest of the polypeptide chain usually denotes as random coil or non-regular structure. Nevertheless, there are some fragements with recurring conformation within coiled structurs. The most frequent one is the β-burn, which account sor nearly 32 percent of all amino acid residues.

β-turn is a poolypeptide fragment comprised of four consecutive amino acid residues in a region where the polypeptide chain change direction roughly 180°. Conformation of the β-turn was first described in a theoretical study of three eptide units (or four neighboring residues) that are stabilized by a hydrogen bond between C=O of residue i and N-H of residues) $i + 3$. Two most stable conformations were revealed, corresponding to type I and type II β-turns, which represent 35 percent and 15 percent, respectively, of all β-turns. In addition, 15 percent of all turns are single turns of the 3_{10}-helix (type III β-turns), and 10 percent of all turns are mirror conformations of the type - I-III turns.

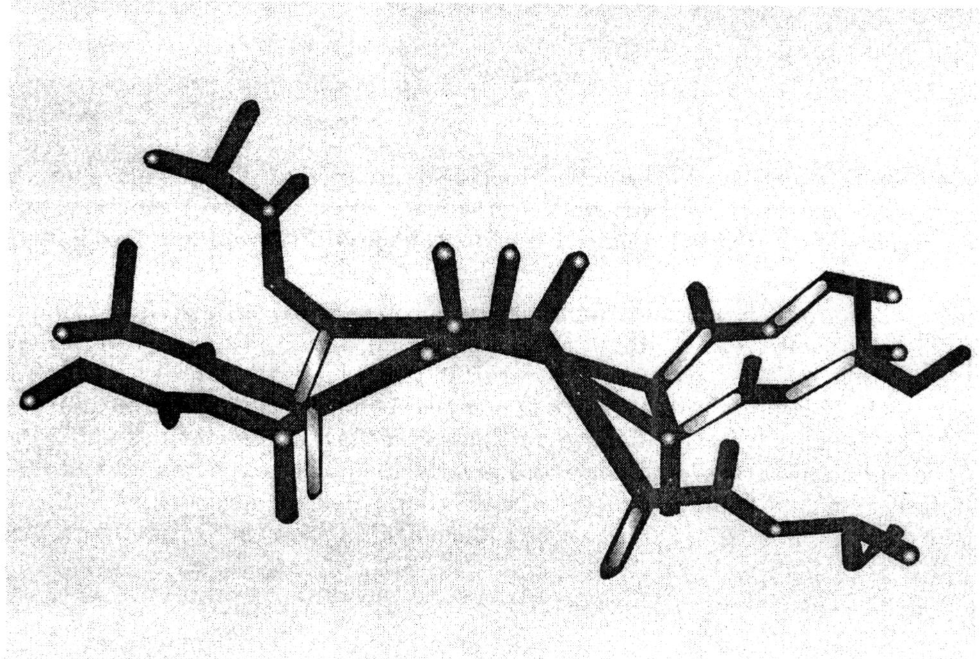

Fig. 10.7. Twist in β-sheets.

Non-standrad turns (or bends) containing no hydrogen bonds occur in 25 percent of cases. They can be recognized if the distance between the C^{β}- atoms of residues i and $i + 3$ is less than 7 Å and those residues are not the part of an α-helix.

β-turns are usually located on the protein surface and contain many polar and chargaed amino acid side chains. Most turns contain glycine in the seond or third postion, where the absence of a side chain in glycine is favorable for the interaction among main chain atoms. Proline often occurs at the second position of turns. About two-thirds of Pro-Gly and Pro-Asp sequences in porteins with known 3D structures are located in the two middle residues of β-turns. Many β-connect neighboring fragements of secondary sturctures (α-α-α-β-, and β-β). In these cases, the β-turn conformation can be stabilized by non-local interactions between the long secondary structurs.

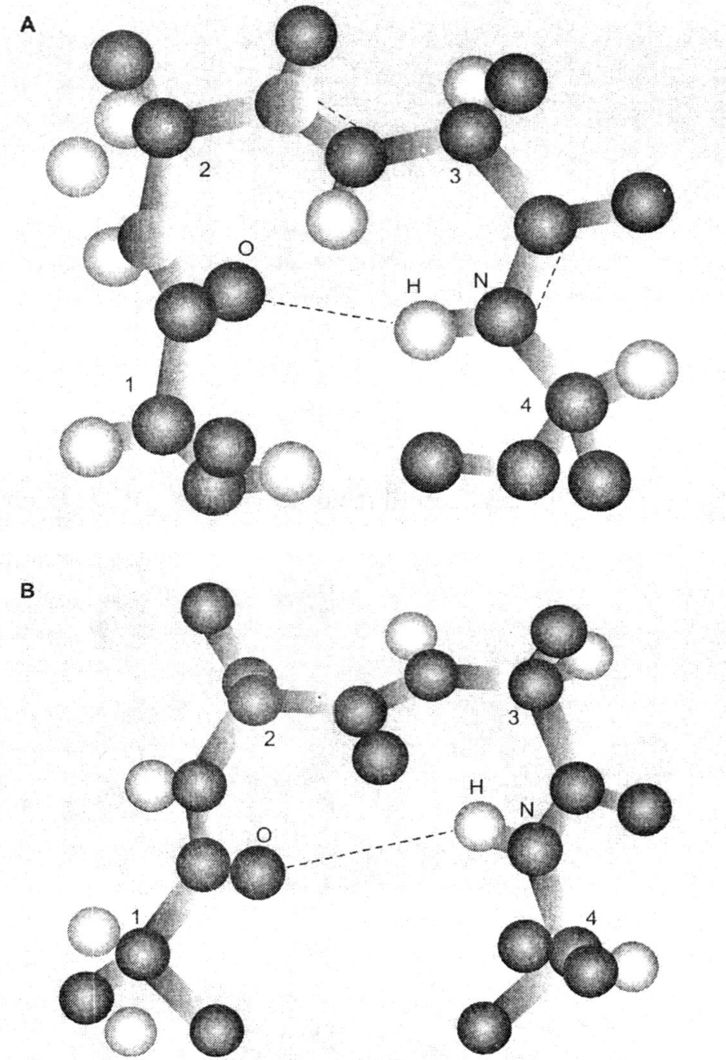

Fig. 10.8. Type I and type II of β-turns. (A) I β-turn in tenascin (PDB ID ITEN) at positions 815-818, length of H-bond 2.77. (b) Type IIβ-turn in mannose permease (PDB ID IPDO) at positions 84-87, length of H-bond 2.99 A.

3D STRUCTURES

Secondary sturcutre assignement can be defined as the detection of regions that belong to the various major types of secondary structure in polypeptide cahins with known 3D *structure.*Major criteria for the detection of secondary structure are the presence of a hydrogen bond between main chain CO- and NH-groups and local geometry. Sevceral approaches have been developed for secondary structure and assignment. We will describe in detail the algorithm proposed by Kabsch and Sander (1983), which is known as *DSSP* by the name of the databse where secondary structure assignments for all stuructues in the Protein Data Bank (PDB) are provided.

The DSSP algorithm takes a single decision parameter-the presence of absence of an H bond. H-bonding patterns are described an "n-turns," which are H-bonds between Co of residues i and NH of residue $i + n$, where $n =$ 3, 4, 5; and "*bridges*" which are H-bonds, between residues not near each other in sequence. In this defintion, α-helices can be described as repeating 4-turns and β-structure as repeating bridges. The results can be easily summarized and presented as a single character string corresponding to the amino acid sequence. it detects the most commonly used representation of a protein secondary structure.

An H-bond is defined based on an electrostatic model with energy calculated as follows:

$$E = q_1 q_2 \,(1/r\,(ON) + 1/r\,(CH) - 1/r\,(OH) - 1/r\,(CN)\,.f$$

with $q_1 = 0.42\,e$ and $q_2 = 0.20e$, where e is the unit electron charge, (r (AB) is the interatomic distance from A to B in Angstroms, $f = 332$ is the dimensional factor, and E is the energy in kcal/mol.

(13IL) Lysozyme_ (E.C.3.2.1.17)_Mutant

**** SECONDARY STRUCTURE DEFINTION BY THE PROGRAM DSSP. VERSION JUNE 1983 *****

#	RESIDUE	AA	STRUCTURE	BP1	BP2	ACC	TCO	KAPPA	ALPHA	PHI	PSI	X-CA	Y-CA
1	1	M		0	0	59	0.0000	0.0	360.0	360.0	150.6	43.4	–1.8
2	2	N	> –	0	0	68	–0.919	0.0	–82.9	–152.1	179.2	40.2	–0.8
3	3	I	H > S+	0	0	26	0.815	124.2	51.2	–58.6	–36.6	38.1	–2.3
4	4	F	H > S+	0	0	75	0.938	113.4	42.6	–68.9	–47.8	40.2	3.6
5	5	E	H > S+	0	0	94	0.840	144.0	54.1	–65.5	–36.1	43.4	3.3
6	6	M	H X S+	0	0	0	0.951	111.9	41.5	–63.1	–53.2	41.8	4.7
7	7	L	H X S+	0	0	0	0.863	109.1	609	–64.6	–31.6	40.5	7.8
8	8	R	H X S+	0	0	91	0.883	107.7	44.5	–63.4	–33.9	43.7	8.2
9	9	I	H < S+	0	0	87	0.915	116.6	46.4	–74.4	–40.4	45.6	8.5
10	10	D	H < S+	0	0	19	0.823	125.0	28.7	–71.1	–34.1	43.0	10.9
11	11	E	H < S–	0	0	50	0.722	89.4	–156.0	–98.3	–29.3	42.7	13.2
12	12	G	< –	0	0	22	–0.242	25.7	–86.5	75.0	–167.5	46.1	13.0
13	13	L	+	0	0	41	–0.690	40.6	174.3	–143.3	125.6	46.6	13.7
14	14	R	E –A	28	0A	146	–0.998	17.7	–160.5	–131.3	133.6	47.1	17.1
15	15	L	E S+	0	0	60	0.454	74.1	58.9	–96.5	–3.9	47.3	17.7
16	16	J	E S-C	57	0B	124	–0895	102.9	–81.4	–22.6	155.4	46.6	21.4
17	17	I	E +	0	0	29	–0.265	58.7	176.7	–49.7	135.1	43.7	23.2
18	18	Y	E –A	26	OA	23	–0.918	34.1	– 108.5	–139.6	162.0	44.1	23.7
19	19	K	E –A	25	OA	122	–0.802	38.5	–139.8	–91.1	132.0	42.2	25.2
20	20	D	> –	0	0	47	–0.094	37.7	–78.1	–78.5	–168.2	40.8	22.5
21	21	T	T 3 S+	0	0	115	0.770	136.6	45.8	–59.5	–27.7	40.7	22.4
22	22	E	T 3 S–	0	0	77	0.373	124.4	–105.6	–94.2	–0.5	37.6	24.8
23	23	G	S < S+	0	0	36	0.643	74.7	138.6	80.5	22.9	39.3	27.0
24	24	Y	–	0	0	75	–0.815	60.7	–97.4	–102.9	155.3	37.3	25.7
25	25	Y	E +AB	19	34A	32	–0.325	56.1	153.1	–65.6	126.1	38.6	25.0
26	26	S	E –AB	18	32A	2	-0.899	23.3	–158.7	–144.2	175.2	39.3	21.3
27	27	I	E > –B	0	31A	0	–0.990	51.6	–1.4	–155.4	164.6	41.4	91.1
28	28	G	E 4 S–A	14	0A	0	–0.344	120.8	–2.3	63.3	–126.3	42.8	15.5

Fig. 10.9. Sample of seconday structure assignment by DSP (Kabsch and Sander 1983).

If the energy is below-0.5 kcal/mol, then an H-bond is assigned. The choice of the energy cutoff takes into account weak or bifurcated H-bonds and errors in the 3D coordinates. Figure 10.10 sumarizes the types of turns and bridges detected and the corresponding H-bonding patterns. In the next step, turns and bridges are analyzed for cooperative H-bonding patterns.

Minimal helices are defined as two consecutive turns of the corresponding type (*e.g.* 3-turn for 3.helix). There are three types of helices defined: 3 Helix, 4-helix,and 5-helix, corresponding to 3_{10}, α-and π-helices, repectively. Minimal 4-helix of length 4(starting at residue i) requires two-4 turns at residues $i-1$ and i Longer helices are built from the overlapping minimal helices. in the summary, the helices are represented with letters G,H, and I for 3-, 4, and 5-helices, respectively, bracketed with H-bonds (shown as ">" and "<").

```
3–turn

      Notation        >       3       3       <
      Residues     –N–C–C– –N–C–C– –N–C–C– –N–C–C–
                    H   O   H   O   H   O   H   O
      H–Bond            \_______________/

4–turn

      Notation        >       4       4       4       <
      Residues     –N–C–C– –N–C–C– –N–C–C– –N–C–C– –N–C–C–
                    H   O   H   O   H   O   H   O   H   O
      H–Bond            \_______________________/

5–turn

      Notation        >       5       5       5       5       <
      Residues     –N–C–C– –N–C–C– –N–C–C– –N–C–C– –N–C–C–N–C–C–
                    H   O   H   O   H   O   H   O   H   O       O
      H–Bond            \_______________________/

Parallel bridge

      Notation                x
      Residues     –N–C–C– –N–C–C– –N–C–C–
                    H   O   H   O   H   O
      H–Bond              \/      \/
                          /\      /\
                    H   O   H   O   H   O
      Residues     –N–C–C– –N–C–C– –N–C–C–
      Notation                x

Antiparallel bridge

      Notation                X
      Residues     –N–C–C– –N–C–C– –N–C–C–
                    H   O   H   O   H   O
      H–Bond            |   |   |   |
                        |   |   |   |
                    O   H   O   H   O   H
      Residues     –C–C–N– –C–C–N– –C–C–N–
      Notation                X
```

Fig. 10.10. Hydrogen bonding patterns.

To describe a β-structure, two types of patterns are intorduce: *"ladder"* and *"sheet"*. A ladder is a set of one or more consecutive bridges of the same type. A sheet is one or more ladders connected by shared residues. Ladders and sheets are labeled alphabetically, parallel ladders by lower case, antiparallel by upper case, and sheets by upper case. Ladders of size 1 are labeled by "B," and longer ladders are labeled by "E" in the summary.

Additional rules were introduced to handle the secondary structuere irregularities. For helices, it follows implicitly form the definition that two overlapping minimal helices offest by two or three residues are joined into one helix. For a β-structure, it was explicitly defined that allowance for β-bulges is such that two perfect ladders or bridges can be connected through a gap of one residue in one strand and four residues in the other strand. In the summary, all bulge-linked ladders are labeled with "E".

In addtion to secondary structure, two geometrical features of polypeptide chains are also incorported into the summar: *"bend"* and *"chiraligy"*. The bend is assigned at sites where the polypeptide chain has an angle of more than 70° between line segments connecting C^{α}-atoms from residue i to $i-2$ and i to $i+2$. The chirality for residue i is defined as a sign of the dihedral angle built on C^{α} atoms from residues $i-1$, i $i+1$, $i+2$.

A similar approach to secondary structure assignment was intorduced by Frishman and Argos (1995) in the STRIDE algorithm, which extends the DSSP definition by adding information on backbone torsion angles. H-bonding criteria were also improved by incorporation of experimental information hydrogen bonds geometry and refining energy fuction. This resulted in better approximation to the assignment of secondary structure provided by experimentalists who determined the 3D structure. For 226 proteins in a test data set, STRIDE assignes 58 percent of the proteins closer to experimental assignment compared to DDSP, whereas DDSP assigns only 31 percent closer than STRIDE. For 11 percent of the proteins, both computational assignments are in the same proximity to the experimental ones.

COMPUTATIONS METHODS

Methods for the prediction of secondary structure from amino acid sequence have been developed for more than 30 years. They can be grouped into the following basic groups: (1) those that are based on stereo-chemical rules, (2) those that use a combination of rules and statistical parameters, (3) those that apply plysical models of secondary structure formation, (4) those that are based on stastistical information and information theory (5) those that analyze sequence patterns, (6) those that are based on discriminant analysis (7) neural-network approaches, (8) analyzing evolutionary conservation in multiple alignment (9) nearest-neighbor algorithsms and, (10) consensus prediction algorithms . There are also methods that combine characteristics and principles from several of these categories.

At the beginning of 1990, these methods had achieved about 65 percent accuracy in predicting secondary structure in three states; α-helices, β-strands, and coil segments. During the last decade, there has been much improvemet in methods based on discriminant analysis, neural network, and nearest-neighbor based approaches; algorithms such as PHD, NNSP , SSPAL, DSC, alignment based nearest-neighbor and PSI-PRED and its modifications have yielded an accuracy generallyranging from 70 to 80 percent. In the following sections, we will describe the major features of these methods and discuss their performance.

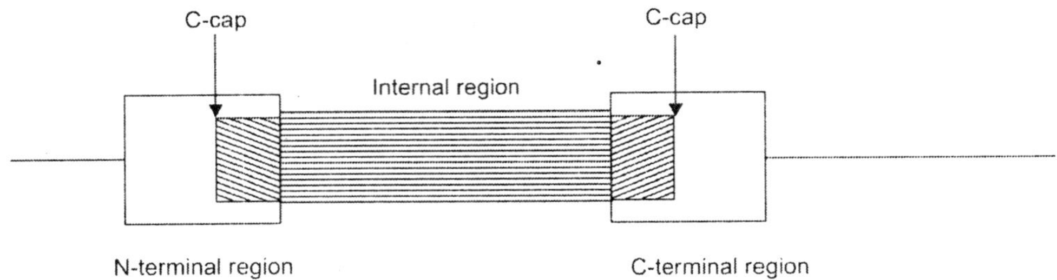

Fig. 10.11. A model of α-helical (β-strand) segments.

Discriminant-analysis

The definition of secondary structure is somewhat imprecise and different authors may arrive at different secondary structure assignments for the same protein by using different assignment algortihms. For the purpose of modeling tertiary structures, it is more important to predict accurately the location of entire α-helix and β-strand segments than to reach higher single residue accuracy. Taking this into account Solovyev and Salamov (1991, 1994) developed the Secondary Structure Preactdiction (SSP) program for the identification of whole α-helices and β-strands. They postulated that a particular conformationof protein segment can be determined by the combined action of three elements: N-terminal internal, and C-terminal regions, because the N-and C-terminal of α-helix or β-strand conformation was estimated on the basis of the corresponding values of linear discriminant functions (LDF), which combined threee characteristic of segmetns; average single-residue, a pair of residues preference parameters, and hydrophobic moment. The LDF were determined using linear discriminant analysis, described in detail in chapter 9 on a database of proteins of pairs of alternative classes: α-helical and β-strand/non-β-strand segments. Because the mechanism of formation of long and short secondary structures may differ during protein folding, and also to increase the discriminating power of LDF, Solovyev and Salamov separately examined the cases of long and short structures. Thus four pairs of alternative classes were considered for short and long α-helical and β-strand segments.

Characteristics Description

Only three simple characteristics were used to assign a given amino acid region to one of the secondary structure classes.

The *singleton characteristic* is an average of single-residue preferences. Using a database of known protein structures, we calculated preferences of all 20 types of amino acids to be in different part of α-helix and β-strand segments N-and C-terminal, internal part, adjacent N- and C-terminal regions). For any segment, the average preference for α-helix or β-strand was computed as the sum of the individual preferences of all residues in the segment (10.1):

$$S=\frac{1}{L}\left(\sum_{i=l_1-m}^{l_1-1} S_i^{N_l} + \sum_{i=l_1}^{l_1+m-1} S_i^{N} + \sum_{i=l_1+m}^{l_2-m} S_i^{in} + \sum_{i=l_2-m+1}^{l_2} S_i^{C} + \sum_{i=l_2+1}^{l_2+m} S_i^{C_r}\right) \quad ...(10.1)$$

Here $L = l_2 - l_1 + 2_m + 1$, l_1 and l_2 are the first and the last position of the segment; $m = 4$ or 3 for long α-helices and β-structures, respectively; m = Int $[(l_2 - l_1)/2]$for short segments; $S_i^{N_l}, S_i^{N}, S_i^{in}, S_i^{C}, S_i^{C_r}$ are the preference parameters of amino acid in position i that is adjacent to N-terminal, N-termianl, internal, C-terminal, or adjacent to the C-terminal part of the segment, respectively (Chou and Fasman 1978):

$$S^k(j) = \frac{P^k(j)}{P(j)} \qquad ...(10.2)$$

Here P (*j*) and P^k (*j*) are the fractions of amino acid of type *j* in the whole database and in the secondary structure of *k* type, respectively (*j* = 1, 20; *k* = N_1, N in C, C_r).

The *doublet characteristic* is simillar to the Chou-fasman singleton characteristic. The pair of residues can be separated by *k* (*k* may be 0, 1, 2, or 3) other residues. Preferences for a particular type of secondary structure *s* (α-helix. β-strand, or their N- or C-terminals of a pair of amino acids of type *i* and *j*, separated by *k* residues (pair of the *ijk* type), are defined as:

$$D^s_{ijk} = \frac{P^s_{ijk}}{P_{ijk}} \qquad ...(10.3)$$

Here P_{ijk} and P^s_{ijk} are the fractions of the *ijk* pair in the whole database and in a given secondary structures *s* , respectively. For N- nad C-terminal regions, only those pairs that have one residue inside and one outside of these secondary structure segments were taken into consideration. The average preference of a segment to be in the *s* conformation (α-helix or (β-strand) was calculated as the sum of all its possible pair characteristics osscurring in the N-terminal, internal segment, and C-terminal of the *s*.

Eisenberg *et al.* (1984) described the *hydrophohic moment characteristic* and periodicity in hydrophobicity of a fragment of amino acid sequences as

$$M = \left[\left(\sum_{k=0}^{l-1} h_k \cos(kw)\right)^2 + \left(\sum_{k=0}^{l-1} h_k \sin(kw)\right)^2\right]^{1/2} \qquad ...(10.4)$$

Here h_k, $k = 0, 1, ..., l$ is the value of the hydrophobicity *(Cornette et al. 1987) of the k-th* residue in the fragment with the length *l*, and *w* is an angle of the amino acid residue hydrophobic moment in the corresponding conformation of the polypeptide chain. The hydrophobic moments with a periodicity corresponding to α-helix (w= 100°) and β-strand (w = 180°) were taken as the characteristics for discriminating α-helix and β-strand segments, respectively.

The Mahalonobis distances, showing the significance of each characteristic and their combined value for each pair of alternative classes, are given in table 10.1. In all four cases (long and short α-helices and β-strands), we observe that the strongest characteristic is the average pair preference parameter, followed by the single preference parameter and hydrophobic moment. AS expected, long α-helix and β-strand segments were discriminated much better than the shorter ones.

SSP Algorithm of Secondary Structure Predication

To predict the secondary structure, a nucleus of potential α-helix is searched for as a region five residues long with an average single characteristic higher than a particular threshold. The value, *d,* of the LDF for short α-helices from non-helical regions), the next (p_{as} is a threshold distinguishing short α-helices is calculated for this segment. If the segment has $d < P_{as}$ (P_{as} is a threshold distinguishing short α-helices from non-helical regions), the next sequnce segment displaced to the right one position was examined. On condition $d > P_{as}$, the segment was expanded in both directions one residue at each step. The LDF value for a short or long α-helix for the corresponding segment was calculated depending on the segment length. A segment with the maximal d was considered a potential α-helix (the maximal extension on both sides was up to 15 positions). The search for other structures was continued after the C-terminus of the last selected helix. A similar procedure was performed for the potential β-strand segments. The result is a set of a potential β-helices and α-strands segments. In the final prediction, over lapping regions were assigned to the structures with the highest LDF value in the region of overlap. Non-overlapping edges of the segments with lower values of LDF were retained or discarded depending on their length (the minimum length for assigning an α-helix was five residues; for the β-strands, three residues).

Table 10.1. Mahalonobis's distances

Segment type	*D*	*S*	*M*	*Total D_2*
Long α-helices	3.97	2.31	1.77	5.19
Long β-strands	4.16	2.91	0.27	4.31
Short α-helices	2.10	1.16	0.98	2.76
Short β-strands	2.69	1.85	0.20	2.74

Significance, of various characteristics of secondary structure segments measured by Mahalonobis disctance D^2. D, doublet preferences; S, singleton preferences; M, hydrophobic moment.

Measures of Prediction Accuracy

The jackknife procedure was used for estimating the prediction accuracy of the method on 126 nonhomologous proteins, the secondary structure of which was assigned by the DSSP program. For assessing single-residue accuracy, several performance measures were used: the percentage of correct residue predications $Q_3 = 1/L\ (P_a + P_b + P_{coil})$. 100% and Sensitivity (Sn), Sepecificity (Sp), as well as Matthew's (1975) correlation coeffcient CCC). Measures for single-residue accuracy do not completely reflect the quality of a prediction. For example, the clear wrongly predicated structure αβαβαβ in the α -helix region would still assign correctly the conformational states of 50 percent of the residues. That is way in addition to single-residue accuracy, a measure reflecting the number of correctly predicted residues.

The prediction accuracy Q_3 for various combinations of the considerad characteristics is shown in table 10.2. Prediction using only a single characteristic gave an overall 58.2 percent prediction accuracy for three states. Therefore, the model used in this method is better than the standard Chou-Fasman method (about 50 percent accurate;), although both methods used similar parameters. Using only pair characteristics gave 62.2 percent accuracy, which is better than analogous characteristics used in the GOR III method, which have 53.5 percent accuracy. It can be seen that the addition of single characteristics only slightly increased overall prediction, from 64.8 percent to 65.1 percent. This can be explained by the high correlation between single and pair characteristics ($r = 0.80$).

Table 10.2. Single-redidue prediction accuracy Q_3 for various combinations of secondary structure characteristics

Characteristics	*Q3 (%)*
Singleton characteristic (S)	58.5
S + hydrophobic moment of the segment (M)	61.4
Doublet characteristic (D)	62.2
D + M	64.8
D + S + M	65.1

Assignment of secondary structure according to the three-dimensional coordinates performed by different methods may differ up to 21 percent. However, a detailed comparison of the secondary structure of proteins belonging to the same structural family showed that a per-reside comparison is not sufficient to assess the presence of segments in a three-dimensional structure of proteins.

Single-residue accuracy measures sometimes poorly reflect the actual predication of secondary structure. For example, assigning coil state to all amino acids in the protein 4sgb gave Q_3 = 76.7 percent, but this protein has several missed β-structures. Conversely, for the protein 3b5c, SSP correctly predicted four our of five real α-helices and three out of five real β-strands, although Q_3 was only 56.5 percent. A simple measure for assessing the quality of predicted secondary structure segment may be the average lengths of the predicated segmetns. The observed and predicted average lengths for α-helices are 10.6 and 10.7 amino acids, and for β-strand segments are 5.1 and 5.8 amino acids, respectively. The predicated values are very close to the real ones.

The segment predictions accuracy can be estimated by a simple measure comparable to that proposed by Taylor (1984): the structures is considered correctly predicated if it has at least two amino acids in common with the real one. It was observed that long segments were predicated much better than short ones: 89 percent of α-helices longer than eight residues and approximately 71 percent of β-strands longer than six residues were correctly located with specificity of correct prediction 0.82 and 0.78 respectively.

An example of SSP secondary structure prediction is given in figure 10.12. The output of the prediction program presents not only the final optimal variant of the secondary structure assignment, but also a set of potential α-helix and β-strand segments that were computed without consideration of their competition. Because the protein secondary structure is finally stabilized during the formation of the tertiary structure, the alternative variants of the α - helix and β-strand segments may important for methods of tertiary structure predction. Occurrence of alternative structure may often point to wrongly predicted regions.

To improve the prediction accuracy of the method, a simple version, which treats multiple sequence alignments used as input in place of single sequences, has been developed. For testing, the alignment of homologous proteins was obtained from the HSSP database. Using the mean values of discriminant functions over the aligened sequences of homologous proteins, a predictions accuracy of 68 percent has been achieved.

Table 10.3. Segment prediction accuracy for short and long α-helices and β-starnds

	Sn (%)	*Sp (%)*
All α-helices	75	78
Long α-helices	89	82
Short α-helices	52	51
Allβ-strands	51	71
Long β-strands	71	79
Short β-strands	45	64

```
Name: >gi | 493758 | pdb | 131L | __Lysozyme_(E.C.3.2.1.17)_Mutant
ssp    Tue   Sep 26   16:25:37    PDT 2000
>>gi | 493758|pdb + 131L | __ Lysozyme_(E.C.3.2.1.17)_Mutant
 pred A:          aaaa                                      aaaaaaaaaaaaaaa
 AA               N 3.C                                     N       2.3       C
 pred B:                      bbbbbbb      bbbbb
 BB                           N 2.7 C      N 2 . C
 predi c          aaaaa       bbbbbbb      bbbbb            aaaaaaaaaaaaaaa
 a/acid         MNIFEMLRIDEGLRLKIYKDTEGYYSIGIGHLLTKSPSLNAAKSELDKAI
                        10          20          30          40          50
 pred A:                 aaaaaaaaa    aaaaaaaaaa     aaaaaaaaaaaaaaaaa
 AA                      N   5.3   C  N    3.3   C   N          4.1          C
 pred B:             bbbb                                                bbb
 BB                  N2  C                                               N 1
 predic              bbbbaaaaaaaaaa   aaaaaaaaaa     aaaaaaaaaaaaaaaaa
 a/acid         GRNTNGVITKDEAEKLFNQDVDAAVRGILRNAKLKPVYDSLDAVRRAALI
                        60          70          80          90          100
 pred A:                      aaaaaaaaaa    aaaaaaaaa            aaaaaaa
 AA                           N    3.9    C  N   4.0    C         N      2.1
 pred B:        bbbb                                                     bbb
 BB             .9 C                                                     N 2
 Predic         bbbb          aaaaaaaaaa    aaaaaaaaa                    bbb
 a/acid         NMVFQMGETGVAGFTNSLRMLQQKRWDEAAVNLAKSRWYNQTPNRAKRVI
                       110         120         130         140          150
 pred A:        aaaa
 AA                C
 pred B:        bbbb
 BB             .5 C
 Predic         bbbb
 a/acid         TTFRTGTWDAYKNL
                       160
```

Fig. 10.12. An example of secondary structure prediction for Lysozyme protein. SSP outputs potential α-helices (pred A) and β -structures (predB) with the final prediction (Predic). The last actual helix is predicted as β-strand, but we can observe potential α -helix here also with approximately the same weight.

The main feature of the SSP method is that it recognizes α-helix and β-strand segments rather than single residues. Another advantage is that even though the method is very simple in nature, it gives results comparable with more elaborated methods.

Table 10.4. lists the results of three state accuracy for many published methods. Due to the different training sets and testing procedurs, direct comparison among methods was rather difficult. Except for the PHD and SSP methods most methods published before 1994 used databases that contained proteins with larger than 30 percent homology. Among algorithms using statistical or neural network approaches, that is algorithms that did not use homology information, SSP, and nearest-neighbour methods give the best results. As mentioned by Rost and Sander (1993a), actual performance of some methods may be even lower, when testing on non-homologous datasets or using a jackknife proedure. For examole, GOR III was originally reported to give 63 percent accuracy, but when tested on the enlarged database gave only 56.7 percent.

DCS Method

The method proposed by King and Sternberg (1996) is also based on linear discriminant functions involving from 10 to 27 protein features. Many of them are similar to those introduced earlier in the GOR method. The features are represented as a set of *amino acid propensities* calculated for 20 residue types in three conformational states at positions $i-8$ to $i+8$ to providing $20 \times 3 \times 17 = 1{,}020$ parameters.

Additionally, several new characteristics were introduced, such as distance from the end of a protein chain, and the moment of hydrophobicity. The distance from the chain end accounts for greater flexibility fo residues at the end of protein chain. Moment of hydrophobicity evaluates possible hydrophobic patches typical for α-helices and β-structures.

Table 10.4. Comparison of prediction

	Type	*Q3 (%)*
Methods using single sequences		
GOR III (Gibrat et at. 1987)	Inform	63
Qian and Sejnowski 1988	NNw	64.3
Gibrat et al. 1987;Holley and Karplus 1989	NNw	63.2
Kneller et al. 1990		63
SM (Zhang et al. 1992)		63.5
Expert-NN (Zhang et al. 1992)	NNw	63.1
Reference network (Rost and Sander 1993b)	NNw	61.7
SSP Segment Prediction	DA	65.1
Combined methods using nearest neighbours		
COMBINE (Biou et 1988)	NNb	65.5
Hydrid (Zhang et at. 1992)	NNb	66.4
PHD (Rost and Sander 1993, 1994)	NNw	62.6
Yi and Lander 1993	NNb	66.5
NNSSP (Salamov and Solovyev 1995)	NNb	67.6
SSPAL (salamov and Solovyev 1997)	NNb	71.0
Methods using homology information		
Levin and gamier 1988	NNb	63
MBR (Zhang et at 1992)	NNb	64.5
Salzberg and Cost 1992)	NNb	65.1
SSP (Solvyev and Salamov 1994)	DA	68.2
Levin et al. 1993	NNb	68.5
SDC (King and Sternberg 1996)	DA	70.1
PHD (Rost and Sander 1994)	NNw	71.6
NNSSP (salamov and Solovyev 1995)	NNb	72.2
SSPAL (Salamov and Solovyev 1997)	NNb	73.2
SSIPRED (Jones 1999)	NNw	76-78

Single-residue prediction accuracy Q_3 for different methods of secondary structure prediction. Whenever available the accuracy data shown for 126 nonhmologo protein data set (Rost and Sander 1993). Inform, information theory-based approach' NNw, neural networks based; NNb, nearest-neigbhour based; DA, discriminant analysis-based approaches.

Information from aligend sequences provides additional features, such as position of deletions and insertions. The moment of conservation measures the degree of variability at a particular position.

The weak part of this approach is that the linear discriminant function per se cannot capture some higher order properties such as auto-correlation, secondary structure feedback effects, and neighborhood constraints on secondary structures. Hence a post-processing step was introduced to take this into account by using some "if then" filtering rules. These rules are based on short patterns of five to seven consecutive positions, and they introduce corrections into secondary structure assignment made on the previous steps. The filtering rules were found using the machine learning approach.

Overall prediction (Q_3) achieved by this algorithm when trained on 126 protein chains (the same set was used forPHD algorithm) was 70.1 percent (an overall per residue three-state accuracy).

Neural Networks

Several aglorthms using neural networks for secondary structure predictions have been developed. One of the most popular is the PHD algorithm by Rost and Sander (1993), which we are going to consider in detail.

The algorithm takes multiple alignments of protein sequences as an input. The sequences for the network training come from the HSSP database, where alignments of similar sequence in the alignment has known structure. For prediction (when the structure is unknown) similar sequences are searched using the BLAST program and then aligned using the Max Hom algorithm.

Prediction of secondary structure makes sense for proteins, which are not homologous to any other proteins with known 3D structure; otherwise, the secondary structure can be predicted by homology with higher accuracy than by any existing secondary structure prediction algorithm .To reproduce this situation in the trining set, all sequences with similarity to any other sequence in the set exceeding the level given by the so-called HSSP curve were excluded. This means exclusion of proteins, which can be aligned at 80 or more residue positions with a sequence identity $\geq$ 25 percent. Thus 130 protein chains were selected from seven hundred protein chains of PDB in 1992.

The multiple alignment is converted into a profile: for each position, the vector of amino acid frequencies is calculated based on the alignment. The neural net is applied sequentially to all protein sequence positions to predict secondary structure in every position as a state from three possible alternatives: helix, strand, and loop Prediction at a given position depends on amino acid frequencies (in the profile) at that position and neighbouring positions within a range defined by the window for which inputs are collected.

The achitecture of the neural net consists of three layers. The first layer takes 21. w inputs, where 21 corresponds to 20 amino acid residues plus one for the missing residues at the beginning or at the end of the sequence; w is the window size, that is how many neighbouring residues are involved in secondary structure calculations for a single position $-w$ = 13 was used (figure 10.13 depicts the w = 7 case). Amino acid frequencies for every positoin as well as for neighbouring positions within the window are few into the first layer of the neural network.

The frequencies are processed by the network according to where J_{ij}^{λ} is the junction

$$s_i^{2,v} = f\left(\sum_{j=1}^{N^{1+1}} J_{ij}^{2}\, f\left(\sum_{k=1}^{N^{0}+1} J'_{jk}\, s_k^{0,v}\right)\right) \quad ...(10.5)$$

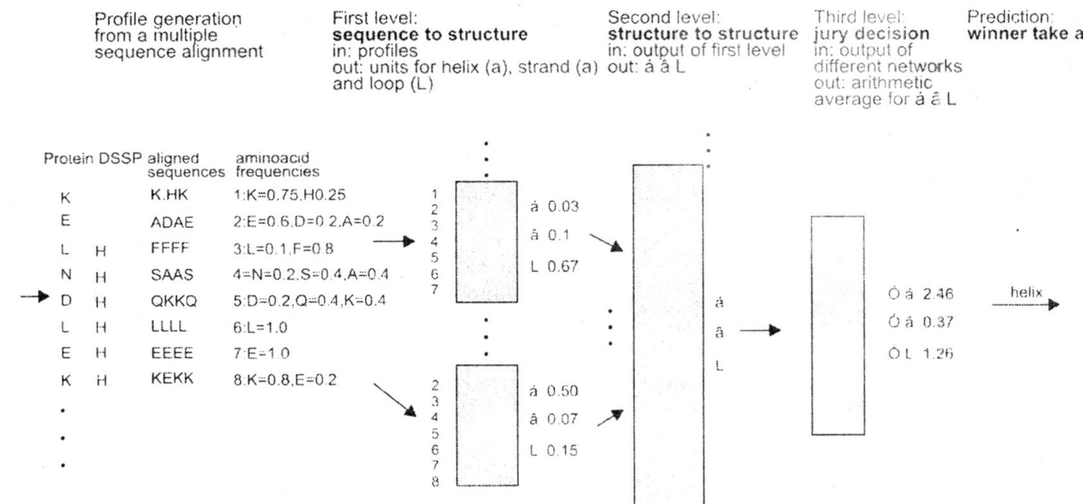

Fig. 10.13. Neural network scheme for PHD secondary structure prediction

(describing the connection between the nodes) between unit j in layer $\lambda-1$ and unit i in layer λ (0 for the input lalyer, 1 for the hidden layer, and 2 for the output layer), N^0 and N^1 are the numbers of input and hidden units, respectively, $S_k^{0,v}$ is the input from kth input unit for a given sample of data v, $f(x) = 1/(1 + e^{-\beta x})$ is the sigmoid trigger function, which describes how the outpur form the units is computed. Training the neural net was done using a strightforward gradient descent method. The first level produces classification of independent segments of residues with respect to the state of the central residue according to three calsses of protein secondary structure: helix (α) strand (β), and loop (L). Therefore, each segment is characterized by three output values.

In the second layer of the network the information form the first layer is modified, providing accounting of correlation between adjacent residues. Inputs are collected from the $w = 17$ units of the first layer, where w is the window size of the second layer, where w is the window size of the second layer. These inputs are converted into four binary units according to the following layer. These inputs are converted into four binary inits according to the following :

$$
\begin{array}{llll}
0000 \text{ for frequency} & & f < 0.02 & \\
0001 \text{ for} & 0.02 \leq & f < 0.33 & \\
0011 \text{ for} & 0.33 \leq & f < 0.66 & \quad ...(10.6) \\
0111 \text{ for} & 0.66 \leq & f < 0.98 & \\
0011 \text{ for} & & f \geq 0.98 &
\end{array}
$$

The output of the second layer is again three values corresponding to the three classes of protein secondary structure.

In the third layer, called "*jury decision*", the integration of outputs from different networks take place. What was described above as layers one and two represents a single instance of one from many networks, which all contribute to the third level. Introduction of many networks addresses the issue of local optimums occurring in the gradient descent training protocol Depending on intial choices of juctions, there will be different optimal solution achieved referred to here as differnt architectures. Combining these architectures together even by the simplest way as arithmetic averages results in improved overall performance.

Overall, a three state predictions accuracy of 70.8 percent was achieved. If we consider the prediction, then the accuracy reaches 82 percent. Further modifications of this approach increased its performance to 72 percent.

Further improvement in overall prediction accuracy was achieved by Jones (1999) in the algorithm called PSIPRED by using a slightly different neural networks desing and essentially diffferent sources of aligned sequences. Instead of HSSP (used in the PHD algorithm), sequence alignments produced by the PSI-BLAST program (Altschul et al. 1997) were considered here for training and prediction (figure 10.14). PSI-BLAST also builds sequence profiles needed for the prediction algorithm. Using PSI-BLAST bring two major advances: (1) significantly faster calculations of predictions; it takes only two minutes on a Silicon Graphics Origin 200 to produce prediction starting form the query sequence; and (2) higher accuracy of prediction, about 76.5 per cent compared to 72 per cent for the PHD algorithm.

Nearest-neighbour

The basic idea of the nearest-neighbour approach is the predictions of the secondary structure state of the central residue of a given segment, based on the secondary structure of homologous segments from the proteins with known three dimensional structure. The predicted type of secondary structure of a test residue is selected as the type of the majority of its nearest neighbours. as max (n_α, n_β n_c), where n_α, n_β and n_c are the numbers of nearest neighbors are selected by comparison of the test window sequence against all n residue windows from the database using the similarity score measure averaged over all window residues.

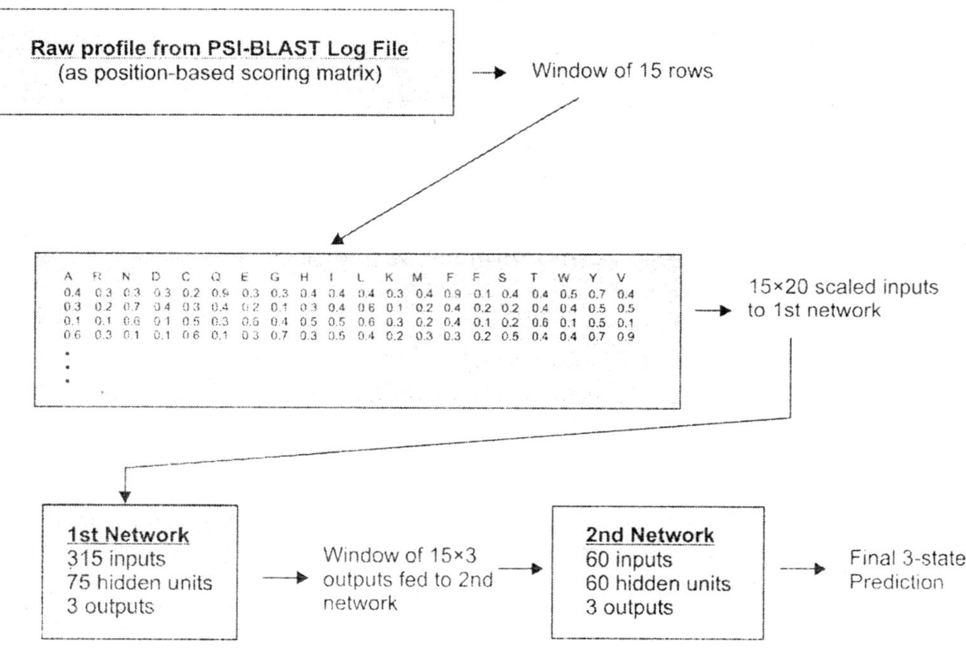

Fig. 10.14. A schematic diagram of the PSIPRED method

A key element in any nerest-neighbour prediction algorithm is the choic of a scoring measure for evaluation of similarity. The local structural environment scoring developed by Eisenberg and coworkers assignes everey residue of a protein with known three-dimensional structure to an *"environment class"* based on the local structur features of the residue position, such as the solvent accessibility, polarity, and secondary structure. Y*i* and Lander (1993) suggested a score of matching a query residue with the database residue as the environment score plus a score, estimated by mutation matrix. The environment score for matching a residue in position *i* of type R_i with the database residue in position *j* having known local structural environment E_j is defined as

$$Env(i,j) = \log_{10}\left(\frac{P(R_i|E_j)}{P(R_i)}\right) \qquad ...(10.7)$$

where P (R_i / E_j) is the probability of finding residue R_i in any enviornment E_j and $P(R_i)$ is the probability of finding residue R_1 many environment. The score of central postition *i* is computed as the average score in a window of length *l*:

$$Score(i,j) = \frac{1}{l}\sum_{k=-1/2}^{k=1/2} Env(i+k, j+k) + \text{SM}(i+k, j+k) \qquad ...(10.8)$$

where Evn (*i,j*) is the score for matching a residue at postion *i* of query sequence with the environment class of postion *j* of the targest protein and SM (*i,j*) is an element of an amino acid substitution matrix for residues at positions *i* and *o* of the query and target protein sequences, respectively. A substitution matrix was calculated based on the multiple sequence alignments for 126 database proteins taken form the HSSP database.

The NNSSP method has created additional environment classes as N- and C-ends of α-helices and β-strands. Also β-turns were considered as a specific class. In this way, 12 classes of secondary structure (five for α-helices: internal, N-and C-caps, the left N- and the right C-adjacent postions; the analogous five for a β-stands, β-turns, and coils) were combined with six categories of solvent accessibility/polarity, giving 72 enviornmental classes. Twelve classes of secondary structure were used only for nearest-neighbors selection (equation 15.8), but the only three-state secondary structure type (α, β, or *c*) of the center residue of nearest-neighbor windows was used for secondary structure assignment by majority rule.

An additional improvement of the predictiong accuracy was reached by reducting the database where we search for amino acid fragments similar to a test protein sequence. We limited the database to a subset of proteins closet to the test protein in some general properties. Distance measure, based on the Chou-Fasman preference parameters for helices, strands, and coils (D_{cf}), is used during selection of the subset.

$$D_{cf} = \sum_{k=1}^{3}\left((a/l_t)\sum_{j=1}^{l_t} f_t^k(j) - (1/l_b)\sum_{j=1}^{l_b} f_b^k(j)\right)^2, \qquad ...(10.9)$$

where $f_t(i)$ and $f_b(i)$ are frequencies of amino acid of type *i*; $f_l^k(j)$ and $f^k{}_b(j)$ are Chou-Fasman co-efficients of the amono acid residue in the *j*-th position for the secondary structure type *k*(α -β-c) and l_t, and *lb* are the lengths of a test and database proteins, respectively.

To exclude small elements, a few simple filtering rules were applied: (1) all helices of length 1 or 2 are converted to coils, except the case of β α β, which is converted to βββ; (2) all starnds of length 1 are converted to coils; and (3) all strands of length 2 surrounded by α-helical residues are converted to α-helices, that is, αββα-to αααα.

The NNSSP method provides about 72 percent of sustained overall three-statr accuracy we use multiple sequence alignment, or about 68 percent accuracy for single sequence input when tested on a benchmark database of 126 nonhomologous proteins.

Table 10.5. A comparison of prediction results using SSPAL method with results using the PHD method and the NNSSP method tested on the same date set of 126 proteins (set) and on a big data set of 461 proteins (set 2)

	Q^a_3	S_n^a	Sp^a	C_a	$\langle L_a \rangle$	$S_n b$	$S_p c$	C_b	$\langle L_b \rangle$
Input: single sequence									
PHD (Set 1)	62.6	57	62	0.42	6.2	42	53	0.35	3.8
NNSSP (Set 1)	67.6	69.1	67.6	0.55	6.2	38.2	66.0	0.41	3.1
SSPAL (Set 1)	71.0	70.5	71.4	0.60	9A	52.6	66.3	0.49	4.4
SSPAL (Set 1)	71.0	72.9	71.8	0.61	9.8	51.4	67.1	0.49	4.4
Input: multiple sequence alignments									
PHD (Set 1)	71.6	70	76	0.61	9.3	62	63	0.52	5.0
NNSSP (Set 1)	72.2	72.4	76.2	0.64	11.5	52.2	67.4	0.50	4.3
SSAL (Set 1)	73.5	75.8	73.6	0.65	9.5	52.7	72.2	0.53	4.3

Pre-residue measure Q_3 residues predicated correctly in three states (helix, strand, loop) divided by all residues; Sn^a correctly predicted residues in helix divided by observed residues in helix; Sn^b the same as helix, but for strand Sp^a correctly predicted residues in helix divided by predicted residues in helix; Sp^b the same as helix, but for stand. Ca and Cb are the Matthews correlation coefficents for helix and strand, respectively (Matthews 1975). (L_a) and (L_b) are the averge segment lengths of the predicted helices and strands, respectively (for observed helices (L_a) = 10.6 and for observed strands $(L_b$ = 5.1).

It is more informative to provide more than one state prediction to computer probability values of three possible states for each residue. Let n_α, n_β, and n_c be numbers of the best selected nearest-neighbors for some positons of our database of proteins with known 3D structure. One can compute the proportion of α-,β-, and *c*-states for these positions and use these data as probability estimation. We scale our n_α, n_β, and n_c values in 1-10 scale and produce a (10 × 10 × 10) matrix with probabilities belonging to a particular state. These probabilities can be matrix with probabilities beloging to a particular state/ These probabilities can be used in scoring the resembence to a tested fold, providing a better accuracy of recogition in comparing with the one-stated fold, providing a better accuracy of recognition in comparing with the one-state secondary strtucture prediction.

Matching any query sequence segment with segments from the database can be viewed as un-gapped alignments between the segments of fixed length. The fixed length of segments and the absence of gaps can significantly decrease the accuracy of nearest-neighbor methods, as the best local alignments usually have different lengths for different sequence regions and often contain deletions and/or insertions. Taking this tnto account, several new approaches for secondary structure prediction using local or global alignments were developed.

The SSPAL method used the Waterman-Eggert algorithm (1987) to compute the *K* best noo-intersecting local alignments of a query sequence with sequece of the subset selected from the database of known 3D structures. Then the prediction is based on the information about secondary structure states of aligend sequence segments. The term "non-intersecting" assumes that each next subpotimal alignment do not share any of the aligned pairs with the preceding alignments. It was shown that if we choose K large enough (in the range of 30-60), it significantly improves the prediction accuracy. For a given query protein, the approach proeduced up to N× K local alignments, variously located along the entire sequence, where N is the number of used database sequences. The score of a local alignment is taken as the score assigned to a given aligned position. The score of each one of three conformational states is computed as the sum of alignment scores with corresponding positions belonging to a particular conformational state. The first 50 alignments with the highest scores were taken into account. The predicted state of position is the state having the highest total score. The best performance (Q_3 = 71.2%) Was observed at K = 30 when the gap-opening penalty equals –20 and the gap-extension penalty equals –10. Another test of the method was performed using a representative list of 461 proteins (PDB_SELECT)

```
nnssp    Tue Sep 16   16:35:14 PDT 2000
>>gi| 493758|Pdb131L|_Lysozyme_(E.C. 3.2.1.17)_Mutant
L = 164 SS content: a⁻     0.63 b = 0.05 c =  0.32
                     10             20               30              40            50
PredSS     aaaaaaaaa      bbbb                 bbb         bb          aaaaaaaaaaaa
AA seq     MNIFEMLRIDEGLRLKIYKDTEGYYSIGIGHLLTKSPSLNAAKSELDKAI
Prob a     88998987744222322321111111111123211101288999999999
Prob b     00000000110000144554211111467742244421000000000000000000000000
                         60             70              80         90             100
PredSS                    aaaaaaaaaaaaaaaaaaaaaa              aaaaaaaaaaaaaaaa
AA seq     GRNTNGVITKDEAEKLFNQDVDAAVRGILRNAKLKPVYDSLDAVRRAALI
Prob a     31111011111888999999886889999998888644232244548899999999998888
Prob b     00111144200000000000000000000000000000111222100000000000
                     110            120             130        140            150
PredSS     aaaaaa            aaaaaaaaaaaaaaaaaaaa                 aaaaaaaaa
AAseq      NMVFQMGETGVAFGTNSLRMLQQKRWEAAVNLAKSRWYNQTPNRAKRVI
Prod a     8888888622112222455799999757799999987553311111134577866
Prob a     00111210122432100000000000000000000000011221111100001123
                     160
PredSS     aaaa    aaaaaa
AA seq     TTFRTGTWDAYKNL
Prob a     55641113577877
Prob b     33221011111011
```

Fig. 10.15. An example of NNNSp secondary structure prediction for lysozime protein NNSPP computes probability of β-and β-structures for each position. The small last α-helix actually is G-helix according to DDSP and we can observe that it has lower probability values for α-helix than the other predicated α-helices.

Table 10.6. Percentage of OVER, UNDER, and WRONG SSPAL predictions on the data set of 126 proteins

Input	*OVER*	*UNDER*	*WRONG*
Single sequences	10.5	15.3	3.3
Multiple sequence alignemts	9.4	14.4	2.6

With less than 25 percent sequence similarity. Four hundred and sixty-one protein sequences were selected from 486 protein chains of this list (PDB_SELECT, excluding all PDB entries with only known C^{α} atom coordinates and membrane proteins acording to the scop. For this dataset, a prediction accuracy of 71 percent was achieved. In addition to the Q_3 score, we also computed several other measures of prediction accuracy that take into account some of the characteristics fo predictions that can be important for building tertiary structural models.

Secondary structure predictions can be evaluated by subcategorizing the incorrect predictions into three categories: OVER (extra α or β) UNDER (unpredicated α or β), and WRONG (α as β or β as α) Table 10.6 presents the values of these categories for SSPAL method. It is assumed that the WRONG predictions can affect the fold recognition more seriously.

The SSPAL method presents an improvement of nearest-neighbor algorithms using local alignments and their scores instead of fixed-length up-gapped segment pairs. It is simple in realization and shows a certain increase in accuracy with using the database constructed from PDB_SELECT list with a 35 percent cutoff value. An additional advantage of the method is obviously a high level of prediction accuracy (up to 100 percent) when we analyze a sequence having some similarity with one of the database sequences.

(a) 50 local alignments:

```
                                243
                                ...GEFDIDCDNLSYMPTVVEEINGKMYPLTPSAYTOSQDQGFCTS
                                ... ccbbbcccccccccbbbbbbccbbbbbbbbbchhhhhhhbbbbccccbbbbbbbbbb
                                                                *
      5er 2E                    ... GGYVGPCSA---TLPSFTFGVGSARIVIPGADYFGPISTGSC
      K = 1 s=3267              ccbbbbbccc--ccbbbbbbccbbbbbbcchhhbbccccbbb
                                                                *
4          erhvl 197            YGITVLNHMGSMAFRIVNE
K = 2      K = 2 S = 384        cccccccccccccbbbbbbbbbccccc
                                                                *
3cd4 1                          KKVVLGKKGDTVELTCTA
K = 3 S= 374                    cbbbbbbbbbccccccccbbbccccc
                                                                *
                                                                ...
           6 cpp 170                        YLTDQMT
           K = 50 S= 50                     hhhhhhhh
                                                                *
```

```
(b)        243

                ...GEFDIDCDNLSYMPTVVFEINGKMYPLTSPAYTSQDQGFCTS
                ... ccbbbcccccccccccbbbbbbbbcccbbbbbbbbchhhhhhhbbbbbbbcbbbb

                                                    *
                                                    ...
                4 rhvl 202        LNJMGSMAFRIVNEHDE
                K = 1 S = 304     ccccbbbbbbbbccccccbbbb

                                                    *
                                                    ...
                5 havpa 5         LWQROLVTIKIGGQLKE
                k = 2 s= 238 cccccbbbbbbbbccbbbbb

                                                    *
                                                    ...
                6 cts 178         IAKLPCAAKIYRNLYR
                K = 3 S = 234     hhhhhhhhhhhhhhhhhc

                                                    *
                                                    ...
             4 pfk 296         IAEALANKHTIDQRMYA
                K = 50 S = 44     hhhhccccccchhhhh
```

Fig. 10.16. Prediction of the secondary structure state of Phe at position 261 of chymosin B (4cm) based on: (*a*) the 50 best local alignments; and (*b*) the 50 nearest-neighbor segments. The PDB identifier, the first postion of matched segment and the score are shown.

In the previous relization of nearest neighbor methods, the score was computed as the score of similarity of two short segments and the best nearest-neighbor score was not significantly higher than the scores of the other nearest-neighbors. In the current method, the prediction is mostly based on the first optimal alignment between a query sequence and homologous targtet, becuase the score of this alignment will significantly dominate the scores of all other alignments.

There are several attempts to create a consensus method of secondary structure prediction by combining the prediction from different approaches. Jpred (combining the NNSSP DSC, PREDATOR, MULPRED, ZPRED, and PHD methods) provides better and probably more stable results then each of the single methods used.

Table 10.7. Web servers for seconday structure prediction

Name of program	*WWW address*
DSC	http://www-bioweb.pasteur.fr/seqanal/interfaces/dsc-simple.html
	http://www-.aber.ac.uk/~phiwww/prof/
Jpred (consensus method)	http://www-jura.ebi.ac.uk:8888/
PSIPRED	http://www- insulin brunel.ac.uk/psipred/
PHD	http://www-cubic.bico.columbia.edu/predictprotein/predictprotein.html
	http://www-.ebi.ac.uk/~rost/predictprotein/
PREDATOR	http://www-.embi-heidelberg.de/argos/predator/run_predator.html
SSPAL, NNSSP, SSP	http://www-dot.imgen.bcm.tmc.edu.1931/psspprediction/pssp.html
	http://www.softberry.com/protein.html

Nearest-neighbor approaches have the potential to improve their performance from about 70_75 percent to 80 percent with an increasing number of known tertiary structures and the number of homologous sequences in protein databases, as we observed recently for neuralnetwork approaches Further progress will probably be difficult due to dependence of small secondary structure elements formaion on the 3D structural environments,as well as on the limitations of seconday structure assignment. However, we can envision some improvement in seconday structruce assignment and further development of approaches that predict the entire secondary structure segments rather than single residue state.

CHAPTER 11

PROTEIN FOLDING

A knowledge of the molecular basis of life is curucial for advances in biomedical and agriculatural research. Proteins are a diverse calss of biomolecules performing vital functions in all living things. Therefore, developing a fundamental understanding of how proteins fold is of immense intellectual and technological significance. A solution to the protein folding problem is often likened to the deciphering of the *"second generic code."* The question is simple to pose: Given a sepcific sequence of amino acids, how do physical and chemical forces determine its myriad properties, especially the essentially unique folded structure of a globular protein/ In principle, this question should be answerable because numerous proteins fold and unfold reversibly in vitro. Protein folding is much more complex in vivo, with the involvement of chaperones and other cellular machineries. Nevertheless, any first-principles account of these complicated processes must begin with an understanding of the considereably simpler folding processes in vitro.

PHYSICS OF PROTEIN FOLDING

Like all natural phenomena, protein folding is a physico-chemical process. Hence, all methods for protein structure prediction require various degrees of understanding of , or assumption about the underlying energetics. Recent protein structure prediction techniques, which include comparative or homology modeling, fold recognition, and "ab inito" approaches, have focused primairly on empirical and statistical analyses of sequence and structure databases. Because of their *inductive* nature, techniques that rely predominantly on correlative parameters derived from collection of existing structural and sequence information are called "knowledge-based." These methods are often constrasted with *"physics-based"* methods that attempt to *deductively* explain or predict protein behaviours form elementary physical forces. At present, knowledge- based approches-some of which are covered in excellent chapters elsewhere in this volume-seem to be more successful than anyh current physics-based technique in predicting native structures from given amino acid sequences. In fact, as Moult et al. (1999) and Shortle (2000) have noted, most current protein structure prediction alogrithms give little consideration to biophysical questions such as balance of forces, energetic components, or folding pathways. Indeed, with the advent of experimental structural genomics, comparative modeling will certainly become incerasingly important for providing structural and functional in-sight into the world's rapidly expanding sequence databases.

Protein folding is complex. To make progress, correlations among observations of all kinds need to be sough at every level, even if the underlying physical reasons are not quite known. Knowledge-based methods for protein structure prediction have contributed tremendously in this regard. But one should never lose sight of the importance of developing physics-based theories.

First, aside from their intrinsic intellectual value, such endeavors are necessary for justifying or improving knowledge based method; example are discussed below. Second, even if technologies for predicting protein native structure turn out to be achievable by purely knowledge-based means, to ascertain how a protein functions or malfunctions often requires biophysical information that cannot be gleaned from a static picture of its native conformation alone. Notably, a proteins's dynamic properties can be crucial for its functions. Conformational distribution and fluctuation in a proteins denatured state are important for native thermodynamic stability, and bear directly on a proteins' tendency to adopt disease-causing misfolded causing misfolded and/or aggregated forms. Physics-based theories, which are the main focus of this chapter, are necessary to address these essential aspects of protein behaviour.

Our purpose here is to provide a broad-stroke panoramic view this area of research. As illustrations of general principles, several physics-based theoretical and computational approaches to protein folding are highlighted. We place special emphasis on critical evaluations of methods and identifying unresolved issues that are crucial for future progress. We assess the strengths and limitations of a a number of approaches, focusing especially on features and assumptions that are important in determining whether a model is protein like or not, but whose ramifications have not been fully appreciated in the literature. The following is an outline of the rest of this chapter: (1) a rough sketch of all-atom simulations and driving forces in protein folding; (2) an introductory discussion of solvation effects in protein energetics and the complexities they entail, using electrostatic and hydrophobic interactions as examples; (3) a delineation of the advantages and limitations of using implified chain representations to study protein folding, and the physical implications of popular modeling approaches; and (4) summaries of recent applications of simple lattice protein models to evolutionary landscapes, protein aggregation, protein calorimetric cooperativity, and folding kinetics.

DRIVING FORCES PROTEIN FOLDING

A protein molecule is a large collection of covalently linked atoms. In principle, its properties should be deducible from these atoms' interactions among themselves and with the atoms of the solvent molecules. Computational approaches using all-atom empirical force fields are predicated on this premise. Typically, the potential energy (force field) V in these studies is a sum of contributions and a function of the spatial position vectors $\boldsymbol{r}_1, \boldsymbol{r}_2, \ldots, \boldsymbol{r}_N$ of all atoms in the system. For example, equation 11.1 in Leach (1996) may be rewritten to make explicit its dependences on atomic coordinates:

$$V(\boldsymbol{r}_1, \boldsymbol{r}_2, \ldots, \boldsymbol{r}_N) = \sum_{[i<j]}^{N} \frac{w_{ij}}{2}\left(|\boldsymbol{r}_i - \boldsymbol{r}_j| - b_{ij}^0\right)^2 + \sum_{\substack{[i,j]\\ [i,k]\\ i<k}}^{N} \frac{v_{ijk}}{2}\left(\cos^{-1}\frac{(\boldsymbol{r}_i - \boldsymbol{r}_j)\cdot(\boldsymbol{r}_j - \boldsymbol{r}_k)}{|\boldsymbol{r}_i - \boldsymbol{r}_k|\,|\boldsymbol{r}_j - \boldsymbol{r}_k|} - \theta_{ijk}^0\right)^2$$

$$+ \sum_{\substack{[i,j]\\ [j<k]\\ [k,l]}}^{N} \sum_{n=0}^{m} \frac{V_{ijkl}^{(n)}}{2}\left[1 + \cos\left(n \cos^{-2}\frac{\left[(\boldsymbol{r}_i - \boldsymbol{r}_j)\times(\boldsymbol{r}_j - \boldsymbol{r}_k)\right]\cdot\left[(\boldsymbol{r}_j - \boldsymbol{r}_k)\times(\boldsymbol{r}_k - \boldsymbol{r}_l)\right]}{|(\boldsymbol{r}_i - \boldsymbol{r}_j)\times(\boldsymbol{r}_j - \boldsymbol{r}_k)|\,|(\boldsymbol{r}_j - \boldsymbol{r}_k)\times(\boldsymbol{r}_k - \boldsymbol{r}_l)|} - \gamma_{ijkl}\right)\right]$$

$$+ \sum_{\substack{i<j\\ \text{nonbonded}}}^{N} \left\{4\varepsilon_{ij}\left[\left(\frac{\sigma_{ij}}{|\boldsymbol{r}_i - \boldsymbol{r}_i|}\right)^{12} - \left(\frac{\sigma_{ij}}{|\boldsymbol{r}_i - \boldsymbol{r}_j|}\right)^{6}\right] + \frac{q_i q_j}{4\pi\varepsilon_0 |\boldsymbol{r}_i - \boldsymbol{r}_j|}\right\} \quad \ldots(11.1)$$

where the square brackets in the expressions $[i, j]$ and $[i < j]$ indicate that a summation is only over atoms i and j that are covalently linked (to form bond $i - j$); w_{ij} and v_{ijk} are force constants for deviations of bond length (between atoms i *and* j) and bond angle (subtended by bonds $i - j$ and j and $j - k$) from their reference (natural) values b_{if}^{0} and θ_{ijk}^{0}, respectively; $V_{ijkl}^{(n)}$'s are "barrier-height" parameters and γ_{ijkl} is the phase factor for the torsion angle between bonds $k - l$ and $i - j$, ε_{ij} and σ_{ij} are, respectively, the well depth and collision diameter of Lennerd-Jones interactions between atoms i and j, q_i is the effective electric charge of atom i ϵ_0 is vacum permittivity,[1] and the last summation over pairwise non-bonded interaction is over atom pairs i, j that are not covalently bonded.

Because of the large number of atoms involed, simulations of proteins using all atom force fields such as equation (11.1) are computationally intensive. The CPU time needed is many orders of magnitude that of the physical time simulated. For an up-to-date review of advanced in relatively-timescale protein and peptide simulations, see Daggett 2000. To date, the longest simulation of any protein folding process has been Duan and Kollamans's 1998 study of the 36-residue villin headpiece (involving 9,295 atoms), tracking a single dynamic trajectory that corresponds to 1 micosecond of physical time. Results from some recent all-atom simultions are encouraging. For exanple, a short (7-unit) β-peptide in methanol has been sucessfully folded by molecular dynamics simulation to its experimentally determined stable conformation. However, no corresponding sucess has yet been reported for globular proteins. A reason may be that even the faster folding proteins known to date take tens of microseconds to fold which is more than one oeder of magnitude longer than the longest simultions per-formed so far.

The functional form in equations (10.1) and others similar to it represent the most detailed atomistic consideration currently feasible for building workable protein folding modles. However, as matter of principle, it is important to realize that they are still drastic simplications of the real physics. In particular, the pairwise potentials in the last sumation do not model the true fundamental interactions between pairs of isolated atoms. Insted, they are empirically parameterized to approximate the effects of many body interactions that are often intrinsically not pairwise additive. A clear example is that the dipole moment of many water models are significantly stronger than that of a single isolated water molecule in the gas phase, because the models are parameterized to reproduce properties of liquied water. if follows that parameters in empirical force fields are not universal, as they need to be optimized for particular sets of applications. Indeed, even for the relatively simple system of argon atoms in the liquid state, quantum mechanical calculations have shown that there is no universal effective pairwise potential that can adequately reproduce three-body dispersion interactions for all applications.

Coarse-Grained Mechanics

All-atom treatments are impractical if broad conformational sampling is required because they are computationally intensive, and general inferences cannot be reliably drawn from a small number of molecular dynamics trajectories. In view of these limitations and the non-universality of empirical force-field parameters, researches have also pursued complementary coarse grained statistical mechanical models of proteins. As for ferromagnetism in condense matter physics, a hierarchy of organizing principles is expected to be needed to bridge our conceptual understanding of physical phenomena of atomic and mesoscopic length scales to which proteins belong. it is natural to use coarse-grained models for low-resolution descripitions in thtis regarded. Furthermore, coarse-grained models are computationally more tractable,and because of their relative simplicity, can often provide insight that would have been obscured otherwise. A case in point is the recent "Gaussian network model" of collective motions and correlations in folded states of globular proteins. At a computational cost approximately three orders of magnitude less than that required by all-atom calculations, the predicated dynamics of these coarse-grained models are similar to that obtained by atom molecular dynamics simulation and normal mode analysis.

In *"big-picture"* conceptual formulations of protein folding and in coarse grained modles, it is customary to classify noncogvlent interactions into a few energetic components or interaction types, namely hydrophobic, hydrogen bonding electrostatic, and dispersion (attractive) and repulsive van der Waals forces . In terms of the atomic interactions presumed by equation 11.1 above, *"hydrophobic interaction"* may be viewed as the combined effect of van der Waals interactions between nonpolar atoms in the protein and their interactions with the surrounding water and cosolvent molecules. Hydrogen bonding interaction is sometimes treated as a type of electrostatic interaction, as provided by the last term in equation 11.1 or sometimes it is implemented by extra terms. As Cooper (1999) noted,the short list of interaction types that are belived to be relevant to protein folding has not changed for the past 40 years. However, even at this coarse-grained level, coherent physcial picture is still lacking. For instance, even the basic question of whether hydrogen bonding stablized or destabilizes native states of proteins remains controversial.

SOLVATION EFFECTS

A major difficulty in accounting for protein energetic is the effect of the aqueous solvent. solvent mediacted interactions between constituent groups of a protein can have more complex properites than the corresponding direct interactions in vacuum. For instance, ions in a solvent can lead to partial screening of the direct electrostatic interactions between two macromolecules and hydrophobic effect in proteins involve protein-water as well as water-water interactions

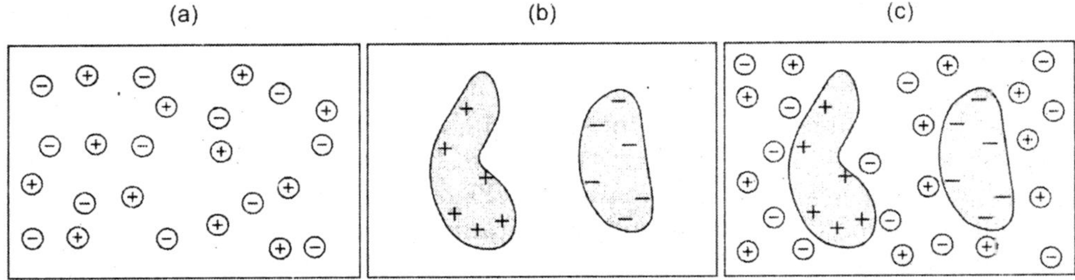

Fig. 11.1. Schematics of solvent mediated electrostatic interactions, (*a*) Solvent with dissiolved ions. (*b*) Direct intraction in Vacuum between two macromolecules (shaded shapes). (*c*) Solvent-mediated interaction between the two macromolecules. Ions in the solvent tend to migrate near oppositely charged macromolecules, partially shielding their direct electrostatic interactions.

Electrostatic Interaction

To illustrate the complexities of solvation effects, we present a brief outline of the Poission-Boltzmann approach, which is an approximate continuum treatment of the solvent-mediated electrostatic interactions. The analysis starts with the poission equation $\nabla \cdot [\epsilon(\boldsymbol{r})\nabla\phi(\boldsymbol{r})] = -\rho(\boldsymbol{r})/\epsilon_o$, which is a re-statment of Gauss' law generalized to include polarizable media. Here $\epsilon(\boldsymbol{r})$ is the free charge density (which excludes the bound charges, whose effects are approximately accounted for by $\epsilon(\boldsymbol{r})$ of the dielectric media), and $\phi(\boldsymbol{r})$ is the electrostatic potential for the corresponding *macroscopic* electric field.

In the presence of mobile ions in the solvent and free charges fixed at a given set of spatial positions (Figure 16.1c), $\rho(\boldsymbol{r})$ may be expressed as a sum of fixed charge density $\rho_{fix}(\boldsymbol{r})$ and controbutions from the mobile ions. Let the discrete charges of the ions be $+\ q(>0)$ and$-\ q$. The standard Poission\Boltzmann argument posits that the eqqilibrium populations of the $\pm\ q$ mobile ions at any position r accessible to the ions are proportional to the Boltzmann factors $\exp\ (\mp\ q\phi(\boldsymbol{r})/k_B\mathrm{T})$, where $k_B\mathrm{T}$ is Boltzmann's constant times absolute temperature. Now let C_q $(r) \geq 0$ be the average (bulk) concentration of either the $+q$ or the $-q$ ions at these positions; $\mathrm{C}_q\ (\mathrm{r}) = 0$ at positions inacessible to mobile ions (in the core of a folded globular protein, for example). The bulk concentrations of the two oppositely charged ions are equal because the solvent is electrically neutral macroscopically. It follows that the mobile-ion contribution to the charge density is equal to $q\mathrm{C}_q(\boldsymbol{r})\ [\exp(-q\phi/k_B\mathrm{T}) - \exp(+q\phi/k_B\mathrm{T})]$.

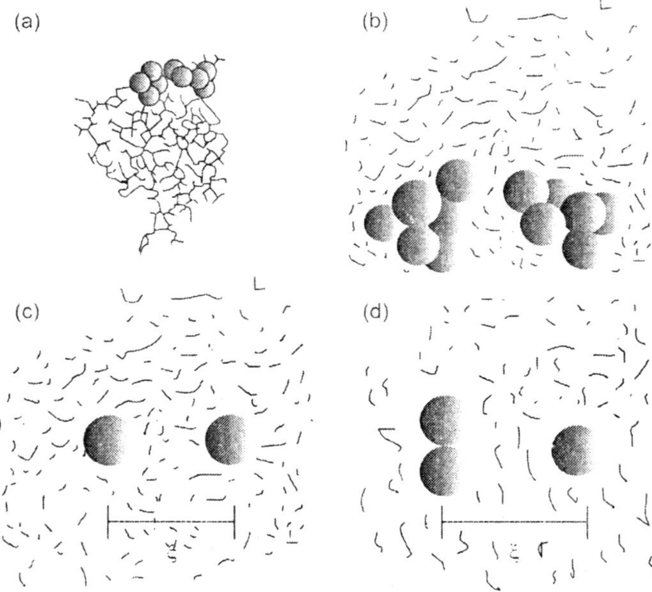

Fig. 11. 2. Schematics of hydrophobic interactions in protein energetics. (*a*) Native structure of chymotrypsin inhibitor 2 (protein databank entry 2C12). As an illustration, two hydrophobic residues-a leucine at postions 12 and an isolecucine at postion 63–are highligheted by space-filling representaions. (*b*) A blown-up view of the two hydrophobic residues in water. Water molecules are depicted by V-shaped representations, whereas the methlyl and mthoylene groups of the residues are shown as spheres. (c , d) Modeling hydrophobic interactions among methyl and methylene groups by water-mediated interactions among methanes. Here each methane (sphere) is treated as a united atom, meaning that its nonpolar hydrogen atoms are not modeled explicitly. The distance ξ defineds the methance (*c*) and three-methane (d) configurations for the potentials of mean force in figure 11.3.

The approximation introduced into this line of agrument is not difficult to discern: The discrete ionic charge density (upon which an earlier step in the argument is premised) has been smeared out in the last expression to a continuous smooth distribution in the solvent. Consequently, posible sharp variations in $\phi(\boldsymbol{r})$ that might have resulted from discrete chargees are precluded. The Poisson-Boltzmann equations is obtained by incorporating this approximate mobile-ione contribution into the Poission equation, viz.,

$$\nabla\cdot\left[\in(\boldsymbol{r})\nabla\phi(\boldsymbol{r})\right]-\frac{2qc_q(\boldsymbol{r})}{\in_0}\sin h\left[\frac{q\phi(\boldsymbol{r})}{k_B\mathrm{T}}\right]+\frac{\rho_{fix}(\boldsymbol{r})}{\in_0}=0 \qquad (11.2)$$

where the value of $C_q(\boldsymbol{r})$ in the solvent is proportional to its ionic strength and the (simpler) direct interactions case in the absence of mobile ions corresponds to $C_q(\boldsymbol{r}) = 0$ for all $\boldsymbol{r}$. Equation 11.2 implies that $\phi(\boldsymbol{r})$ can be solved for any system with a given set of fixed charges, a solvent ionic strength, and a posion dependent dielectric constant. This continuum treatment of solvent charge effects has been applied to complex-shaped biomolecules to provide useful physical insight into, and graphical visualization of the role of electrostatic interactions in a wide range of biological phenomena.

Hydrophobic Interactions

Solvent-mediated interactions can be modeled directly by empirical force fiedl simulations with explicit water and other solvent molecules. Explicit solvent simulations take into account the shape of discrete solvent molecules, giving predications that are often more physical than that form continuum solvent models Figure 11.3 shows recent exppicit-solvent simulation results that bear on our under standing of hydrophobic effects.

Plotted in this figure are potential of mean force[2] (PMF) among methane molecules in water. These systems have been investigated extensively as models for hydrophobic interactions in proteins because methane is chemically similar to the nonpolar merthyl and methylene groups in hydrophobic amino acid residues. Each PMF in figure 11.3 is a free energy function obtained from averagine over water configurations. it represents the sum total of solvent effects plus the direct interaction between the methanes, and it determines the relative probabilities of different methane configurations. Methanewater and water-water interactions lead to PMFs with spatial dependences that is significantly more favorable to methane-methane association than that are more structured than that of the direct two-methane interaction. The sailent features of two-methane PMFs are: (1) a contact mininum at $\xi \approx 3.8$ Å that is significantly more favorable to methane-methane association than that contributed by the attractive direct methane-methane interaction alone, underscoring the role of water in promoting hydrophobic association; (2) a free energy barrier (desolvation peak) at $\xi \approx 5.7$ Å and (3) a second, shallow (solvent-separated) minimum near $\xi \approx 7.0$ Å. Implications of these features on protein folding have recently been discussed.

In protein folding energetics, a crucial question is to what extent interactions can be teated as additive. This issue is pertinent to the question as to whether and appropriate scoring function for protein structure predication can be devised using only pairwise addivite amino-acid based contact energies. Whether hydrophobic and other iteractions are additive is also relevant to understanding protein calorimetric data. Potential of mean force simulations can help address these issues. Here, figure 11.3*b*, *c* demonstrates that even if the underlying non-bonded atomic interactions of a model are *assumed* to be pairwise additive—as in the case for the force field used in these simulations—the resulting PMFs, which are *effective* interactions, are not necessarily additive. From the results of Rank and Baker (1997) in figure 11.2*b*, effective interactions among methanes in water appear to be anti-cooperative at least for the interaction at the contact minimum. Free energy for binging three methanse from infinty to contact is predicated by their simulation to be less favorable (less negative) than the sum of bringing three paries of methanes together spearately. On the other had, the results in figure 11.3*c* from a different simulation of the same system by Czaplewski *et. al.*, (200) implies the opposite-that effective contact interactions among methanes in water are slighly cooperative; in other words, the free energy for bringing three methanes from intogether separately. More recently, the discrepancy between these two studies has apparently been resolved. Using the same water model as that in Rank ansd Baker (1997) and Czaplewski et al (2000) but a more reliable technique of analysis, simulation data from our groups indicates that multiple methane hydrophobic interactions are largely anti-cooperative under ambient conditions.

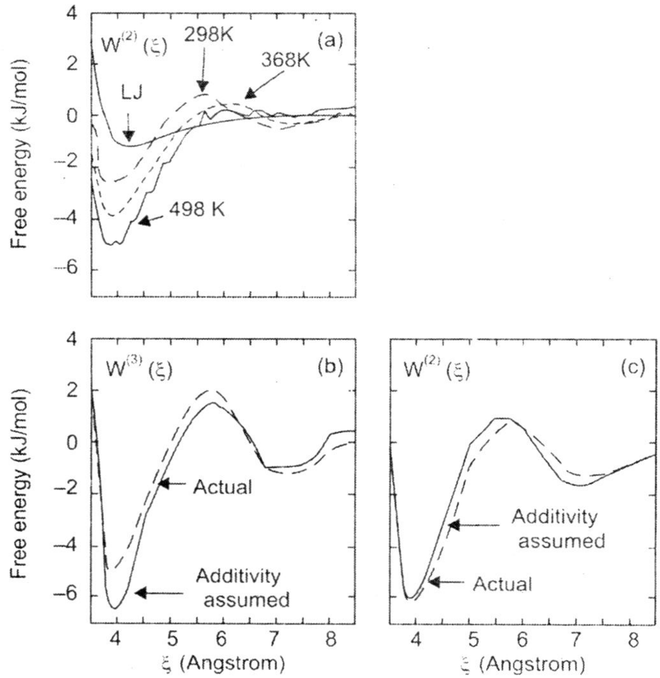

Fig. 11.3. Hydrophobic potentials of mean force (PMF). The variable ξ in (*a*) is defined in figure 11.2*c*, whereas ζ in (*b*) and (*c*) is defined in figure 11.*d*. (*a*) Two-methane PMFs, PMFs, $W^{(2)}$ ξ, simulated under atmospheric presure at 298 k (25° C) and 368 k (95° C) Included for comparison are the direct Lennard-Jones (LJ) model potejntial between the two methanes, and a W^2 (ξ) simulated at an average water density of 829 kg/m^3 at 498K (225°C) under constant volume conditions. Computational details are given in Shimizu and Chan200. (b) Three-methane PMF results from Rank and Baker (1997), adapted from their figure 4a. The free energy $W^{(3)}$ (ξ) plotted here is for bringing a methane from infinity to position ξ relative to the other two methanes that are already in contact. The "Actual" curve is the simulated three-methane PMF, $W^{(3)}$ (ζ) For each three-methane configuration specified by ξ the sum of two-methane PMF values for the two pairs of methans (between one of the contacting methanes and the third methane) is plotted as the "Additivity assumed" curve. 9c) Corresponding three-methane PMF results from Czaplewski et al. 2000 for the same three-methane system as in (*b*) adapted from the 12 window plots of their figure 8*d*.

Explicit-water simulations show that PMFs can be temperature depednent even when the underlyig non-bonded atomic potential function of the model is temperature independent. Take two methanes i water as an example. The effective independent. Take two methanes in water as an example. The effective interaction $W^{(2)}$ (ξ) at each separation ζ between the two methanes involes an average over a huge number of differnt configurations of water molecules. The relative populations of these configurations are temperature dependent because not all configuraions have the same engergy. Hence $W^{(2)}$ (ξ) depends on temperature. This fact is equivalent to characterizing the effective interaction as having in *intropic* part, whereas temperature-independent interaction may be refereed to as being *purley enthalpic.*"

Temperature dependences of potentials of mean force have ramifications for recent all-atom simulations of protein unfolding kinetics. In an attempt to circumvent current computational limitations of all-atom simulations of protein folding (see above), all-atom *unfolding* molecular dynamics have been performed at a high simulation temperature of 498k (225 °C) to speed up the unfolding process, using a significantly reduced average water density of 829 k/gm^3 for water under atmospheric pressure between 0° and 100 °C. It has been asserted that kinetic protein folding pathways are temperature independent but this assertion is not valid in general. PMFs in aqueous solutions are sensitive to temperature and average water density. Just as the PMF at each position involves aver agin many solvent configurations, the probability for a protein molecule taking any *microscopic* folding pathway (as defined by the trajectories of all atoms in the protein but *not* the trajectories of solvent degrees of freedom) is a function of the ptobabilities of many possible trajectories of the solvent molecues consistent with the given kinetic pathway of the protein. Because the relative probabilities of solvent trajectories can shift as temperature is varied (because they can involve different energy barriers) the relative favorabilities of different portein folding and unfolding pathways can change with temperature. Figure 11.3 3a shows a two-methane PMF simulated under conditions that have been used for high-temperature unfolding molecular dynamics simulation. It has features significantly different from PMFs at ambient temperatures and pressure. The high temperature and low water density lead to a much more favorable contact minimum, and the desolvation peak and solvent-separated minimum are all but abrogated. These observations suggest that, through insight can be gained from high temperature unfolding simultions caution should be exercised in their interpretation.

SELF-CONTAINED POLYMER MODELS

Proteins are chain molecules. Chain connectivity, stiffness, and excluded volume impose significant constraints on protein behaviour. In addressing the physical forces in protein folding, self-contained polymer models are indispensable. By self-contained polymer models, we refer to theoretical constructs in which the conformational distribution of a chain molecule is determined solely by the energetic components that are being considered explicitly. This is the most intuitive and straightforward approach to physical modeling. Our only reason for emphasizing it here is because the traditional discurse of protein energetics often involves non-selfcontained constructs. Typically, denatured states in non-self contained constructs are simply assumed to have certain a priori conformational distribution (partiaially unfolded, random-coil-like) that are not derived from the elementary interactions under consideration. As a result, logical connections between elementray interactions and predicated properties cannot be unequivocally eastablished using non-self contained constructs.

Obviously, all-atom protein models based on empirical force field such as equation 11.1 are self-contained polymer models. The ability to use these models to simulate biochemical processes is expected to be steadily enhanced by advances in computer technologies, improvements in simulation techniques and the development of efficient conformational sampling algorithms Nevertheless, brute-force all-atom simulation of folding is still beyond our. More fundamentally, it is not always straightforward to clarify the essential physics of a a complex system from a huge amount of detailed simulation data generated by a large number of parameters. Therefore, to address general principles in protein folding, it is necessary to also construct and analyze self-contained polymer models with simplified representations of protein chain geometries and intrachain interactions, as we have argued above. Research in the past decade has demonstrated that simplified models can lead to novel concepts and provide useful insight. But they also raise new questions, especially with regard to their relations to real proteins.

Lattice Representations of Chain Geometries

Simple protein models use reduced representations the polypeptide chain. Chains in most of the recent simple models are configured on regular lattices. Simple off-lattice protein models with chains configured in the continuum and off-lattice discretized conformational spaces have also been investigated. Owing to space limitations, only lattice models are discussed below.

The basic components of typical lattice chain models are shown in figure 11.4 Lattice models are more tractable because the number of possible chain conformations is restricted by lattice regularities, allowing for faster searches and broad coverage of the discretized conformational space. Many recent lattice protein chain models are configured on three-dimensional simple cubic lattices. A further simplification is to configure them on two-dimensional square lattices. Both of these extremely simplified representations are introduced to capture chain connectivity, excluded volume, and certain general features of intrachain interactions (see below). They do

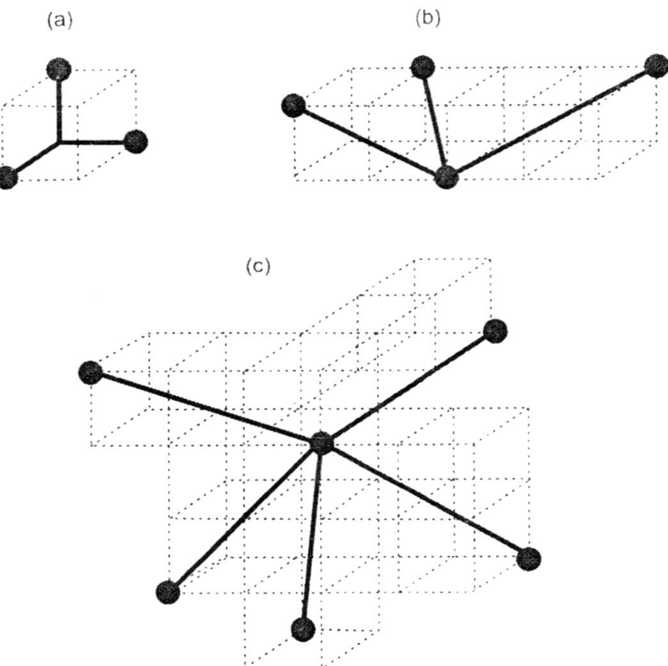

Fig. 11.4. Lattice representations of protein chains. Example bond constructions are drawn as thick lines joining two solid circles. The dotted lines show the underlying mesh. (*a*) Simple cubic lattic models, (*b*) The hydrid-210 model. (*c*) The ultra-310 model. Bonds in the high-coordination lattice models (*b*) and (*c*) correspond to protein C_α-C_α virtual connections (Godzik et al 1993)

not attempt to model geometric details of polypeptide chains. In most applications, the nodes along these lattice chains are not specified to correspond to any particular position of a real polypeptide; hence, they are not used as models for specific proteins, but rather as theoretical tools to address general properties of generic proteins. The rationale for this modeling approach is discussed in dill et al 1995.

To model protein structure at higher resolutions, extensive investigations have been undertaken by Kolinski, Skolinick, and coworkers to design lattice mimics of polypeptide geometries. This effort began more than a decade ago with protein chain models on diamond lattices. Two of their models are shown in figure 11.4*b* *c*, both of which use an underlying simple cubic lattice mesh to construct model polypetide $C_\alpha - C_\alpha$ virtual bond vectors. The unit length of the Cartesian coordinate system (*i.e.*, the magnitude of vectors [± 1, 0 ,0] and their permutaions) is equated with a length scale *a* to match model dimensions to real proteins. Different bond-vectorconstruction schemes require diffrent values of *a*. Inan early *"knight's walk"* model, $C_\alpha - C_\alpha$ virtual bonds were constructeed from the 24 permutations of the vectors (±, 2, ± 1, 0), and *a* was set to 1.70 Å so that the model virtual bond length $\sqrt{5}a$ = 3.80 Å matches that of the predominant *trans* peptide bond; and amino acid side chains were treated as on lattice entities.

In a subsequent *"hybrid"* –210 system $C_\alpha - C_\alpha$ virtual bonds are constructed from the 56 permutations of the vectors in the set {(± 2, ± 1, 0), (±, 2, ±1, ±1), (±, ± 1, ±1)}, and *a* = 1.70 Å. Virtual. This was followed by a finer (less coarse-grained) "ultra" –310 system that constructs C_α–C_α bonds from the 90 permutations of the vectors in the set {(± 3, + 1, 0), (± 3, ± 1, ± 1), (± 3, 0, 0), (± 2, ± 2, ± 1), (± 2, ± 2, ± 1), (± 2, ± 2, 0)}, with *a* = 1.22 Å. In more recent applications, amino acid side chains were incorporated as off-lattice entities. To conform to polypeptide geometries, restictions are placed on the directions of consecutive virtual bonds to exculdue some actue and open bond angles. Virtual bond angles are confined to the range of 78.5 – 143.1° and 72.5 – 154° for hybrid-210 and ultra-310 models, respectively. These *"hybrid"* and *"ultra"* systems are *"fluctuating bond"* models, in that $C_\alpha - C_\alpha$ virtual bond lengths are not fixed within a model. In units of *a* possible virtual bond lengths are $\sqrt{3}$, $\sqrt{5}$, $\sqrt{6}$, for the hybrid-210 model, and $\sqrt{8}$, 3, $\sqrt{10}$, $\sqrt{11}$ for the ultra-310 model. The value of *a* were chosen so that the predominant virtual bond length of 3.80 Å for real proteins corresponds roughly to the average virtual bond lengths in these models. Bond fluctuations models have been used to model polypeptides in other contexts of particular intersest is a recent fluctuationg-bond "side-chain-only" lattice protein model, in which 646 possible model virtual bonds with lengths ranging form 3.8 Å to approximately 10 Å are used to connect side-chain centers of mass instead of C_α's.

High-coordination lattices are quite flexible. Because many bond anlges are allowed, conformations constructed on these lattices can match quit closely any polypetide conformation at the C_α level. Average root-mean-square deviation of fitted lattice conformations from C_α traces of Protein Data bank structure is ~ 1.0 Å for the hybrid-210 model, and ~ 0.8 Å for the ultra -310 model and the recent side-chain-only model.

High-Coordination Lattice Models

Interactions schemes for high-coordination lattice protein models use knowledge based parameters statistically derived from protin native structure databases as well as postulated potentail functions. These schemes have evolved over the years. Typically, their potential functions include energetic contributions form: (1) local orientation correlations between amino acid side chain rotamers: (2) lattice-defined hydrogen bonding interactions: (3) amino-acid-specific one body terms (here a "body" refers to an amino acid residue) that depend on a rsidues' distance from the center of mass of a given protein confromation, or the number of contact a residue made with other residues; (4) pairwise (two-body) terms; and (5) four-body "tertiary interactions" terms to promote certain preferred side chain packing patterns Parameters for the assumed functional forsm of (1) – (4) were derived from protein databases, whereas (5) was postulated "by hand". Other additional features have aso been incorporated.

Even with their rather complex chain representation and interaction schemes, these lattice models are computationally more tarctable than all-atom models, and have provided insight into folding thermodynamics and kinetics many of which cannot be addressed by other current models at comparably high levels of structural resoultiin. Physical interpretation of high coordination lattice model results, however,is not always straight forward. This is because of the large number of tunbale knowledge based and postulated parameters involved, whose relationships with physical forces are sometimes not entirely clear. For instance, an "one-body" term described above postulates a gravitation-like pull on some residues toward a certain attractive center in a protein. But physically the collapse of a protein chain must originate from the solvent mediated interactions among the residues, and therefore should be describable by atomic interactions, or two- and higher body interactions at the residue level. It is not certain whether a universal high-coordination lattice interaction scheme that can describe a broad range of protein protein properties would ultmately emerge.

Low-coordination Lattice Protein Models

A complementary route to physical understanding is to adopt an incremental approach that as a first step, seeds to establish a workable conceptual framework to account for general properties of proteins. For this purpose, highly coarse-grained chain representations, such as those on simple cubic and square lattices, are used to capture only the rundimentary chain nature of protein but not their structural detaisl. The *postulated* interactions in these modles are simple, but can nonetheless be based on physical considerations, in a spirt very much akin to that of Ising models for ferromagnetism and related phenomena (pathria 1980). These models have an important logical advantage because they are readitly falsifiable. Their results are essentially direct deductions from the basic axioms of a given model's energetics, derived without intervening approximations and additional assumptions. Most of these models consider only contact interactions (see below), but orientation-dependent interactions can also be incroporated. Because of their models consider only contact interactions (see below), but orientations-dependent interactions can aslo be incorporated. Because of their higly simplified nature, extra care is needed to assess whether the basic interaction scheme of a given model warrants certain conclusions about real proteins, especially with regard to structure specifics and microscopic mechanisms of protein folding. Several such question are raised below. Nonetheless, provided that both the advantages and shortcomings of simplified modeling are taken seriously, much can be learned about protein energetics from these exercises. This is underscored by the fact that although there can be many designs for higly simplified chain modles, obtaining predictions consistent with experimental observations is nontrivial . with proper applications of experimental constraints, simple protein model cfonstruction is not as arbitray as it might seem.

Interaction Potentials

Go and HP + Model

The "Go potential" is one of the first interaction schemes used in simple lattice models. Starting with a target native structure, a Go model assigns equal favorable energies (< o) to all contacts between a pair of residues (monomers) that occur in the given target structure, and assigns neutral energies (= 0) to all non-native contacts that do not delong to the target structure. Recent varioants include an "HP +" model that assings repulsive (> o) in-stead of neutral energies to nonnative contacts. From a physical perspective, such *teleological* constructs may appear trivialbecause they do not seek to acount for the energetic favorability of the native state in terms of universal elementary physical interactions but, instead, fabricate a different interaction scheme for each different target structure. Because of their non-umiversality, Go and Go like potentials do not provide a modelsolution to the most basic question in protein folding, which is how the amino acid sequence of a protein determines its native structure.

Having said that, in view of our severely limited knowledge at present, it is profitable to test all kinds of assumptions and ask *"what if"* questions whenever possible. Such exercises are useful because they allow us to partially sort out the many logical relationships that may lead to further progress, even though the physical origin of the assumptions remians to be elucidated. Pursuing nontrivial implications of Go like interactions can be fruitful within such a conceptual framework. in fact, currently some generic protein properites can only be qualitatively reproduced by such highly artificail constructs and results from some Go model studies can provide remarkable and unexpected information. Go like native-centric approaches have also been applied to modelprotein folding in non-lattice contexts. However, it should always be borne in mind that these approaches by themselves do not tell us what physical interactions can conspire to create the highly specific " molecular recognition" features they assume. Answer to such basic questions have be sought in studies that attempt to model the physical driving forces in proteins.

HP Models

A widely applied simple lattice protein potential desigend to capture physical driving forces is that of the HP model. It used an extremely coarse-grained two-letter folding alphabet: sequences are strings of monomers that are either H (hydrophobic) or P (polar). (These model monomer names should *not* be confused with the one-letter codes for histidine and prolien!) To mimic hydrophobic effects, each nearest neighbor contact between two H monomers not consecutive along the sequence is assigned a favorable energy, irrespective of whether the contact is in the native (ground state) structure or not; all other contacts are modelled as neutral. Thus, ground-state structures of HP sequences are determined by a universal model potential. This contrasts with the Go model approach, which uses a particular potentail for each target native structure. Example HP sequence and ground-state structures are given in figure 10.5 *a d*. Other two-and three-letter potentials have also been investigated.

The HP model potential is quite nonspecfic and interaction heterogeneity is not high. As a result ground state degenearcy (number of conforamtons with the same lowest energy) is generally high for HP sequences. The fraction of short HP sequences with up to approximately 20 monomers that have a unique ground state conformatoin is $\approx$ 2.5 percent on two-dimensional square lattices. Degeneracy is higher in three dimesnions. Determining the ground-state conformations of long HP sequences is a computational challenge because in general it is an NP-complete problem. No HP sequence configured on simple cubic lattices has been found to possess a unique ground-state conformation recently, Buchler and Goldstein (2000) suggested that certain structural conclusions from HP model strudies may not be general because these features are sensitive to the size of the alphabets tó smaller ones . These observations raise a fundamental modeling question: Two what extent are HP and other simple lattice potentials proteinlike?

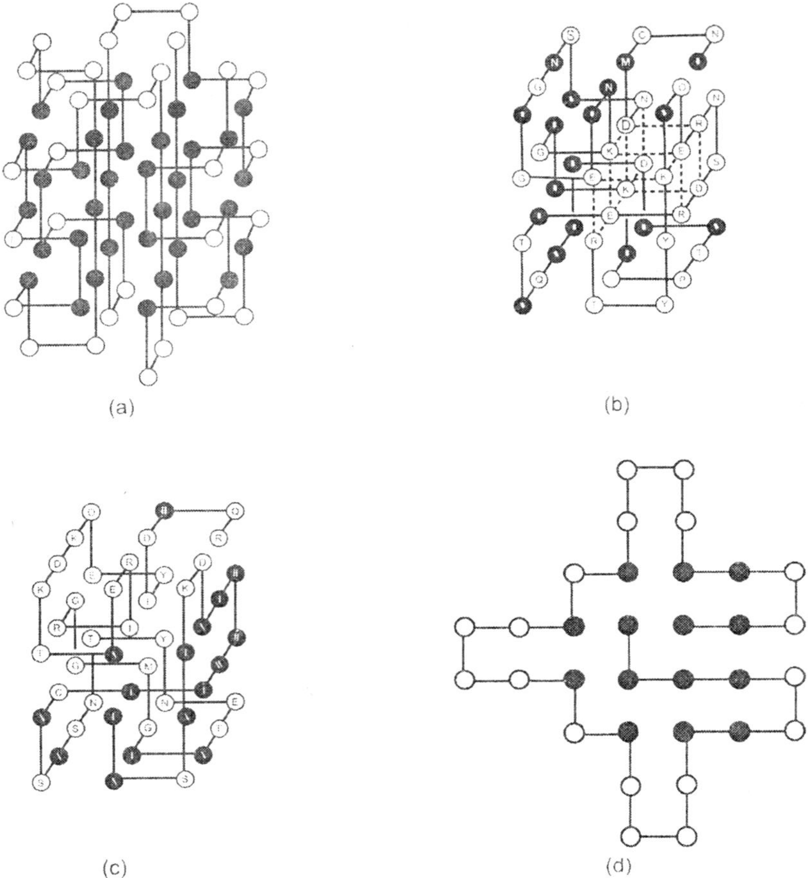

Fig. 11.5. Simple lattice protein models (*a*) One of three ground-state (lowest-energy) conformations of a67-monomer (67 mer) three-dimensional HP model sequence (Y ue and Dill 1995). H and P monomers are depicted as filled and open circles, respectively. (*b*) The conformation with the lowest simulated energy jfor the 20-letter 48 mer sequence shown (as given by the one-letter codes for the amino acids), computed with a particulr set of interaction parameters modified by Shaknovich *et. al.*, (1996) from table VI of Miyazawa and Jernigan 1985. Black circles are used for monomers corresponding to eight amino acid types (A, V, L, I, M, P, F, W) that are customarily considered to be hydrophobic. White circles are sued for monomeres corresoonding to the other 12 amino acid types, including D, E, K, R, and H that are customarily referred to as charged. Each dotted line indicates a salt-bridge-like contact between a pair of monomers with opposite charges *c*. Same conformations as in (*b*) but for a different 20-letter 48 mer sequence governed by a different set of interaction parameters. This conformation has the lowest simulated energy for the given sequence according to a set of contact energies modified from the two-body in teraction parameters of Kolinski et al (1993). (*d*) A 32 mer two-dimensional HP model sequence in the conformation with the lowest simulated energe.

Despite its obvious limitations, the HP poteitial is useful in the study of proteins, for the following reasons.

First, in light of a host of simplifications in any simple model, a model folding alphabet is not necessarily less proteinlike solely because it has two instead of 20 letters. Other physical issues beside alphabet size can be equally, if not more critical. These include the nature of a model's packing forces, namely the model interactions' specificity and stablizing mechanisms (see below), the discriminating role of repulsive interactions, and whether the model conformaional ensemble is unphysically resticted to only the macimally compact conformations.

Second, HP modles should be useful for expliring the mapping between protein sequences and their native structurs inasmuch as different energetic components conftibuting to protein folding (hydrophobic and other interactions) have no significant conflict with one another in the native conformation, that is, insamuch as the situation envisioned by the consistency principle (Go 1983) or principle of minimal frustration is valid, In that case we may adopt the working assumption that for a sequence to be proteinlike, it must have only a unique most-favored conformaion or a very small set of near-unique most-favored conformations based on its hydrophobic-polar pattern alone, notwithstanding the fact that contributions from other energetic components have to be incorporated to fully account for its thermodynamic and other properties. This formulation has been motivated by the observation that symmetries exahibited by many HP model ground state structures are intuitively more proteinlike. Than the ground state conformations favored by other model potentials, including some with large alphabets.

In this view the unique model ground state conformations determined by an HP or HP-like potential are adopted as coarse-grained models of protein native conformations, but with the undersatanding that additioal *consistent stabilizing* driving forces may have to be invoked in applications that address more refined features of protein energetics . The viability of the present physical interpretation in buttressed by the recent finding that thydrophobicity pattersn in two-dimensioanl HP model proteins as characterized by mean- square block fuuctuation are indeed qualitatively similar to that found in real protein sequences.

Twenty-Letter Models

Simple lattice models that use 20-letter alphabets have aso been extensively studies. As for the HH model, ground state confrimations in these models are determined by a universal set of interaction parameters. Typically, the interaction parameters are modified from knowledge-based pairwise contact energies between amino acid residues which are often represented as a symmetric 20 × 20 energy matrix with 210 independent elements. In general knowledge-based statistical potentials can be quite different form physical interactions, although under cerain restrictive conditions the former statistically derived parameters can provide a reasonable description of the underlying physicas. Figure 16.6 provides one such example.

Here we analyze the 1985 version of the Miyazawa-Jernigan (MJ) contact energies by a matrix diagonalization technique. This method is a variation of an earlier analysis of L_i et al. (1997). Shown on the left in figure 16.6 *a* is the specturam of eigenvalues of this particular MJ matrix. One of the eigenvalues is favorable (< 0) and dominant (= – 64.7); others have much smaller magnitudes, between – 1.93 and + 3.42. In this formultion, the eigenvector of a given eigenvalue correspind to a hypothetical linear combination of amono acid resides that has only one nonzero interaction, namely with itself; whereas its interactions with all other eigenvectors are zero. It follows that when a model system has a dominat eigenvalue, which may be inter preted as a dominant mode of interaction, the interaction of the system may be approximated by the interaction involving only the eigenvector of the dominant eigenvalues.

Figure 11.6b strongly suggests that the main physical interactions captured by the MJ parameters are the "hydrophobic" interactions, as has been pointed out previously. This is because the magnitudes of the amino acids' projections on the jdominant eigenvector correlate quite well with amino acid hydrophobicity scales determined from water/ oil solute transfer experimetns, one of which is included here. In general residues that are more hydrophobic (*i, e*,.those with lower water- to-oil transfer free energies) have large component along the dominate eigenvector.

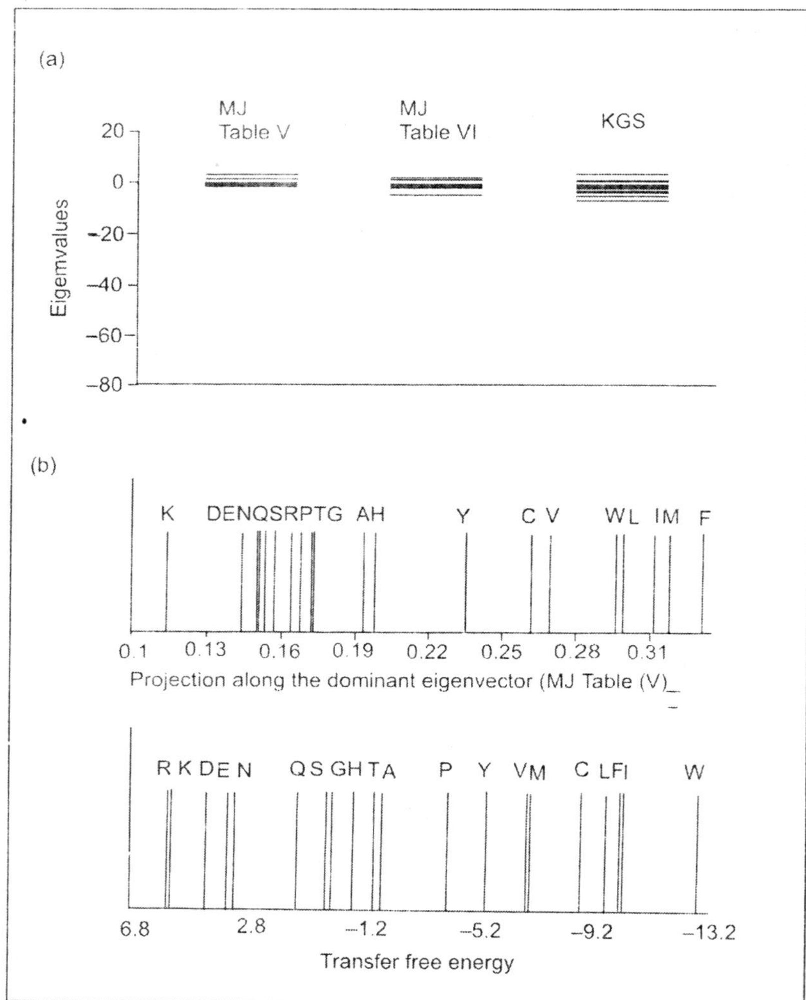

Fig. 11.6. Matrix analyses of statistical contact energies " MJ table V" and "MJ table VI" here and in the text referes to the upper half of table V, and table VI of Miyazawa and Jernigan 1985, respectively. The KGS" pairwise interaction parameters are from table III of Kolinski et al 1993. The eigenvalues of these three matrices are given in (a). The upper plot in (b) shows the amino-acid components of the eigenvector of the dominat eigenvalue of KJ table V. for details. The lower plot in (b) is an hydrophobicity scale, in which an aminoacied's exaperimetal preference to be in octanol rather than in water is given by a transfer free energy in units of Kj/mol.

These results are very similr to that obtained earlier for the 1996 version of this MJ matrix the latter is the physcial basis for a recent novel algorthim for desgning optimal sets of reduced amino acid alphabets with fewer than 20 letters. The finding are consistent with the original interpretation, based on a quasichemical approximate physcial description of solvent mediated interactions between amino acid parirs. In view of this, its is noteworthy that the MJ paramete in Shakhnovich and cowrker's 20 leteer model are not from KN table V but from MJ tabnl VI.[4] The latter was derived from MJ table V by shifting contact energies by a certain amount. Physically, this procedure may by be viewed as the adoption of a new solvent or *"refeerence state"* different from that of the original system.

In general, just like changing the solvent in a protein experiment can change the protein's conformational distribution, "shifting" a set of energy parameters can change the physicas of a model system. This is celarly demonstrated by the fact that, unlike that for MJ table V, the specturm of eigenvalues for MJ table VI lacks a dominant eigenvalue. Because the 20 letter model of Shaknovich et al. is the foundation of a large body intersting work we now take a closer look at the physical imlications of its interactions parameters. One conspicuous condequence of using Mj table VI is the prevealence of slat bridge-like charge interactions in the cores of model native conformations, as in figure 11.5b. Instead of a proteinlike hydrophobic core, the core of this model sequence is made up entirely of chared monomers.

All hydrophobic monomers are on the sufrace, non is inside. In short, compared to real proteins, the native sttucture of this model has an "Inside-out" hydrophobicity distributions. Does this matter/ Whether this ostensively protenlike feature is a cause for concern depends on what knowledge about proteins we are seeking from these models. If our interestis restricted to how higly coarse garined protein properties may arise from general properties of heteropolymers, we may view the use of MJ table Vi just as a conveinent way of implementing a model interaction scheme with sufficent heterogerity, without regard as to whether tis details are proteinlike or not. The calss of 20-letter models in question can be instrumental in such investigations. In that case, however, we should refrain from, or at least be very conservative about indentifying detailed features of the model iwth structural or mechanistic aspects of real proteins.

By contrast,if we are interested in how physical interations in proteins affect folding mechanism, it is only logical to inquire to what degree the model interaction scheme conciders with the driving forces in proteins. If this is ones' goal, it is necessary to ascertain whether the very nature of the physcial inteactions has been changed by using MJ table VI instead of MJ table V. To get to the root of this matter, we look beyond the nominal identities of the model monomers and focus on the mathematical properites, namely the signs and magnitudes, of the contact energies themselves. The native core of the 20 letter model in figure 11.5b is stablized by a network of 15 contact interactions between nominally charged monomers (dotted lines), all but one of which are contacts between negatively charged monomer types DK and Positively charged monomer, types K, R. There are 6 E-K, -3 D-K, 3 D-K, 3 E-R, and 2 D-R contacts. According to MJ table VI, the energies of these contact types are –0.97, –0.76, – 0.74, and – 0.72 respectively.[5] Second only to the –1.06 value for a pair of cysteines, these are the next four most favorable interactions in MJ tableVI. The (unweghted) average of these four strongly attractive contact types in the antive state equals – 0.80, on the other hand, the corresponding average for the six contact types among D,E R, and K that *do not* appear in the native structue is + 0.16, as they are either intermediately jto strongly repulsive or essentially neutral to sligthly attaractive: according to MJ table VI, the D-D, D-E, E-E, R-R, R-K, and K-K contact energies are 0.04, – 0.15, – 0.03, + 0.11, + 0.75, and + 0.25, respectively.

This implies that the interactions among these four monomer types in this model do have features similar to that of electrostatic interactions, although MJ table VI provides for stronger attractions between opposite charges than the repulsions between like charges. This analysis reveals the mechanism by which alternate nonnative core packing arrangements are energetically disfavored in this model. Native stability is the outcome of a competition between native and nonnative configurations. Structural specifictiy and thermodynamic stability of the native structure increase if compactnonnativ econfigurations are disfavored. The above observation means that this modle's relatively high degree of interaction specificity originates from an ionic-crystal-like interation pattern in its native core. Interactions with high specificties are likely to be need to account for real protein behaviors.

The currently accepted physical picture is that nonlocal electrostatic interactions, whose features the model in figure 11.5 b appears to be capturing, do not play a significant role in stablizing the protein core. The interaction patterns of some other 20 letter lattice models are more prtotein-like, at least ostensively. Examples include models based on KGS parameters.The two-body KGS interaction parameters of Kolinski et at. (1993) were not intended by the origianl authors to be used as the sole contribution to the energy of a model protein, because there are many other terms in their potential. The distibution of KGS eienvalues is quite similar to that for MJ table VI but not the hydrophobicity-dominant MJ table V presumably because hydrophobic effects are described mainly by the one-body term in the KGS formaulation. Nonetheless, the core of the conformation in figure 16.5 C is made up of hydrophobic monioners, with 21 contacts among A, V, L, I, M, F, , and W, encompassing 10 of the 28 possible pairings among these monomers. The KGS two-body parameters stipultes athat all these 28 contact tyupes are favorable except one (the slightly unfavorable + 0.1 for A and M).[6] The (unweighted) average energy of the 10 pairwise interaction types that occur in the nativer structure is –0.87, which is only slightly more favorable than the average of –0.68 for the 18 pairwise interaction types that do not appear in the native conformation.

The interaction pattern stablizing the native core of this model may be characterized as low-specificity huydrophobic interactions, which are belived to be a significatn contributing factor to protein native stability A similar example is a 24 mer model sequence of Betancourt and Thirumalai (1999b), whose antive core is stablized by eight contacts amsong V, L, and A mononers. Consistent with the usual notionof hydrophobicity, their interaction parameter set stipulates that contact energies between all possible pairs among V, L, and A are favorable. These include contact pairs that do and do not occur in the model native conformation in question. These observations suggest that the stablizition mechanisms of the latter two 20 letter models are quite different from that of Shankhnovich and coworkers. Besides figure 11.5b, other native structures in the model of Shamhnovich et al. also tend to have noninally, charged monomers buried in their cores, though it is sometimes possibnle to a chieve similar results with less stable model sequences that bury nominally hydrophobic monomers.

The above analysis shows that the issur at hand is not a trivial semantic quibble about the nominal identities of the model monomers, because in essence it is about the underlying mathemetical and physicl properites of the model potential function that determine folding mechanism and a model's ability or inability to predict real protein properties. The 20 letter model in question has provi\ded insight in many applications but apparently this model fails to capture some key features of protein energetics. An indications is that although the inportance of native-structure topology to protein folding is generally recongized by its authors this model has predicted a correlation trend between folding rate and relative contact order (which is a measure of the average sequence separation of contact resideues) that is *opposite* to the observed experimental trend. Significant advances have been made in the past decade in simple lattice protein modeling, but much work is still needed to achieve better matches between theory and experiment.

APPLICATIONS

We nowturn to more recent results, Because conmprehensive review is beyond the scope of this cahpter, the topic covered below are necessarily selective and obviosuly baised by our own research interests. A major purpose in presenting them here is to illustrate how physical insight can be gained from modeling.

Evolutionary Landscapes

Simple lattice protein models are useful for addressing conceptual issues in evolution. In general, a model of evolution requires a mapping from sequence (genotype) to fitness. Such a relation is often referred to picturesequely as a given model's evolutionary landscpe. Biologiclly, a sequences' fitness is derived from its funcion, and function is intimately related to the structure(s) phenotype) the given sequence enodes. Analytical models of sequence-fitness mapping are instructive and useful in many respects. But these models do not address the underlying physical mechanisms by which sequences are mapped onto structures. To tackel these question many theoreticians have adopted computational approaches, focusing on cosntructing models of sequences-stucture mapping that aer mootivated by various aspects of polymere physics.

To our knowledge , the first such effort was the seminal work of Fontana and Schuster (1987) on RNA secondary structures, which has led to an elucidation of ocntinuous versus discontinuous evolutionalry changes. For globular proteins, mutations and evolutionary issues were addressed using the HP model which was applied early on by Lipman and Wilbur (1991) to investigate whether non-lethal mutations form a connected network. Since then, protein evolution has also been investigated suing other simplified chain modles. These include approaches that employ sampling and/ or enumeration of the full ensemble of chain conformations in on-lattice and off-lattice models, and analyses based on enumeration of macimally conpact lattice conformations.

Neutral Nets

"Neutral net" is an evolutionary concept whose properties can be explored using these simple physical models. A neutral net is a collection of sequences interconnected via single-point mutations and encoding for the same ground state structure. An interesting idea emerging from these studies is that neutral net *topology* the pattern of the sequences' interconnections in a neutral net-can have an effect on the distribution of steady-state evolutionary populations among the sequences in the net. This hypothessis posits that, when all else is equal, sequences that have more connections to other sequences in the same neutral net are expected to be more populated in evolution simply by virtue of their mutational stabilities. As a determinant fo evolutionary population, simple lattice model studies show that this neutral net topology factor often acts in concert– as in the *superfunnel* scenario with a *prototype sequence*– but can also oppose the evolutionary force that arises from the relative functional fitness of individual sequences.

Van Nimwegen et al. (1999) have independently arrived at a similr conclusion via a different modeling approach. In principle, this topology-population hypothesis should be testable by extensive mutagensis exeriments. Simple lattice model studies also show that sequences in a neutral net are often homologous. Results from these investigations have been applied to provided theroetical underpinning for a recent technique that uses "averaged" information from a set of homologous sequences to reduce the statistical noise in knowledge-based potentials. To illustrate the methods used in these studies, figure 11.7 lists the sequences of two HP model netutal nets obtined by exhaustive enumerations of all possible conformations.

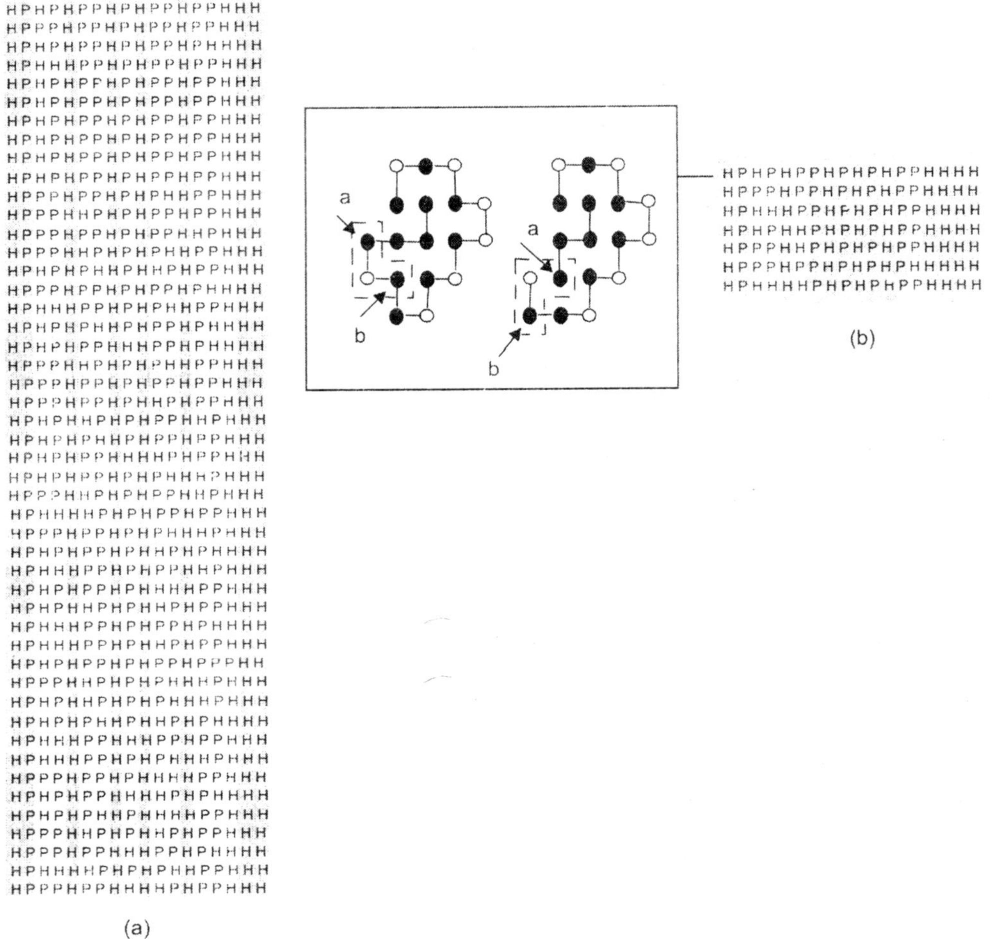

Fig. 11.7. A switch between two neutral nets. Sequences in an HP model neutral net are connected vai a network of single site H → P or P → H mutations (a) The complete list of all two-dimensional HP sequences that constitute a neutral net Each of the 48 sequences encodes *uniquely* for the left conformation in the center box. (b) A smaller neutral net with severn sequences; each encodes *uniquely* for the right conformation box. (b) A small neutral net with seven sequences; each encodes *uniquely* for the right conformation in the center box. In both the 9a) and (b) columns, sequences are listed from top to bottom in decreasing order of native stability. In each neutral net, the top sequence is the most stable; and conserved H or P monomers are shaded.Shown in the center box is a doubly degenerate sequnce that encodes for both conformations and serves as a 'bridge' between the two unique sequnces connected by horizontal lines to the box. The bridge sequnce differs from either of these two unique sequnces in (a) and (b) by only a single site substitution, at the specific sites a and b marked by arrows, respectively. thus it may be viewed as an evolutionary link between the two neutral nets.

As in real proteins, the HP model protein cores are made up of conserved hydrophobic monomers. The *"biride"* sequence in the center box serves as an intermediate step along a possible mutational path by which a sequence changes from encoding uniquely for one conformaion to another. Remarkably, some features of these HP model predictions are reminiscent of the elegant experiments recrntly performed by Sauer and coworkers. Cordes et al. found that interchanging a hydrophobic core residue and a surface polar residue in each of the two subunits of the Arc repressor homodimer changes a surface inter-subunit two-stand β-sheet in the wild-type native conformation to a pair of α-helices in the native conformation of what they called a *"switch"* mutant. Analogously, in figure 11.7, a pair of mutations that inter changes the sequence postions of an H and a P results in a similar effect: starting with the unique sequence connected to the left side of the center box, the pair of substitutions P → H at site a and H → P at site b leads to a different local fold on the surface (dashed boxes) of the HP model protein, namely from the "L" shape on the left to the "Γ" shape on the right.

More recently, Cordes et al (2000) discovered that a sequence that is a single substitution mutant of both the wild-type Arc repressor and the "switch" mutant have approximate equal populations in the ewild-type Arc repressor and the "swithc" mutant have approximated equal populations in the wild-type native conformations and the "switch" mutant's native conformation. The role of this new Arc repressor mutant is thus analogous to that of the model *"bridge"* sequence in figure 11.7. This set of experiments has important implications for evolution because it demonstrated the possiblilty that for some proteins a very small number of mutations can lead to a new fold. It show that such processes can be determined in large measure by the evolution of the hydrophobicity pattern of the protein sequence. In this regard, as we have argued above, even a minimalist notion of hydrophobicity as inplemented in the HP model can be used to rationalize coarse-grained poroperties of the mapping between protein sequences and their native structures.

Conformational Propagation

Because of their conputational tractability and broad coverage of conformational space, simple lattice models can be used to investigate protein misfolding and aggregation. For instance, lattice models have been applied to expore the feasibility of a proposed iterative annelaing meachanism of chaperonin action, which hyporthesizes that a chaperonin functions by partially unfolding a misfolded protein–multiple times if neceassary–so as to give the protein more chances to fold correctly. Lattice kinetics models of the proposed process have been used to study the physical factors involved .

More recently, simple lattice models have been applied to protein aggregation. In most cases, high-resolution modeling is not currently feasible for these multiple-chain processes. The thermodynamics cand kinetics of two-chain systems have been studied using lattice models in both two dimensions Kinetics simulations of three-chain systems in two dimensions and a concentrated solution of up to 40 two-dimensional HP 20 mer sequences have been conducted. Based on a presumed idealized packing geomety for an extensive two-dimensional aggregate state, Giugliarelli et al. (2000) investigated how aggregated chains may adopt a stable fold different from that of its single-chain native conformations. Many of these studies have been motivated by "*protein misfolding*" diseases, including Alzheimer's, systematic amyloidoses, and prion diseases.

Prion diseases of humans and other mammals are bnelived to be cause by a misfolded from PrP^{sc} of the normal cellular prion protein Prp^{c}. Other prions in lower organisms have also been discovered. Here we highlight several recent finding of Harrison et al. (1999-2001). Their modeling identifies PrP^{c} with the ground-state native) conformation of s single isolated model chain, whereas multiple-chain aggregates with two or more model chains in alternate folds different from that of the single-chain conformation are identified with PrP^{sc}. Consistent with the proposed prion disease mechanism, they found that a non-negligible fraction of HP sequences adopt theromodynamically more stable alternate folds upon dimerization and that model proteins whose singel-chain native states are less stable are moe susceptible to this form of misfolding.

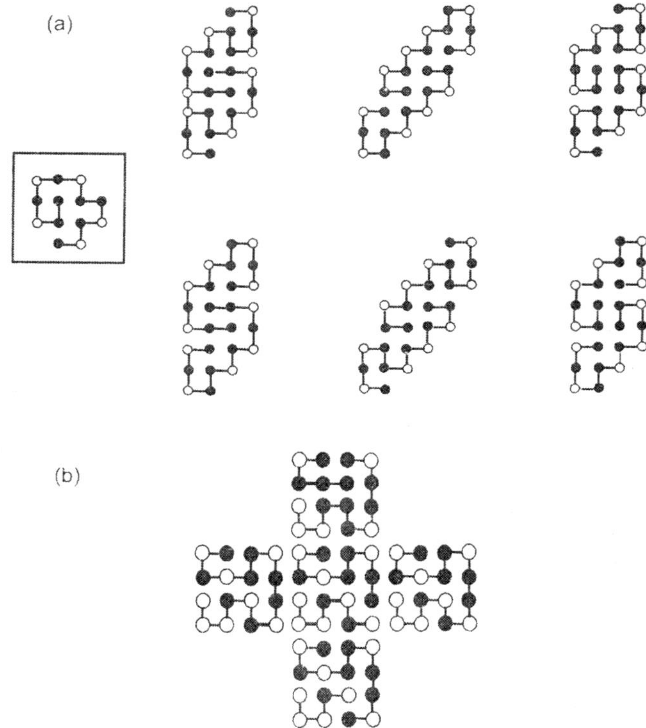

Fig. 11.8. Models of protein aggregation. (a) In the box on the left is the unique ground-state (native) confromation with six HH contacts of a single isolated HP model chain. The same sequence gives rise to six different (two-chain) homodimers shown, each of which has 14 HH contacts. These energetically equally favorable homodimer configurations are obtained from extensive conformational enumerations (Harrishon et al. 1999). They are thermodynamically more stable than any docking of two single-chain native conformations because the latter can at most have a total of 13 Hh contacts. (b) A "super-lattice" of aggregated model proteins.

Multiple alternate folds that are equally stable can arise upon dimerization of the same sequence (figure 16.8 a). This lattic scenario offeres a persective for undrstanding PrP^{sc} strains, which have distinct self-propagating properties that are apparently encrypted in their different tertiary structures. A characteristic of the prion diseases is that appraently the disease-causing form PrP^{sc} can serve as a template to lower the kinetic barriers that normally prevent PrP^{c} from connverting to PrP^{sc}. The lattice scenario in figure 16.9 suggests that multiple-chain kinetics with these postulated fratures are feasible. This perspective suggests that, vis-a vis the onset of prion diseases, the normal healthy condition corresponds to a situation in which the normal form of the prion protein is kinetically prevented from reaching theromodynamically more stable disease causing forms. Model simulation suggests that molecuar croweding effects may be a mian reason for the extremely slow kinetics. In other words, in normal situations, the conversion of PrP^{c} to PrP^{sc} may have been arrested by kinetic traps, the physics of which is reminiscent of the sluggish dynamics in glassy materials.

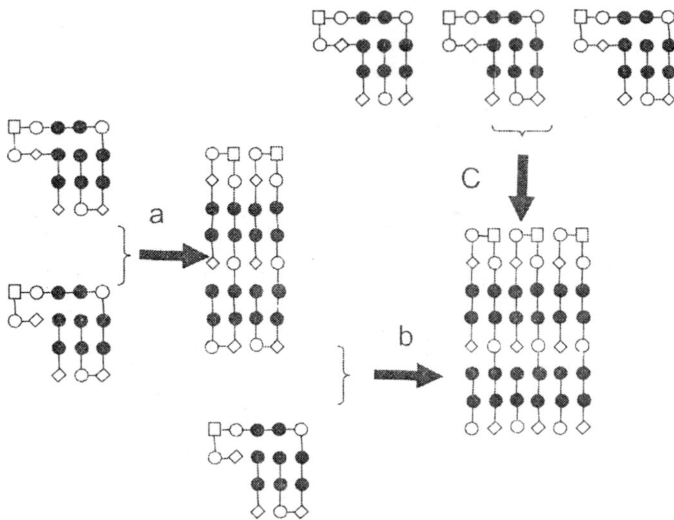

Fig. 11.9. A simple 4-letter chain model of conformational propagation kinetics. Kinetic process a isa *sponataneous* two-chain conversion from two single chain native conformations (on the left) to a more stable homodimer in which the chains adopt an alternate fold different form the single chain native conformation. b is a *templated* three-chain conversion wherby a chain in the single-chain native conforamtion is placed near an already convetted homodimer, then all three chains are allowed to refold in close proximity with one another to the more stable three-chain aggregate shown. c is a *spontaneuos* three -cahin conversion process whereby all three chains initially in the single-chain native conformation are converted to the same three-chain aggregate ias inb. Simulated templated process b proceeds at approximately five times the rate of spontaneous process.

Caorimetric Cooperativity

One of the generic theromdynamic properties of small single-chain proteins is their calrimetric two-state cooperativity, that the ratio of their van't Hoff to calorimetric entahalpirs $\Delta H_{vH}/ H_{cal} \approx 1$. Recently, it has been pointed out that this criterion,in conjunction with other experimental protein properties from small-angle X-ray scattering and NMR measurements, can be used to ascertain the extent to which a self-contained polymer model is proteinlike. The main ideas are illustated in figure 11.10 which shows that to satisfy the calorimetric two state cirterion ($\Delta H_{vH}/ H_{cal} = K_0 \approx 1.$), the denatured desnity of states has to be very narrow. The heat capacity function C_p of a claorimetrically more coopertive model protein is sharper, whereas that of a less cooperative one has a long tail of appreciable non-zero contribnution at high temperatures.

The physical origion of the latter behavior is that the average enthalpy of the denatured population of a less cooperative or non-cooperative model protein under-goes a significant shift as tempeature is raised,whereas the enthaply distribution of acalrimetrically cooperative model's denatured population does not shift significantly over a wide range of temperature. Chan (2000) and Kaya and Chan (2000 b) offer deatiled analyses. Nearly all popular contact energy based models, except three-dimensional go models, were found to fall short of meeting the calorimetric two-state cirterion to vrious degrees. This suggests that some key energetic ingredients are missing in existing simple protein models. Therefore, the calorimetric two-state criterion should be applied as a constraint to facilitate the development of more proteinlike models.

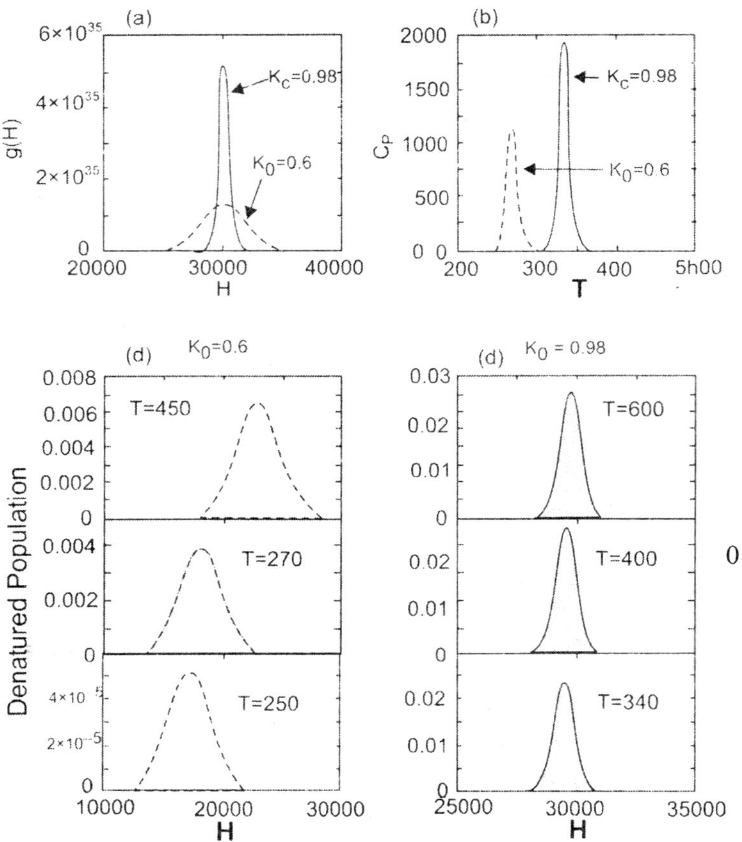

0

Fig. 11.10. Density of states determines calorimetric copperativity. Results are obtained using the random emergy Gaussian model which are based on experimental parameters from chymotrypsin inhibitior. 2 Here g (H), H, C_p, and T denote density of states, enthalpy in units of k_B, heat capacity in units of k_B, and absolute temperature, respectively; the native enthaply is equal to zero; k_0 is the population-based van't Hoff to calorimetric enthalpy ration $\Delta H_{vH}/\Delta hcal$. Here a model protein's density of states is its number of conformations as a function of enthalpy. Results for a high-cooperativity model (k_o = 0.98, solid curves) and a low-cooperativity model (k_o = 0.6, dashed curves) are compared. The curves in are the denatured parts of the densities of states of the two models; and (b) shows their heat capacity functions. The two lower plots hsow the distribution of denatured state enthaply at the temperatures indicated, for the model with k_o = 0.6 (c) and k_0 = 0.98 (d), respectively The shaded areas under the curves are proportional to the total fractional denatured populations. Note that different vertical scales are used for different temperatures.

Preliminary analyses suggest that both local and nonlocl interactions and their cooperative interplay are necessary for calorimetric two-state cooperativity. This investigation has provided novel insight into theroretical and experimental aspects of native-state conformational diversity, including a chlrificationof the physical meaning of the multiple conformation native state defined in some 20 letter models. Moreover, because the calorimetric criterion addresses energetic distribution among the unfolded non-native confromations, it also bears on the Z-score used in empirical protein structure predication.

SIMPLE LATTICE MODELING

Recent years witness extensive effforts in using simple lattice models to study protein folding kinetics. We do not repeat their many finding here, as there is no shortage of comprehensive studies and reviews on the subject, including perspectives on a proposed nucleation folding mechanism. Here we only focus on modeling issues that we belive are basic, but that have received relatively little attention. For protein folding kinetics, in-depth analyses are often necessary to match simple lattice model results to experimental data. One noteworthy example is that temperature dependences have to be introduced in to both the model intraching interaction and the *"Monte Carlo clock"* of the model dynamics in order to provide a self-fconsistent account of the experimentally obsrved chevron plots and non-Arrhenius kinetics. Extra caution has to be used in Kinetic than in thermodynamic applications of lattice models. The main reason is that, unlike continuum molecular dynamics, lattice dynamics are not governed by Newton's equation of motion.

To mimic reality, model daynamics on lattice are designed with ingredients that are untuitively recognized to be physical features of real dynamics. It has long been realized that liattice protein folding kinetics models with local chain moves alone may have serious artifacts . As a result, authors of high-coordination models have been careful in choosing moves and assigning physically plausible weight to the moves. To ensure that various possible meachanisms of protein assembly are not *a priorit* excluded, thye have desigen their lattice moves in such a way that the model dynamics allow for the slow diffusion of assembled fragments of seconday and supersecondary structure, but at the same time the assembled fragments can also dissolve and reassemble at a different location. On the other hand, the move set in many recent simple lattice model simulations of folding kinetics are limited to local moves.

These studies are still useful because they can demonstrated the posibility of a particular folding path or a folding mecahnism, for example, how the conformational search problem can be solved. But obviously they cannot be used to exclude physically plausible folding meachanisms their move set a priori precludes. Significant move-set dependences have been observed in the folding time of two dimensional Hp model sequnces by comaring two move stes, MSI and MS2.

Figure 11.11 shows a similar comparison, now over a continuous class of dynamic models that have diffrent relative weights (probabilities) for the moves in the two move sets. In this particular example, folding times in different dynamics models can differ by more than six orders of magnitude . The underlying physical reason is pictorically illustrated in figure 16.11 aA a rigid rotation (In MS2) –even at a low diffusion rate. Can bring two assembled fragments of a nonative conformation together to form the native structure,and can do so without encountering any energy barrier. However, a move set limited to local moves (MSI) must take may steps and break at least two existing HH contacts along the way to arrive at the same end point.

This implies that the kinetics with the two move sets are very different when breaking of existing contacts are energetically costly, as is the case in this calculations. It is clear from this example, as heve been pointed out before that rotation and diffusion of intact assembled fragments of a protein, as envisioned by Karplus and Weaver (1976, 1979) in their diffusion collision-adhesion modle, are precluded by lattice move sets with only local moves. Figure 11.11

Suggests that different folding meachanism would be predicted if different move sets are assumed, even though the underlying interaction potential reamains unchanged. Therefore, to again a clear picture of the relationship between protein energetics and the predicted folding mechanisms in a lattice model, the role of move sets must be taken into account.

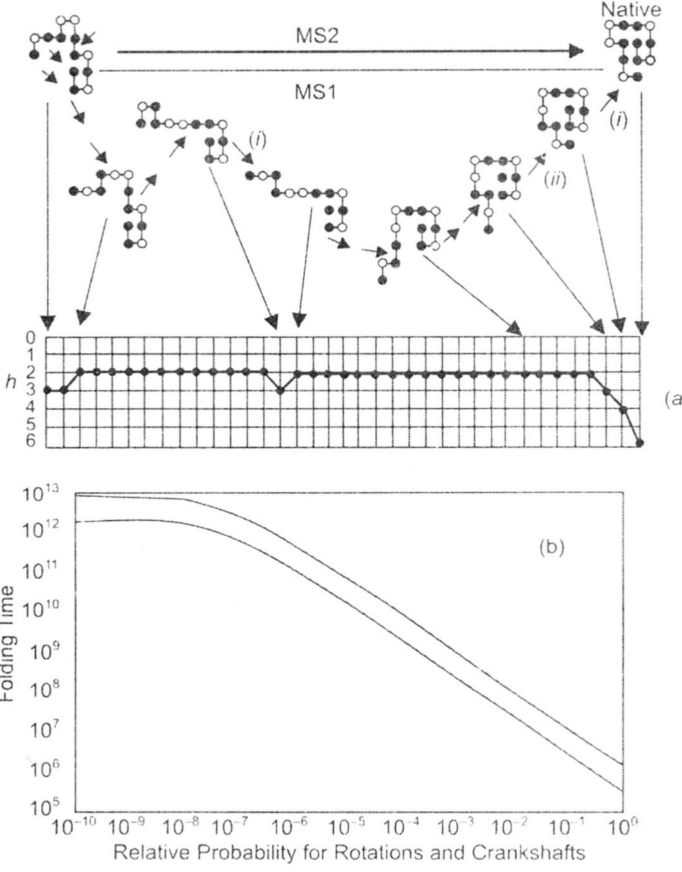

Fig. 11.11. Folding kinetics of lattice protein models depend on move set. Two move sets MSI and MS2 are compared suing a two-dimensional HP model sequence. MSI contains only three point flips and end flips, whereas MS2 includes also crankshafts and rigid rotations. Folding paths from any nonnative conformation to the enative conformation (upper right corner) can be very different for the two move sets. For example, the conformation at the upper left corner can reach the native state via one single rigid rotaions step in MS2 (top). But at least 36 MSI steps are needed if the number of energetically unfavorable breaking of HH contacts in minimized along the folding path (*a*) is the energy landscape along one such folding path, where *h* is the number of HH contacts. Selected conformations along this path are shown, chain moves (*i*) and (*ii*) are examples of three-point and end flips, respectively. (*b*) is the folding time of the model HP sequnce at HH interaction energy = – 9.3 k_BT, computed by a reduced transition matrix method for a ragne cf probabilities of the additional movs in MS1. The upper and lower curves are the time needed to achieve 95 percent and 50 percent native popultions, respectively, from an initial ensemble of open conformatins. Even relative probabilities as low as 10^{-6} speed up folding considerably in this case. him or her evalute results reported in the literature.

CONCLUDING REMARKS

We have emphasized physics-based approaches in this survey of computational methods for protein folding, from very brief sketches of all-atom simulation, continuum elctrostatics, and high-coordination lattice models to an expostion of simplified statistical mechaniscs models. This is not a comphehensive review; in fact we have devoted more effort to raising questions than to providing answer. Nevertheless, it is hoped that the brief summaries, analyses, and cited reference in this chapter would make recent development more accessible to the interseted reader, and help him or her evaluate results reported in the literature.

A few example have been analyzed and discussed in some detail to highligh issues we believe are importnat, especially with regard to physical interpretations of simple lattice model results. Protein foldingis a vast fields: many approaces are complementary to one another. Protein folding is a many faceted problem; its elucidation requires theories at many level, and models that account for different degrees of complexity. To move forward in physical understanding, it is necessary to critically evalute theoretical predictions against experiments, to clarify the relationship between the basic modeling assumption and the physico-chemical forces in real proteins,and to eastablish rigorous logical connections from a model's basic assumptions to its predictions. Clearly, the degree to which one in successful in these endeavors would determine the amount of physical knowledge and the quality of information that one can gain from any modeling efforts.

CHAPTER

12 DOCKING AND APPLICATIONS TO DRUG

Several ingredients are needed in order to efficiently and successfully search a library of inhibitiors, or drugs, with the goal of optimally docking them onto a specific target recaptor: first, an adequate molecular surface representation; second, efficient docking techniques; third, a practical way of accounting for molecular surface variability; and fourth, providing for molecular flexibility. These four ingredients yield the candidate molecules. The fifth critical component is a fast, empirical way of scoring the large number of obtained solution and ranking them. Currently, although there exist a variety of computational docking approaches, the scoring step has proven to be the most difficult hurdle. Prediction of the docked conformation of a recaptor-ligand molecule pair without any additional knowledge as to their binding sites is an extremely complex problem. The problem can be defined as follows: Given the atomic coordinates of the two molecules, predict their native bound association. Clearly, in principle, every portion of surface of one molecule should be matched with every portion of the other, in all rotations and translations. The number of possible matched configurations is immense. Further, even in the binding site can be predicted, there is no guarantee that the corresponding candidates trial ligand will not bind at other, alternate sites.

A particularly severe complication is the molecular surface variability. In solution, the surfaces of the molecules are in constant motion. Movements of side chains and surface atoms implicitly forces taking account of intermolecular penetrations of the docked molecule pair. In solution, such surface of the molecules are in constant motion. Movements of side chains and surface atoms implicitly forces taking account of intermolecular penetrations of thee docked molecule pair. In solution, such surface penetrations are alleviated by the movements of the groups of atoms on the molecular surface. By contrast, obviously, allowing full-fledged molecular flexibility, that is, allowing every two atoms to move with respect to each other while we search through databases of molecules is entirely infeasible. Thus, we need to devise some practical approaches in which some degree of flexibility will still be permitted. Moreover, in addition to molecular surface variability, one need to consider domain motions. Docking rigid molecules may miss the correct solutions altogether. Below we describe efficient computational approaches to the docking problem.

Computational techniques to be focussed. The first is a rigid body docking technique, the second allows conformational flexibility of molecular parts, through hinge-bending motions We note the geometrical representation of the molecular surface they currently empioy and pattern matching algorithms to detect geometric surface complementarily. These techniques derive from computer vision and robotics. They are highly efficient, yielding candidate solutions in very short CPU matching times. They are straight forward to run, suing an SGI workstation or a PC. However, despite all of these encouraging attributes, we are still faced with very serious difficulties, namely, the scoring and ranking of the obtained, predicted, docked configurations. Are there then ways to quickly sift through potential docked configurations and rank the correct native configurations at the top of the obtained list?

Analyses of protein-protein interfaces have illustrated that the binding interfaces do not necessarily have the largest extent of buried surface areas. Furthemore, native -like bound conformations do not manifest the largest nonpolar buried surface areas as compared to other potentially feasible docked solutions. They do not contain the largest number of hydrogen bonds or the smallest number of unsatisfied buried polar groups. In solution, these are likely to be handle by surface motions, eliminating such unfavourable energy contributions. Hence, the problem is how, despite these hurdles, to still conceivably be able to detected candidate lead compounds/ protein inhibitors. Here we outline one potential way of handling this problem, namely, through utilization of a *library of functional epitopes*. Such a library can be generated efficiently using techniques based on the same principles. We further consider the pros and cons of their utilization. This chapter is divided into two parts. In the first, we describe computer vision based rigid and hinge-bending docking algorithms, and the generation of potential solutions, In the second, we focus on the generation of functional epitopes, and some attributes of binding epitopes that we have recently obtained.

Docking is an alternate, complementary tool to that of drug design. In drug design, the molecule is built normally in the binding site of the receptor, by ligating groups of atoms, in a step-by-step, trial-and-error calculation. Here we focus on docking algorithms. In such algorithms the three-dimensional structures of the ligands are taken as single entities from the database.

RIGID-BODY DOCKING

We have devised two rigid-body computer vision based docking techniques. In both the representation and the matching are 3D rotation and translation invariant. The principles of both algorithms are similar. Here we describe briefly only one. To align the surfaces of two molecules in a complementary manner, we need to computer rigid transformation that superimposes the surfaces without allowing one molecule to penetrate or overlap the other. The obtain hypotheses for such transformations it suffices to align a triplet of ordered non-collinear points (congruent triangles) from both molecules. However, it may happen that there are no three independent matching point-pairs between the receptor and the ligand. For docking, the points we utilize are those describing the molecular surfaces. These are computed to accurately represent the maxima (*holes*) and minima (*knobs*) of the shape function. We also compute the surface normal, associated with the point. Below, these points are dubbed *critical points*. The docking strategy utilized only Paris of matching critical points with the additional geometric information on their normals.

In order to computer a candidate rigid transformation, we need to detect a pair of critical points in both molecules that share the same internal distance, and if superimposed, have opposing surface normals. This reduces the number of potential docked configurations, and concomitantly reduces the run-time complexity of the program. For each pair of critical points form each of the molecules (any two critical points combination for the protein-protein docking; two holes for the rejector and two knobs for the ligand in the protein-drug cases) we computer transformation invariant *signature*. The signature includes the distance between the two critical points (d); the two angles formed by the line between these critical points and their respective normals (α_1, α_2) ; and the torsion angle defined by the two normals and the line segment between then (ω). If the signatures of the ligand and of the recaptor are compatible, the best rigid transformation between the two pairs of the critical points is computed. In compatible pairs (1) a knob in one molecule matches a hole in the other; (2) the difference between the two distances (*d*) in the two molecules in less than a predefined threshold; and 93) the difference between the corresponding $\alpha's$ and the two $\omega's$ is also within the allowed limits. We impose stronger cumulative constraints on the alignments of the surface normals: the sums of the difference of the two $\alpha's$ and of the three angles should not exceed certain limits. In order to qualify as a candidate match, if one of the normals is not well aligned, the other must compensate and have a reasonable alignment.

There are several advantages in employing the surface normals in the signature, in addition to the critical points: (1) we need only two critical points in the interface area; (2) the combinations of finding correct matches is lower (3) the orientation of the normal is used for fast rejection of a large number of wrong solutions. Because the matching is computed for local patches of the surface, it is essential to verify that the solution is vable for the entire protein. A problem may arise when the entire ligand molecules is brought to dock onto the entire recaptor. It is conceivable that although there is a good complimentary at the matching interface, there could be an overlap between two molecules elsewhere.

To verify that the docked configurations do not seriously interpenetrate each other, we compute a scoring function. The function is based on geometric features: surface contact is awarded, and overlaps where ligand atom centers invade the outer shell of the molecular representation of the receptor are penalized, but retained. However, potential solutions where ligand atoms fall into the "core" of the recaptor are rejected. By allowing a certain extent of intermolecular penetrations, we implicitly take into account molecular surface variability. The details of the scoring and ranking function along with a sample hydrophobicity function utilized in the docking of protein-protein pairs are described elsewhere. Inspection of the obtained complexed conformations immediately reveals that many molecular associations are relatively similar to each other. These represent virtually the same docked solution.

Clustering similar solutions both reduces the number of solutions and allows focusing on alternate configurations. A good clustering scheme should group similar solutions. However, at the same time it should properly distinguish between alternate binding modes. The relative rotation between two conformations is computed form their individual rotations against the initial conformation. This computer vision based docking algorithm is highly efficient. Its fast CPU matching times–on the order of minutes on a PC, even for large protein-protein cases– allows large-scale docking trials of large and small molecules.

We have carried out such experiments on about 230 receptor-ligand molecule pairs, including 26 protein-protein "*bound*" and 19_n "*unbound*" protein cases, where the structures have been determined separately. These have variable, imperfectly matching surfaces. We have further docked about 160 cases of protein-drugs, 13 cases of protein-DNA, and 14 cases of DNA-drugs (Norel et at. 1999b). The molecules vary substantially both in size and in the hydrophobic/polar chemical nature of their surfaces. The location of the active site, or the identity of particular residues/atoms in either the receptor of the ligand, which participate in the binding, has not been taken into account in any of these.

As can be seen in Norel *et al.* The quality of the results (low mods of the docked configurations as compared to the crystal-complex) and the speed of our techniques are highly attractive. Thus, based on this examination of the docking and the results it has obtained, we conclude that shape complementarity almost always enables attaining correct docked configurations. Nevertheless, it is insufficient for ranking. Above, we have outlined a computer vision based algorithm for *rigid* docking. On the technical side, rigid docking algorithms are faster and obtain fewer candidate solutions than those allowing molecular flexibility. If the flexibility is limited, treating the molecules as rigid bodies will still detect the native bound conformations. Such case are handle simply by the "error thresholds." Below we extend the docking repertoire , by describing an efficient method for handling a special kind of flexibility, namely, allowing *hinge-bending* motions.

HINGE-BENDING FLEXIBLE MATCHING

Here we dock a ligand onto a receptor surface, allowing hunger-bending movements of domains, subdomains, or any structural parts. The advantage of our algorithm is that we allow all angular rotations, although we still avoid a conformational space search. We pick a hinge point, to divide the molecule into two parts. However, we do not dock each of the molecular parts separately, necessitating a subsequent reconstruction of the consistently docked molecules. Instead, we dock all parts simultaneously. In particular, we utilize the position of the hinge from the starts. Like pliers closing on a screw, in an automated fashion, the receptor closes on its ligand. Movements are allowed either in the ligand or in the large recaptor.

The algorithm mimicks the so-called "*induced*" molecular fit. More than one hinge can be allowed in the docking. Interestingly, contrary to intuitive expectation, there is no increase in the *matching* (docking) times as the number of hinges increases. However, the ranking (scoring) times increase. This is the outcome of the necessity to examine two types of pentrations, both intramolecular part penetration and intermolecular as in rigid-body docking. Hinge-bending movements are frequently related to molecular association. Such motions can involve domains, subdomains, loops, secondary structure elements, or be between any portions of the molecules connected by flexible joints.

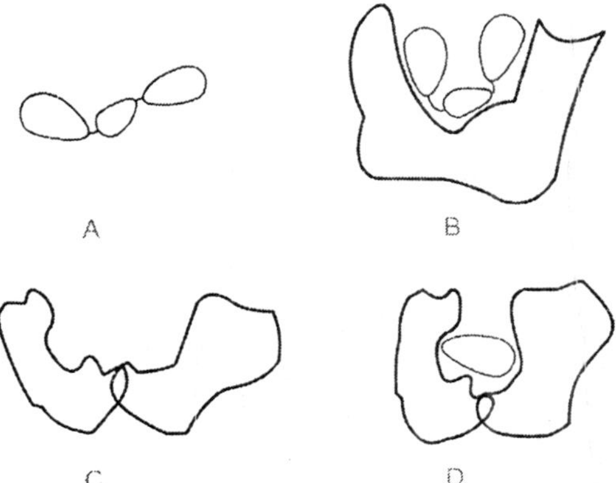

Fig. 12.1. A schematic illustration of the molecular hinges allowed in ligands and recaptors in the bending docking. (A) The hinges in the ligands. (B) The ligand in hinge bent to optimally fit in the recaptor active sit. (C) The hinge is in the receptor. (D) The reeptor is hinger-bent, closing on its ligand.

Currently, we have implemented the hinges at points and at bonds. Clearly, in reality, the only molecular movements that are possible are rotations around a covalent bond. Hence, in principle, we could have limited the matching to the simpler, one degree of freedom rotation. Nevertheless, by enabling full three dimensional rotations around a *point,* rather than around a bond, we can implicitly take into account several rotations about consecutive or nearby bonds. If we so wish, the point rotation can be restricted to bond rotation. We can also permit the molecule to rotate around several, simultaneous hinges.

In enabling several hinge motions to occur *mimulta-neously,* in effect we are simulating the cumulative effect of multiple mutations, each introducing a limited motion. In practice, to date, the algorithm has been implemented to enable two simultaneous hinges although initial prototypes for multiple hinges are already tested. We have successfully applied our algorithm to a number of *bond* and unbound molecular configurations, achieving fast recognitions times of their surfaces. As in the rigid-body case, the atomic coordinates are extracted form the PDB.

The location of the hinges has been pre-defined, through a comparison of similar structures in different, that is, "*open*" and "closed" conformations, in cases where both types of structures exist. When we applied our method to molecules taken from bound conformations, we were able to reproduce the molecular association, in excellent agreement with the experimental crystal structure. Because the ligand and the recptor have been picked in their complexed from, near-native geometrical solutions correspond to those with near zero rotations and translations. The binding modes our algorithm has obtained have small many compared to the native crystal complexes. The average rams of a correct solutionis 1.4 Å, with average run-time for each complex around 1 min (on a SGI-Challenge R 8000 machine). Additionally, for the bound cases, "correct" bound configurations typically rank high. In addition, geometrically well-fitting, alternate binding modes have also been generated.

THE ALGORITHM

For simplicity and clarity, here we outline description of this robotics based method for the single hinge case, where the hinge is positioned in the ligand molecule and the receptor is assumed to be rigid. Nevertheless, although we describe the algorithm for flexible *ligands,* the roles of the ligands and the receptors are interchangeable. That is, the hinges can be positioned in the receptors as well. Elsewhere we describe further enhancements that we have implemented for the multiple hinges case. The algorithm is based on a *voting scheme* designed to find the most suitable ligands (out of a library of ligands), and for the respective transformations for matching each of their parts to a given receptor.

Both the ligand and the receptor molecular surfaces are described by their 3D sets of "interest points." This molecular surface representation is essentially similar to the one described above for the rigid body docking. Details are given in Lin et al refs. 36. 37. Here we proceed to outline the hinge-bending algorithm and its two phases, the *preprocessing* and the *recognition.*

An Overview of the Algorithm

The ligands may undergo translations and rotations of their parts in order to optimally dock to the surface of the receptor. The ligand information is stored in a look-up table, which is invariant to this type of transformations. The table is generated in the *preprocessing* phase of the hinge-bending algorithm. The position of the hinge in each of the ligands is pre-determined by considerations of its more flexible joints. Nevertheless, because this robotics-based algorithm is fast, numerous other, alternate, hinge locations may be tried. The structure of the receptor is presented to the system in the second, *recognition* phase of the algorithm. If a ligand has an interest point configuration (*i. e., portion* of the molecular surface) similar to the receptor interest point configuration, the algorithm scores a match. This is done by casting a vote for this ligand, together with the computed location of its hinge... This hinge location is computed from the transformation between the corresponding receptor and ligand interest point surface configurations.

The highest scoring (voted for) hinge locations of candidate ligands are sought. No knowledge of the binding site, or of the hinge locations relative to the receptor, is assumed. In the *preprocessing* step, the ligand molecule (model) is described as a set of interest points. The (predetermined) hinge location is positioned at the origin of a 3D Cartesian coordinate frame. This frame is the *ligand frame.* The orientation of this frame is set arbitrarily. For each non-collinear triplet of interest points, in each of the ligands' parts, a unique triplet-based Cartesian frame is defined. This is the triplet *frame.*

One way in which this can be done is by defining the origin at the first triplet point, the x-axis in the direction of the vector from the first point to the second, the Z-axis, which is the normal to the triangle plane in the direction of the cross-product of the two vectors originating from the first triangle point, and the y-axis in the direction of the cross product of the x and z unit vectors. However, in practice, in order to not overlook any potential solution, we define three such frames, one for each triplet point. The *sphep signature* of each triplet of points is the ordered triangle side lengths. This geometric shape signature of the triplet constitutes an address to a look-up hash table. The information that is stored at this entry at this address is the lined identification , the part number, and the transformations between the *triplet frames* and the *ligand frame.* In the *recognition* step, the molecular surface of the receptor is similarly described by its set of intersect points.

All non-collinear triplets of the interest points of the receptor are considered in the docking stage. For each of these triplets, the triplet-based Cartesian frames are computed. Each is the *receptor triplet frame.* The mode of calculation is as above. This calculation is invariant under rotation and translation. Thus (almost) congruent ligand triangles will result in similar values. The look-up table calculated in the preprocessing phase is entered using as an address of the currently computed ordered triplet of triangle side lengths of the receptor. For each ligand-record present at that entry in the table, a *candidate ligand frame* is computed, calculated by applying the pre-recorded ligand transformation at that (hash table) entry to the current *receptor triplet frame.* The origin of the candidate ligand frame is the candidate *hinge location.* We vote for the identity of the ligand molecule, along with the location and orientation of the *candidate, hinge-centered, ligand frame.* At the end of this procedure, we seek high scoring pairs of (*ligand, hinge location*). This hinge location defines the 3D translation that the ligand would need to undergo in *candidate* docking.

The appropriate rotations are calculated separately for each part only at a later scoring and filtering stage. In the current step, hinges that have received a large number of votes are selected.

This hinge-bending algorithm exploits the fact that both parts of the molecule share the same hinge. The essential point here is that the way it is taken into account is by locating the origin of the reference frame of the ligand *at the hinge.* In this way, both parts contribute votes to a reference frame at the same location, although the orientations of both parts with respect to each other may differ. Most importantly, by picking up votes from both molecular parts, a ligand, which might otherwise have only a small portion of surface complementary to the receptor surface in each of the parts, may still score high.

Although each of the individual parts of the ligand can obtain an insignificant score, the sum of the votes obtained form both of the ligand's parts may yield an overall acceptable match, which can be automatically detected. This algorithm is general. It can handle the rigid docking as a particular case.

For the rigid-body docking, the ligand, reference frame is located arbitrarily. Above, we have outlined the algorithms for the case of a single hinge. However, it may be extended to

multiple hinges. There, rather than pick a single ligand frame, multiple ligand frames are **defined**. Each of these frames is centered at a different, pre-defined hinge. Next, in the **preprocessing** step, for each ligand triplet in a single part, we encode the transformation of its **triplet frame** to all ligand frames of that part. Hence, if a part has two hinges (as might be the **case if the** part is the middle of a protein), we will store two transformations for each triplet. **On the** other hand, if a part has only one hinge (*i.e.*, if it is an edge part), we store only one **transformation**. (A drugs-ligand with a branched structure may have more than two hinges per part). The recognition phase is unchanged. The exception is that each receptor triplet participates in the voting for as many frames as the number of different transformations stored in its table entry. Run time at the different steps of the algorithm on several examples are given by Sandak et at. (1999).

THE RANKING PROBLEM

Above we have described two algorithms for docking ligands to the surfaces of their respective receptors. The two methods are very efficient, producing high-quality results for cases where the complementary between the surfaces of the molecules is relatively good. Nevertheless, despite the advantages of these techniques, they face considerable difficulties as the fit deteriorates. This problem is uniformly encountered by all current methodologies.

If docking initiates from entire molecular surfaces where the structures of the molecules have been determined separately, that is, when they are in their uncomplexed conformation, a range of solutions is obtained.

The problem of how to rank the solutions, with near-native solutions ranking near the top of the list, is still a major hurdle. Because the number of candidate solutions may be large, detailed chemical calculations are unrealistic, and fast empirical approaches have yielded unsatisfactory results. Among the many schemes that have been tested over the last years we find derivation and utilization of pairwise, knowledge-based potentials, either residue-based or atom-based; calculations of the nonpolar buried surface area, or of the total buried surface area; inspection and counts of hydrogen bonds and salt bridges; minimization of the number of buried polar groups; and minimization of the energies. Any of these may be satisfactory for some case; however, they fail for the others. Thus, even if the docking programs are good and can scan quickly large databases of drugs or inhibitors, the ranking problems till constitutes a huge stumbling block

BINDING EPITOPES

The definitions of binding epitopes vary. In general, the term refers to a recurring pattern of molecular surface in a family (or, families) of proteins, where at least in one family member the site is known to be a binding site. A similar binding epitope can also be observed across family boundaries. To construct a library of binding epitopes we may start by picking known site as a *pattern* , and search for similar portions of surface, or of arrangement of residues at/ near the surface, in other molecules.

Characteristics of Binding Sites

Locating a priori unknown active (binding) sites on the surfaces of enzymes or receptors in important for a number of reasons. First, detection of binding sites is the initial step in the design of drugs. Second, being able able to identify actives sites on the surfaces of proteins enables their modification to enhance their activity toward specific ligands and functions. Knowledge of likely binding sites enables focusing on these locations in the docking simulations. Finally, they can further be used to filter and rank the obtained solutions at later stages.

Although the former is more efficient, we may nevertheless also wish to obtain alternate geometrically feasible configurations. In an insightful review, Ring (1995) surveys the definition of "what makes a binding site a binding site." She describes some guidelines, and proposes that binding sites are in general depressions in the protein surface, "in which there is greater than average degree of exposure of hydrophobic groups." These frequently contain disordered, easily displaced water molecules. Ringe further suggests that conformationally flexible residues at the binding site may be particularly useful in replacing the disordered water by the competing ligand Laskowski et al. (1996) have also examined the active sites. Their extensive study indicates that active sites of enzymes typically consist of large clefts. Their study shows that in the majority of single Chian enzymes, the ligand binds in the largest cleft. Hence, active sites of enzymes may well be identified using geometrical criteria alone. Peters et al. (1996) utilized alpha-shapes a computational geometry tool, in an automated search for ligand binding sites on protein surfaces. This strictly geometry-based algorithm has also found a correlation between the depth of the cleft and enzyme active sites. In addition, a correlation between patches of hydrophobic surfaces and binding sites has also been noted. Both the results of Peters et al. (1996), and of Laskowski et al (1996), confirm an older observation that protein-binding sites are characteristically more shallow, unlike the enzyme active sites. A straightforward approach to locate an active site in proteins if their structures are available in to select a known active site of a (model) protein and use it as a template, searching for similar ones in target proteins. This can be done by initially describing the molecular surfaces of the proteins and picking the points that faithfully represent the surfaces.

Alternatively, we can use the location of the atoms that line the surface, or are near it. In the latter case, we may either select the coordinates of surfaces atoms or of C_α- atoms. C_α-atoms may be those whose atoms line the molecular surface, or more likely, they will be in a certain shell form the surface, and will belong to functionally important residues, such as those of the catalytic trade of the serine proteses. Such search methods need to be flexible enough to allow both inexact and partial matching of the model and target molecules. Molecular surfaces are flexible. In particular, that may hold for active site regions.

Flexibility may be particular, advatageous for proteins with a broad range of ligand/substrates. The less pecific the receptor-ligand (enzyme-substrate) interactions, and the broder the range of binding the more flexible the binding site is likely to be. Further, enzymes and proteins haveing a broad range of bound substrates/ligand may also make substantial use of water molecules to mediate binding to their ligands. Ladbury (1996) outlines the rationale and the consistent evidence, arguing that water is likely to have a dual role in binding. Water molecules may increase the promiscuity of the binding; however, water might also increase the specificity and affnity. In general bridging water molecules are not accounted for in molecular surface matching algorithms. That is largely due to the difficulty in predictiong their locations on the surfaces of the proteins.

A further difficulty is to pinpoint the residues that are critical for binding . Residues may be observed to be in contact with their cognate ligands in a number of different members of the family. This, however, does not necessarily imply that these residues are critical for binding. There is a growing body of examples showing that mutating such residues still enables molecular association. What happens is that owing to side chain flexibility, side chains that are near by move, taking over the role played by the residue that was originally at that location. A binding epitope should therefore always be taken with a grain of salt. Still, as initial guesses, especially when a certian sloppiness in the stipulated requirements is allowed, they are useful to have.

Conserved Residues

A method such as Geometric Hashing enables carrying out a comprehensive and systematic structural analysis of protein-protein interfaces. Beacause the crystal structures of the families of these interfaces are geometrically similar the dataset can be used to study structural charactersitics of protein-protein interfaces. In particular, such a dataset may be used to address questions such as, given a geometric similarity among members of a protein-protein interfaces family, how similar are their binding furfaces? Which residues at the interface are conserved? What are the determinants the binding (hydrophybicity, electrostatic, etc)/ Whic residues are prefereed, avioded, or neutral at the interface/ Which resideus are poor interface formes/ These questions have important consequences for both protein-protein binding prediction and protein ligand design.

An insight into such equstions should enable searches for homologous, potential sites in other structures, where the existence/location of these sites are unknown. A number of studies addressed the question of which are the critical residues at protein binding sites. The studies examined either a single or a few protein-protein interfaces. Bogan and Thorn (1998) carried out an extensive analysis of alanine scanning mutagenesis.

Although the total number of mutations was large, the number of protein interfaces was small, with some of the interfaces closely related. Recently, we have shown that although binding sites are hydrophobic, they are seeded with conserved poor residues at specific locations, possibly serving as energy "*hot spots.*" Our results have confirmed and generalized the alanin scanning data analysis. In that earlier study, Trp, Arg, and Tyr were observed to constitute energetic hot spots. They were rationalized by their polar interactions and by their surrounding rings of hydrophobic residues. Nevertheless, there was no compelling reason to explain why it is specifically these residues. Out study has illustrated that other polar residues are similarly conserved. Conserved residues that are in contact across the protein-protein interfaces have been obseved in all the examined families. These results are based on 11 clustered interface families, comprising a total of 97 crystal structures. The families have at least five members, with sequence similarity between the members in the range of 20-90 percent.

The matching was carried out by the Geomertic Hashing structural comparison algorithm, the conserved residues are at spaticlly similar environments. Additionally, for the enzyme-inhibitors, we have observed that residues are more conserved at the interfaces than at other locations. On the other hand, antibody-protein interfaces have similar surface conservation as compared to their corre-sponding linear sequence alignment, consistent with the suggestion that evolution has optimized protein interfaces for function. Protein-protein interfaces are largely hydrophobic, with hydrophobicity being a dominant force in protein-protein interactions.

Analysis of protein-protein interfaces has shown that the extent of hydro-phobicity,as mesaured by the nonpolar buried surface area between the two chains, may vary to a large extent between the interfaces. Consistently, there are also indications of the existence of hydrophlic the importance of hydrophilic interactions in biological function. With regard to specific residues, the reported preferences vary form case to case, mainly due to the change in the systems studied. In studies of a specific system, it is difficult to differentiate between residue conservation conferring binding specificity and conservation owing to the role of the residues in constrituting energy hot spots.

Interface Residues

We have examined the distributions of all interface residus, of identically matched residues between the representative of the family and each of the family members, and of conserved residues in the whole family. For individula families, there are significant preferences of certain residues to be at the interfaces. Howerver, when examining the counts of indentically matched residues, the preferences become much less significant. Although the distribution of all interface residues yields simple statistics, the propensities of conserved residues provide clues to the more important driving forces.

Table 12.1. Propensities in terms of hydrophobic or hydrophilic residues, classification for all residue identical residues, and conserved residues.

Residue classifications[a]	*Total interface residues*	*Identically matched interface residues*	*Conserved interface residue*
Hydrophobic	5.9	4.6	5.7
Aromatic	6.5	3.7	5.4
Hydrophobic + Aromatic	12.4	8.3	11.1
Polar	9.6	7.0	14.9

*Hydrophobic residues include Ala,Vla, Pro, Met, I l e, and Leu; polar residues inclued Asp, Glu, Lys, Arg, Ser, Thr, Asn, and Gln; aromatic residues include Phe, Tyr, His, and Trp.

Table 12.1 summarizes the propensities in terms of hydrophobic or hydrophilic residue classifications. Here aromatic residues are considered hydrophobic. The table shows that with respect to the total number of interface residues, hydrophobic residue have higher propensities than polar residues (12.4:9.6), consistent with previous, statistically based studies. However, remarkabley, the preference of hydrophobic residues decreass when analyzed in terms of the propersities of identically matched residues (8.3: 7.0). In terms of conserved residues, polar residues are observed to dominate (11.2: 14.9). Hence, we observe a trend of preferred conservation of polar residues in protein-protein interfaces. A comparison of our conserved interface residues with experimentally indentified energetical hot sports illustrates that the two most conserved residues are cysteine and argining.

The high propensity of cysteine stems from the disclfide bond conservation, either to maintain the interface structure or to chemically bind interfaces together. The indentificaion of cysterin evalidates our conserved residue propensity computations. Table 12.2 and figure 12.3 show that our selection of conserved residues corresponds to experimentally identified hot sport. There is a surprising match between the propensities of most of our conserved resieues and the residue enrichment in hot spots complied form the database of alanine scanning mutagenesis. Except for a few outliers, the correlations coefficient of the experimentally determined amino acid enrichement and our computed conservation propensity is 0.72

Of the unmatched residues, alanine is clearly incomparable. Cysteine is favoured in our database owing to its forming covalent bonds. Tryptophan has a lower propensity in our analysis, probably because of tis rareness. Serious disagreement is found for lysine. Lysine is well represented in our interface database. However, althourh in our analysis we failed to identify conserved lysine residues, lysine has a good enrichment in the alnine scanning data. A possible reason for this failure may relate to our high criteria of cutoff percentage (80 percent) for a conserved interface resideu. Alternatively, lysine is a highly flexible surface residue, projecting into the solvent.

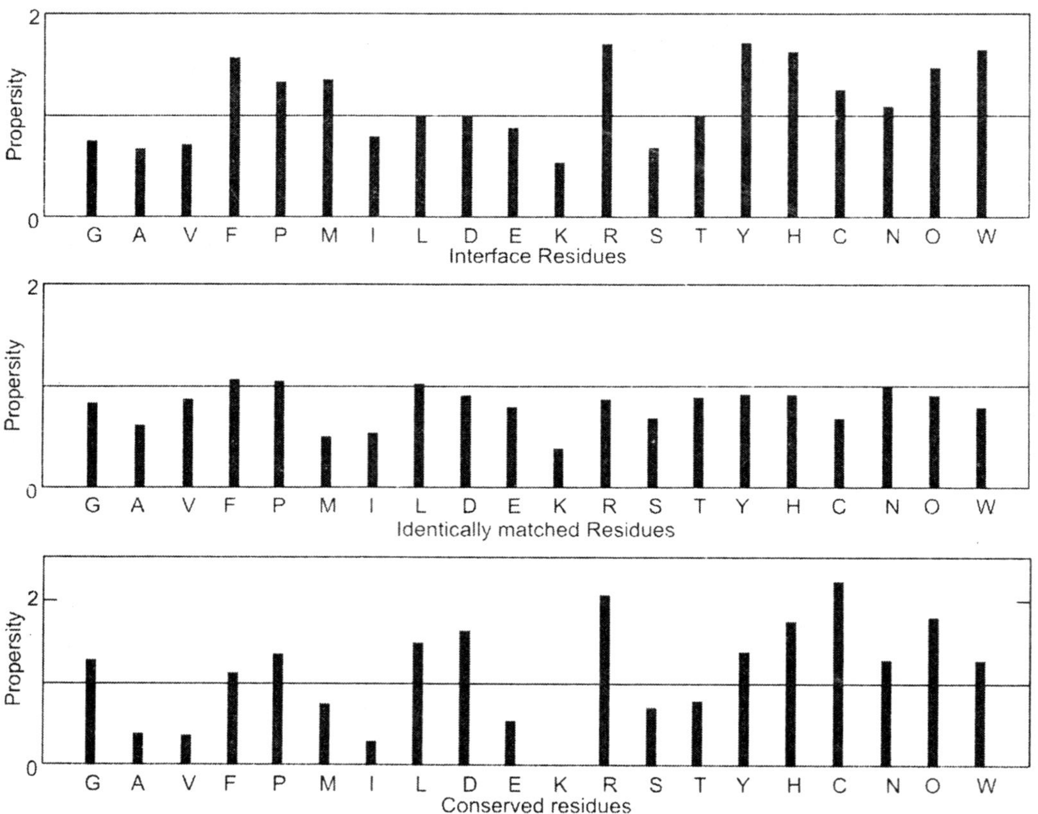

Fig. 12.2. Propensities of all interface residues (A), of identically matched residues (B) and of conserved residues (C).

It is quite possible that it is not detected as a matched residue parir in the super-positioning owing to slightly larger shift in its position. At first sight, it appears there is also a serious disagreement between our results and the alanine scanning with respect to leucine. However, if we combine the contributions of lecine and isoleucine, we obtain good agreement (1.80 from alanine scanning, 1.74 from conservation propensity). This agreement further addresses Bogan and Thorn's question as to why isoleucine has a higher frequencey than leucine in their database. Taken together, the solution may be that one should consider the sum of leucine and isoleucine, as we did in figure 12.3. There are also disagreements in the lower third of

Table 12.2. Conserved residue propernsities vs. experimentrally determined enrichment in hot spots.

Residue	*Propensity to conservation*[a]	*Enrichment in hot spots*[b]
Val	0.36	0
Ser	0.67	0.21
Thr	0.75	0.28
Gly	1.29	0.45
Met	0.74	0.54
Phe	1.12	0.56
Gln	1.73	0.58
Glu	0.52	0.68
Asn	0.24	0.93
Lsy	0	1.17
Pro	1.33	1.25
His	1.68	1.49
Asp	0.59	1.67
Ile	0.29	1.79
Leu	1.45	0.01
Leu + Ile	1.74	1.80
Tyr	1.35	2.29
Arg	2.01	2.47
Trp	1.22	3.91

[a] The propensity of conservation (P_k) of a resude to occur at the interface is calculated as the fraction of the count of residue k in the interface as compare with its fraction in the chains. $P_k = (n_k/n)/(N_k/N)$, where N_k is the number of conserved reisdues of typr k at the interface, n is the number of residues at the interface, N_k is the number of residues of type k in the chainn and N is the total number of resudues in the chain.

[b] Enrichment in hot spot gives fold enrichemnt of that resudue type in hot spots ($\Delta\Delta G \geq$ 2kcal/mol in alanine mutations) over whole database of 2,325 alanine mutations.

the table, particularly in Gln, Phe, and Gly athough the difference in Gly can be related to flecibility, we have no explanation for the more frequent occurrence in Gln and Phe in our interface families as compared to the ananine scanning, apart from the limited number of dissimilar interfaces examined by the alanine scanning.

In general, we see a higher preponderance in conserved polar residues (his, asn,Gln, Thr, Ser), or partially polar (Phe, Met), as compared to the alanine scaning. Further, Bogan, and Thorn (1998) have examined the location of the hot spots accross interfaces. They found that the hot spots are usually located around the center of the interfaces, and hence protected from bulk solvent. This is also observed in our location of conserved interface residues. We note, however, that the number of datapoints, particularly the number of conserved interface residues, in small. Some of the outliers may be due to poor statistics. Another point worthy of note is with regard to the residue conservation and portein function. Althouth these residues recur in all interfaces, with the interfaces belonging to different families and fulfilling different functions, it is still possible that some conservation is due to protein function.

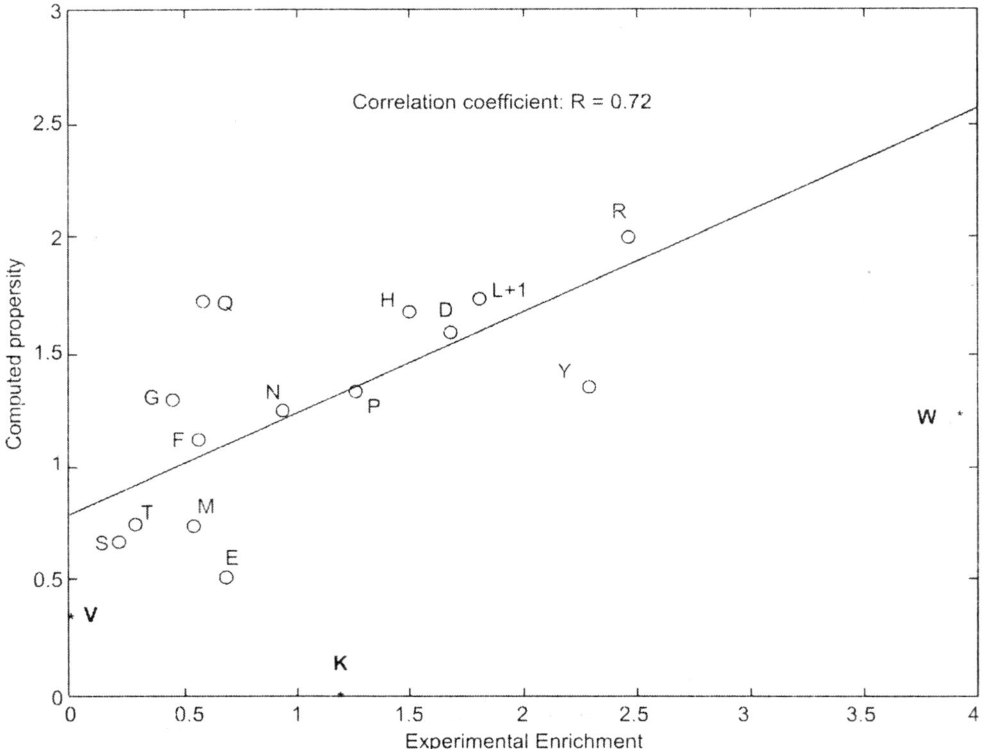

Fig. 12.3. A correlation of experimentally determined anino acid enrichemnt and our computed conservation pro-pensity. Residues are indicated by their one-letter code. Outliners are indicated. They are not included in the least square fitting. Note that Val is classified as an outliner due to its zero value in the experimental results.

Hence, although overall interfaces manifest a higher frequency or occurrence of hydrophobic residues, the polar residues are those that are preferentially conserved. Within these, Bogan and Thorn find a special enrichment of Trp, Arg, Tyr, Asp, Pro, and His. Analysis of the residue conservation in a large number of interfaces shows Arg, Gln, His, Asp, Pro, and Asn to be especially enriched.

CONCLUDING REMARKS

Our goal was to detect binding epitopes shared by members of the same family. In particular, we sought to proble the potential exstence of common principles that hold in general for protein-protein binding sites, and as such can be used in docking. Several observations, have been made in the literature. First, binding sites are generally hydrophobic, indicating that the hydrophobic effect plays a major role in protein-protein interactions. Second, the extent of the hydrophobicity is variable, with some interfaces considerably more hydrophobic than otheres.. Consistently, hydrophilic side cahins play a more important role in binding than in folding. Third, for some interfaces, it has been shown that electrostatics plays a major role, sterring the ligand onto the binding sits of the receptor. Fourth, the availability of the large number of clustered interfaces, combined with the Geomertic Hashing structural comparisin too, which enables superimposing the interfaces despite the fact that they are composed of dis- continuous pieces of the cahins, provide powerful tools.

The interface families that we have examined show a preference for consrervation of polar residues at their interfaces.

In particular, conserved interface residues are strongly correlated with the experimentally indentified "hot spots" complied from the databased or exaperimental alanine scanning mutagenesis. Specifically, the more highly conserved residues in our analysis are Arg, Gln, His, Asp, Pro, and Asn Bogan and Thron (1998) argue that energetic hot spots are cirtically importnat for the affinity of a protein interface. The fact that these residues tend to be conserved at specific locations indicates that they may constitute binding epitopes (functional epitopes). As such, they may be utilized in searches for potential unknow binding sites. They may further be engineered inbinding sites design. Conserved intrface residues may play a dual role in binding, bnoth thermodynaic and kinetic. Additionally, the surface alignment identified conserved residues in the eleven clustered protein-protein interfaces.

In enzyme-inhibitors, we find that the residues are more conserved at the interfaces than in other locations in the proteins, as shown by multiple sequence alignments. In contrast, antibody-proteins, interfaces illustrate similar surface conservation as compared to linear sequnce alignment, consistent with the proposition that evolution has optimized protein interfaces to achieve optimal functions, as conserved residues are more likely to be optimized. The main advantage in the utilization of binding epitpes in docking is that they are inherently resilient to the atomic conformaional detail and provide a practical (albeit statistically based) way of treating surface variability and flexibility. They further enable the utilization of modeled structure, in addition to high- resolution ones.

INDEX